Genome **Biology**

Vol 7 S~~...~~ ~~...~~ ~~gust~~ 2006

EGASP '05: ENCODE Genome Annotation Assessment Project

Supplement Editors:

Roderic Guigó
Centre de Regulació Genòmica, Institut
Municipal d'Investigació Mèdica-Universitat
Pompeu Fabra, Barcelona, Catalonia, Spain

Martin G Reese
Omicia Inc., Christie Ave, Emeryville,
CA 94608, USA

Supported by the National Human Genome Research Institute (NHGRI) and the Wellcome Trust Sanger Institute.

Cover image

The lower part of the cover figure shows a genomic view of the human CFTR loci neighbourhood on chromosome 7. The upper part indicates the GENCODE annotations for that loci region, within the ENCODE ENm001 sequence interval; it also illustrates the results from the computational assessments used in the EGASP experiment. The annotation plots were both generated using the gff2ps tool [1], the original boxplots using the R package [2], and the final composition was made with The Gimp [3].

1. Abril JF, Guigo R: **gff2ps: visualizing genomic annotations.** *Bioinformatics.* 2000, 16:743-744.
2. **The Comprehensive R Archive Network** [http://cran.r-project.org]
3. **GIMP** [http://www.gimp.org]

Published by BioMed Central Ltd, Middlesex House, 34–42 Cleveland Street, London WIT 4LB, UK.

British Library Cataloguing-in-Publication Data.
A catalogue record for this book is available from the British Library.

ISBN 978-0-9540278-2-7

Genome **Biology**

The reviewing and alerting service provided by Genome **Biology** is available on a subscription basis; access to all primary research is free.

Genome **Biology** (Print ISSN 1465-6906; online ISSN 1465-6914) is published by BioMed Central Ltd
Middlesex House
34-42 Cleveland Street
London WIT 4LB, UK
Tel +44(0)20 7323 0323
Fax +44(0)20 7631 9961
E-mail info@genomebiology.com

Genome Biology can be found on the web under the following address:
http://genomebiology.com

Open access
All research articles in Genome **Biology** are published under BioMed Central's Open Access Charter (http://www.biomedcentral.com/info/about/charter): they are freely available online and archived in full in PubMed Central (http://www.pubmedcentral.nih.gov/), and all users may download, store, or redistribute them in their original form as long as the original citation and bibliographic details remain with each article. BioMed Central is committed to maintaining open access for all research articles that it publishes, both retrospectively and prospectively, in all eventualities.

For further information about the journal, please see the description available on the website (**http://genomebiology.com**).

Advertising information
Advertising is accepted by GenomeBiology.com. Further information regarding advertising, including rate cards, specifications, etc, can be obtained from the Advertising Department:
Tel +44(0)20 7323 0323
Fax +44(0)20 7580 1938
E-mail advertising@biomedcentral.com

Disclaimer: Whilst every effort is made by the publishers and editors to see that no inaccurate or misleading data, opinions or statements appear in this publication, they wish to make it clear that the data and opinions appearing in the articles and advertisements herein are the responsibility of the contributor or advertiser concerned. Accordingly, the publishers, the editor and editorial board and their respective employees, officers and agents accept no liability whatsoever for the consequences of any such inaccurate or misleading data, opinion or statement.

Subscription information
Subscriptions are charged on an annual basis.

Indexing services
Genome **Biology** is indexed in Science Citation Index, Science Citation Index Expanded, ISI Alerting Services, MEDLINE, PubMed, PubMed Central, Biotechnology Citation Index and Chemical Abstracts Service.

2006 Subscription Rates (Volume 7, 12 issues, 2006). All rates include full-text access to all articles online.

	North/South America[1]	Rest of World[2]
Personal subscription (online only)	$120	£71.50/€107
Student subscription[3] (online only)	$64	£38/€58

[1]**Canadian** subscribers add GST. [2]**EU** subscribers may be liable to European sales tax. [3]**Students** must give the name of their institution or school, plus the name of their department chairman to qualify for the student rate. The student rate is available for a maximum of 2 years per student. For online **Institutional subscriptions** and **Lab subscriptions**, please e-mail info@genomebiology.com

US Postmaster

Send address corrections to:
BioMed Central Ltd
c/o Current Science Inc.
400 Market Street
Suite 700
Philadelphia
PA 19106-2514
USA

Orders

All orders for Genome **Biology** should be placed with a bookseller or subscription agent, made online at **http://genomebiology.com/subscriptions/** or sent to:

BioMed Central Customer Services
Current Science Group
Middlesex House
34-42 Cleveland Street
London WIT 4LB, UK
Tel: +44(0)20 7631 9131
Fax: +44(0)20 7631 9926
E-mail: info@genomebiology.com

Subscription queries

All enquiries should be addressed to:

BioMed Central Customer Services
Current Science Group
Middlesex House
34-42 Cleveland Street
London WIT 4LB, UK
Tel: +44(0)20 7631 9131
Fax: +44(0)20 7631 9926
E-mail: info@genomebiology.com

CONTENTS

COMMENT

EGASP: Introduction

MARTIN G REESE AND RODERIC GUIGO

http://genomebiology.com/2006/7/S1/S1

REVIEWS

EGASP: the human ENCODE Genome Annotation Assessment Project

RODERIC GUIGO, PAUL FLICEK, JOSEP F ABRIL, ALEXANDRE REYMOND, JULIEN LAGARDE, FRANCE DENOEUD, STYLIANOS ANTONARAKIS, MICHAEL ASHBURNER, VLADIMIR B BAJIC, EWAN BIRNEY, ROBERT CASTELO, EDUARDO EYRAS, CATHERINE UCLA, THOMAS R GINGERAS, JENNIFER HARROW, TIM HUBBARD, SUZANNA E LEWIS AND MARTIN G REESE

http://genomebiology.com/2006/7/S1/S2

Performance assessment of promoter predictions on ENCODE regions in the EGASP experiment

VLADIMIR B BAJIC, MICHAEL R BRENT, RANDALL H BROWN, ADAM FRANKISH, JENNIFER HARROW, UWE OHLER, VICTOR V SOLOVYEV AND SIN LAM TAN

http://genomebiology.com/2006/7/S1/S3

REFEREED RESEARCH

GENCODE: producing a reference annotation for ENCODE

JENNIFER HARROW, FRANCE DENOEUD, ADAM FRANKISH, ALEXANDRE REYMOND, CHAO-KUNG CHEN, JACQUELINE CHRAST, JULIEN LAGARDE, JAMES GR GILBERT, ROY STOREY, DAVID SWARBRECK, COLETTE ROSSIER, CATHERINE UCLA, TIM HUBBARD, STYLIANOS E ANTONARAKIS AND RODERIC GUIGO

http://genomebiology.com/2006/7/S1/S4

Pairagon+N-SCAN_EST: a model-based gene annotation pipeline

MANIMOZHIYAN ARUMUGAM, CHAOCHUN WEI, RANDALL H BROWN AND MICHAEL R BRENT

http://genomebiology.com/2006/7/S1/S5

Vertebrate gene finding from multiple-species alignments using a two-level strategy

DAVID CARTER AND RICHARD DURBIN

http://genomebiology.com/2006/7/S1/S6

Exogean: a framework for annotating protein-coding genes in eukaryotic genomic DNA

SARAH DJEBALI, FRANCK DELAPLACE AND HUGUES ROEST CROLLIUS

http://genomebiology.com/2006/7/S1/S7

Using several pair-wise informant sequences for de novo prediction of alternatively spliced transcripts

PAUL FLICEK AND MICHAEL R BRENT

http://genomebiology.com/2006/7/S1/S8

JIGSAW, GeneZilla, and GlimmerHMM: puzzling out the features of human genes in the ENCODE regions

JONATHAN E ALLEN, WILLIAM H MAJOROS, MIHAELA PERTEA AND STEVEN L SALZBERG

http://genomebiology.com/2006/7/S1/S9

Automatic annotation of eukaryotic genes, pseudogenes and promoters

VICTOR SOLOVYEV, PETER KOSAREV, IGOR SELEDSOV AND DENIS VOROBYEV

http://genomebiology.com/2006/7/S1/S10

AUGUSTUS at EGASP: using EST, protein and genomic alignments for improved gene prediction in the human genome

MARIO STANKE, ANA TZVETKOVA AND BURKHARD MORGENSTERN

http://genomebiology.com/2006/7/S1/S11

AceView: a comprehensive cDNA-supported gene and transcripts annotation

DANIELLE THIERRY-MIEG AND JEAN THIERRY-MIEG

http://genomebiology.com/2006/7/S1/S12

A computational approach for identifying pseudogenes in the ENCODE regions

DEYOU ZHENG AND MARK B GERSTEIN

http://genomebiology.com/2006/7/S1/S13

Genome **Biology**

Editorial Office

Genome **Biology**,
BioMed Central Ltd.,
Middlesex House,
34-42 Cleveland Street,
London WIT 4LB, UK

Telephone +44 (0)20 7323 0323
Facsimile +44 (0)20 7631 9961
E-mail editorial@genomebiology.com

Editorial Director (Biology),
BioMed Central Ltd Theodora Bloom
Editor Michaela Torkar
Assistant Editors Clare Hinkley,
Maruxa Martinez-Campos
Editorial Coordinator Ruth Rowland
Assistant Manager – Editorial
Production Caroline Hering
Editorial Production Assistants
James Heywood, Frances Mulvany
Supplements Jo Baker, Isobel Peters,
Enitan Sawyerr

Introduction

EGASP: Introduction

Martin G Reese* and Roderic Guigó[†]

Addresses: *Omicia Inc., Christie Ave, Emeryville, CA 94608, USA. [†]Centre de Regulació Genòmica, Institut Municipal d'Investigació Mèdica-Universitat Pompeu Fabra, B08003 Barcelona, Catalonia, Spain.

Published: 7 August 2006

*Genome **Biology** 2006, **7**(Suppl 1):S1

The electronic version of this article is the complete one and can be found online at http://genomebiology.com/2006/7/S1/S1

Science is about building causal relations between natural phenomena (for instance, between a mutation in a gene and a disease). The development of instruments to increase our capacity to observe natural phenomena has, therefore, played a crucial role in the development of science - the microscope being the paradigmatic example in biology. With the human genome, the natural world takes an unprecedented turn: it is better described as a sequence of symbols. Besides high-throughput machines such as sequencers and DNA chip readers, the computer and the associated software becomes the instrument to observe it, and the discipline of bioinformatics flourishes. However, as the separation between us (the observers) and the phenomena observed increases (from organism to cell to genome, for instance), instruments may capture phenomena only indirectly, through the footprints they leave. Instruments therefore need to be calibrated: the distance between the reality and the observation (through the instrument) needs to be accounted for. This issue of *Genome Biology* is about calibrating instruments to observe gene sequences; more specifically, computer programs to identify human genes in the sequence of the human genome.

After nearly 25 years of research in the area of computational gene finding, and genome annotation, and after the completion of the human genome sequence in 2003, it became important to assess the current state-of-the-art in this discipline because in the future the success of many genomic and systems biology projects will depend on the quality of genome annotations. In this endeavor we built on the efforts by the NIH initiated ENCODE (for ENCyclopedia of DNA Elements) project, the goal of which, in its first phase, is the development and assessment of methods to identify all functional elements in 1% of the human genome across 44 regions, so that these methods can later be applied to the entire human genome. Within this project, the GENCODE consortium has produced a high quality annotation of the protein coding content of the ENCODE regions. We have used this annotation as the 'golden standard' against which to measure the performance of the computational methods. Developing such a standard has been a difficult task and the paper by Harrow *et al.* in this issue is dedicated to describing the process by means of which the GENCODE standard annotation was obtained.

Scientists working in the field of computational genome annotation were asked to submit predictions on the ENCODE regions. Eighteen groups worldwide participated in the experiment - which we named EGASP (for ENCODE Genome Annotation Assessment Project), the second of its kind after GASP1 [1] - and submitted 30 prediction sets using state-of-the-art methods.

Predictions were compared to the golden standard in a workshop organized at the Sanger Center on May 6 and 7, 2005, and sponsored by the National Human Genome Research Institute (NHGRI) at the NIH. The gene finding evaluation experiment is described in detail in Guigó *et al.* in this issue and the promoter evaluation experiment is described in Bajic *et al.* Many of the computational gene finding methods applied are also described in this issue: Allen *et al.*, Arumugam and Brent, Carter and Durbin, Djebali *et al.*, Flicek and Brent, Solovyev *et al.*, Stanke *et al.* and Thierry-Mieg and Thierry-Mieg. The paper by Zheng and Gerstein describes the analysis of the pseudogenes.

The willingness of the scientists within the gene finding community to provide their computational annotation for public comparison and blind evaluation against one and the same standard annotation set allowed us to really identify pluses and minuses in the various methodical approaches. Too often superior performance is claimed for new datasets without a careful analysis of potential biases within them. Therefore, we hope that with this experiment we have again laid out a test bed for performance enhancement within the field of gene finding. The difficulty of a repeated experiment in the future will be that it is hard to distinguish whether performance has improved due to the more and better auxiliary data (for example better cDNA sequences) or due to algorithmic improvements. Therefore, we suggest repeating the experiment in the future with the same genomic sequence and using the same, 'frozen' auxiliary sequence databases. Furthermore, we expect that a future experiment would include a higher focus on multiple mRNA transcript evaluation, including 5' and 3' untranslated region transcript predictions besides the classic coding sequence evaluations.

EGASP highlighted the recent progress in computational gene finding. Computer programs are increasingly sophisticated, efficient and accurate in mapping cDNA and protein sequences onto the genome sequence, as well as in using genome comparisons to other organisms. Despite the progress, however, programs are still not able to replace the insight of human annotators. Difficulties arise not only from the quality of the source data, but also because of the complexities of biology and the complex structure of human genes: the bulk of cDNA sequences are partial, and contain many sequence errors, and genome sequences are often incomplete and errors may exist in the assemblies; mapping of cDNA sequences onto the genome is compounded by the presence of pseudogenes and recent duplicates, which occasionally makes it very difficult to identify the exact genomic locus for a given cDNA sequence; and alternative splicing, for instance, is widespread, and involves, more often than until very recently expected, exons from apparently different loci. The diversity of the human proteome may be much higher than that derived simply from the total number of genes.

Computational methods at EGASP also predicted many exons and genes that were not included in the standard GENCODE annotation. While predictions mapped within annotated loci could correspond to novel alternative splice forms of known genes, predictions in intergenic regions might reveal novel genes. However, only a handful of such predictions could be verified by RT-PCR experiments using 24 human tissue libraries. This certainly seems to suggest that the standard GENCODE annotation is quite complete, and that, in general, not many novel gene loci remain to be discovered in the sequence of the human genome. This is, however, in contrast with results from recent large-scale surveys of the transcriptional activity of the human genome using high-throughput sequencing and hybridization based technologies [2].These reveal, also in the ENCODE regions,

a wealth of sites of transcription that are neither included in the standard annotation or predicted by the programs. Whether these correspond to real, novel, protein coding genes or to non-coding RNAs could not be answered within the project described here.

In summary, from the EGASP project we have learnt that the current human genome annotation is almost complete in terms of novel protein coding loci. Nevertheless, the annotation of the exact structure and the transcriptional organization of a gene is still nowhere near completed. Three years after the completion of the human genome sequence and after all the human chromosomes have been published it seems that the gene locus annotations are still in flux. Therefore, we believe that efforts towards annotating the human genome should be extended. Almost correct gene annotations are simply not good enough as the blueprint of human biology. These errors can mislead many follow-on projects such as genetic variation experiments, mRNA expression profiling as well as proteomic experiments. Efforts to systematically and continuously sequence high-quality cDNA libraries to obtain full-length cDNA sequences, such as those at the Mammalian Gene Collection [3] need to be continued, although increasingly aggressive sequencing of cDNA libraries appears to have reached a plateau and is yielding only a fraction (which could be small) of lowly or rarely expressed transcripts. Hybridization based techniques, such as high density genome tiling micro-arrays, could constitute, in this regard, a complementary approach.

The ultimate goal in human genome annotation should be to map onto the genome sequence all primary and processed RNA molecules that exist in a given cell type at a given time - and ideally measure their relative abundance. This is a task that will likely take at least a decade to achieve completely. While the biological roles of non-coding RNAs are increasingly appreciated, EGASP focused on protein coding genes. In this regard, EGASP has shown that computational methods do not provide evidence for many additional, still un-annotated protein coding genes in the human genome, and, therefore, there is no need to drastically re-evaluate the current estimations of the total number of human genes. Current annotation efforts within the ENCODE project, in which RACE reactions are hybridized into genome tiling arrays, have, however, uncovered a wealth of additional transcripts mapping onto annotated protein coding loci. Often these transcripts reach upstream genes and include exons from intervening loci. These transcriptional continuums, in which boundaries between loci seem to fade away, challenge our very concept of what a gene is, and makes estimating the total number of human genes almost a futile task.

Acknowledgements
We thank the NHGRI for providing funding for the meeting, the Sanger Institute (Wellcome Trust) for help in its organization.

This article has been published as part of *Genome Biology* Volume 7, Supplement 1, 2006: EGASP '05. The full contents of the supplement are available online at http://genomebiology.com/supplements/7/S1.

References

1. Reese MG, Hartzell G, Harris NL, Ohler U, Abril JF, Lewis SE: **Genome annotation assessment in *Drosophila melanogaster*.** *Genome Res* 2000, **10**:483-501.
2. Cheng J, Kapranov P, Drenkow J, Dike S, Brubaker S, Patel S, Long J, Stern D, Tammana H, Helt G, *et al*.: **Transcriptional maps of 10 human chromosomes at 5-nucleotide resolution.** *Science* 2005, **308**:1149-1154.
3. Gerhard DS, Wagner L, Feingold EA, Shenmen CM, Grouse LH, Schuler G, Klein SL, Old S, Rasooly R, Good P, *et al*.: **The status, quality, and expansion of the NIH full-length cDNA project: the Mammalian Gene Collection (MGC).** *Genome Res* 2004, **14**:2121-2127.

Review

EGASP: the human ENCODE Genome Annotation Assessment Project

Roderic Guigó[*,1,11], Paul Flicek[*,2], Josep F Abril[*,1], Alexandre Reymond[3], Julien Lagarde[1], France Denoeud[1], Stylianos Antonarakis[4], Michael Ashburner[5,12], Vladimir B Bajic[6,12], Ewan Birney[2,11], Robert Castelo[1], Eduardo Eyras[1], Catherine Ucla[4], Thomas R Gingeras[7,12], Jennifer Harrow[8,11], Tim Hubbard[8,11], Suzanna E Lewis[9,12] and Martin G Reese[*,10,12]

Addresses: [1]Centre de Regulació Genòmica, Institut Municipal d'Investigació Mèdica-Universitat Pompeu Fabra, E08003 Barcelona, Catalonia, Spain. [2]European Bioinformatics Institute, Wellcome Trust Genome Campus, Hinxton, Cambridge CB10 1SD, United Kingdom. [3]Center for Integrative Genomics, University of Lausanne, Switzerland. [4]University of Geneva Medical School and University Hospitals of Geneva, 1211 Geneva, Switzerland. [5]Department of Genetics, University of Cambridge, Cambridge CB3 2EH, United Kingdom. [6]South African National Bioinformatics Institute (SANBI), University of Western Cape, Bellville 7535, South Africa. [7]Affymetrix Inc., Santa Clara, California 95051, USA. [8]Wellcome Trust Sanger Institute, Wellcome Trust Genome Campus, Hinxton, Cambridge CB10 1SA, United Kingdom. [9]Department of Molecular and Cellular Biology, University of California, Berkeley, California 94792, USA. [10]Omicia Inc., Christie Ave., Emeryville, California 94608, USA. [11]Member of the EGASP Organizing Committee. [12]Member of the EGASP Advisory Board.

*These authors contributed equally to this work.

Correspondence: Roderic Guigo. Email: rguigo@imim.es; Martin G Reese. Email: mreese@omicia.com

Published: 7 August 2006

Genome Biology 2006, **7(Suppl 1)**:S2

The electronic version of this article is the complete one and can be found online at http://genomebiology.com/2006/7/S1/S2

Abstract

Background: We present the results of EGASP, a community experiment to assess the state-of-the-art in genome annotation within the ENCODE regions, which span 1% of the human genome sequence. The experiment had two major goals: the assessment of the accuracy of computational methods to predict protein coding genes; and the overall assessment of the completeness of the current human genome annotations as represented in the ENCODE regions. For the computational prediction assessment, eighteen groups contributed gene predictions. We evaluated these submissions against each other based on a 'reference set' of annotations generated as part of the GENCODE project. These annotations were not available to the prediction groups prior to the submission deadline, so that their predictions were blind and an external advisory committee could perform a fair assessment.

Results: The best methods had at least one gene transcript correctly predicted for close to 70% of the annotated genes. Nevertheless, the multiple transcript accuracy, taking into account alternative splicing, reached only approximately 40% to 50% accuracy. At the coding nucleotide level, the best programs reached an accuracy of 90% in both sensitivity and specificity. Programs relying on mRNA and protein sequences were the most accurate in reproducing the manually curated annotations. Experimental validation shows that only a very small percentage (3.2%) of the selected 221 computationally predicted exons outside of the existing annotation could be verified.

Conclusions: This is the first such experiment in human DNA, and we have followed the standards established in a similar experiment, GASP1, in *Drosophila melanogaster*. We believe the results presented here contribute to the value of ongoing large-scale annotation projects and should guide further experimental methods when being scaled up to the entire human genome sequence.

Background

During the first decade of the 21st century the sequencing of whole genomes has become a routine biological practice. The list of chordates with assembled genome sequences now numbers nearly two dozen, while the total number of sequenced bacteria, archea, and eukaryota is approaching 2,000. The genome sequence is said to be an organism's blueprint: the set of instructions dictating its biological traits. In higher eukaryotic organisms, however, these traits are apparently encoded by only a small fraction of the genome sequence that is functional (possibly less than 5% in the case of the human genome). The genes are a major component of this functional sequence. While there is growing evidence for many functional non-protein coding RNA genes, such as miRNAs and snoRNAs, the largest and best studied subset of the human genes comprise the protein coding genes, genes specifying the amino acid sequence of the proteins. Thus, locating the genes in a newly sequenced genome is a first, essential step toward understanding how the organism translates its genome sequence into biological function. This paper focuses on the identification of protein coding genes, if not otherwise noted.

Maybe to the surprise of many, five years after the first drafts of the human genome sequence became available [1,2], and nearly three years after the announcement of the completion of the sequencing [3], a complete set of protein coding genes encoded in the human genome does not exist. One reason for the lack of a complete gene set is that an appropriately rigorous standard has been set for the human genome: every gene, exactly correct. And as shown in this paper, only very few of the human genes seem to be missing from the computational predictions, but the exact genomic structure of these genes is estimated to be correct for only 50% of the predicted genes. In other words, only very few protein coding genes appear to have been totally missed today. Nevertheless, getting the entire genomic structure of a protein coding gene right is still a very difficult task, compounded by the large amount of alternative splicing characterizing human genes. Our assessment here tries to quantify the status of these differences in the current human genome annotations and computational prediction programs.

Automatic genome annotation methods

To date, accurate automatic annotation of the human genome (and of other genomes with significant cDNA libraries)

strongly relies on an elaborate mapping of these known gene sequences onto the genome sequence. This method of genome annotation requires high quality and a nearly complete set of cDNA sequences. Datasets trying to achieve this goal, but are still works in progress, are the RefSeq database [4] and those currently being produced by the Mammalian Gene Collection (MGC) [5]. As the MGC project - and similar efforts to deepen the coverage of the fraction of the human genome being transcribed - continues, cDNA mapping based gene identification methods are becoming increasingly accurate. While few organisms will have the rich cDNA libraries that are currently being developed for the human genome, the availability of protein sequence data from evolutionarily close relatives has been effectively used in addition to cDNA data for automatic gene prediction across many of the currently sequenced mammals. The most commonly used annotation pipelines are the ENSEMBL pipeline [6], the UCSC genome browser's [7] Known Genes (KG II) pipeline, and the Gnomon pipeline at the NCBI [8]. It remains unclear, however, what fraction of the low and specifically expressed transcripts and of alternatively spliced isoforms can be effectively recovered from cDNA libraries. Additionally, orthologous proteins from other species may not align genes that are rapidly evolving. For these reasons, current cDNA and protein-based methods are likely to provide an incomplete picture of the protein coding gene content of the human genome. These methods will be less accurate for genomes with fewer expressed sequences and comparative options.

For automatic annotation of genomes without deep expressed sequence libraries, any available cDNA or expressed sequence tag (EST) based annotation is often complemented by dual (or multiple) genome comparative predictions. These predictions are obtained by means of the analysis of the patterns of sequence conservation between genome sequences of evolutionarily related organisms. As examples, programs such SGP2 [9], SLAM [10,11] and TWINSCAN [12,13] have contributed efficiently to the annotation of a number of vertebrate genomes, including mouse [14], rat [15], and chicken [16]. This type of comparative-based automatic gene prediction can produce highly accurate gene sets when the sequence of related species is available, but few ESTs have been sequenced, such as the case with the fungus *Cryptococcus neoformans* [17].

Occasionally, the so-called single genome *ab initio* predictors - programs that use statistical sequence patterns, such as the coding reading frame, codon usage or splice site consensus sequences, for gene identification - are also used to complement cDNA and comparative based methods. When no genome exists at the appropriate phylogenetic distance, and the cDNA or EST coverage of the transcriptome is shallow, single genome *ab initio* predictions play an important role in genome annotation, such as those obtained, for example, by the programs GENSCAN [18] and GENEID [19] in the initial annotation of the genome of the fish *Tetraodon nigroviridis* [20].

In summary, despite substantial progress in the past decade and the existence of highly accurate gene sets in a number of organisms, current gene identification methods are, as yet, not able to produce a complete catalogue of the set of protein coding genes in higher eukaryotic genomes (see [21] for a recent review).

Assessing the accuracy of automatic genome annotation

Over the past quarter century, a large number of automated gene prediction algorithms have been introduced, which can be loosely grouped based on the general strategies described above. These methods vary widely in the details of their implementation and in the number and location of predicted protein coding genes. Thus, the issue of evaluating the accuracy of the predictive methods has been recurrent within the field of computational gene prediction. The early work of Burset and Guigó [22], and the subsequent analysis of Bajic [23], Baldi *et al.* [24], Guigó *et al.* [25] , Rogic *et al.* [26] and others, provide a framework - a set of metrics and a protocol - to consistently evaluate gene prediction methods. Essentially, a set of well-annotated sequences are used as a test set. The gene prediction programs are run on these sequences, and the predictions obtained are compared with the annotations. A number of measures are computed to evaluate how well the predictions reproduce the annotation. Typically, predictions are evaluated at nucleotide, exon and gene levels. At all three levels, two basic measures are computed: sensitivity, the proportion of annotated features (nucleotide, exon, gene) that have been predicted; and specificity, the proportion of predicted features that is annotated. One problem with this approach is that, until recently, very few large genomic sequences were well annotated and only the coordinates of the coding exons within a gene could be considered. Moreover, because methods did not exist to predict alternative splicing, the test sets used to evaluate computational gene predictions consisted of a few hundred short sequences encoding single genes from which alternatively spliced isoforms had been removed. This led to an oversimplification of the problem and, in turn, to an overestimation of the real accuracy of the programs [25]. Furthermore, many programs were developed in-house and were, therefore, not accessible for independent evaluation.

To address the problem of independent, objective assessment of the state-of-the-art in automated tools and techniques for annotating large contiguous genomic DNA regions and eventually complete genomes, a first Genome Annotation Assessment Project (GASP1) was organized in 1999 [27]. In many ways, GASP1 was set up similarly to CASP (Critical Assessment of Techniques for Protein Structure Prediction) [28]. In short, at GASP1, a genomic region in *Drosophila melanogaster*, including auxiliary training data, was provided to the community and gene finding experts were invited to send the annotation files they had generated to the organizers before a fixed deadline. Then, a set of standards were developed to evaluate submissions against the later published annotations [29], which had been withheld until after the submission stage. Next, the evaluation results were assessed by an independent advisory team and publicly presented at a workshop at the Intelligent Systems in Molecular Biology (ISMB) 1999 meeting. This community experiment was then published as a collection of methods and evaluation papers in *Genome Research* [27].

The ENCODE Genome Annotation Assessment Project

Inspired by GASP1, and within the context of the ENCyclopedia Of DNA Elements (ENCODE) project, we organized the ENCODE GASP (EGASP) community experiment, which followed closely the model of its predecessor, GASP1 [27]. The ENCODE project was launched two years ago by the National Human Genome Research Institute (NHGRI) with the aim of identifying all functional elements in the genome sequence through the collaborative effort of computational and laboratory-based scientists [30]. The pilot phase of the project is focused on a selected 30 Mb of sequence within 44 selected regions (Table 1) across the human genome, which represents approximately 1% of the genome sequence.

Within the ENCODE project, the GENCODE consortium [31] was set up. This group, in collaboration with the HAVANA team [32] at the Sanger Institute, has produced a high quality annotation of the gene content of the ENCODE regions through a combined manual, computational and experimental strategy [33]. The EGASP experiment was organized with the main goal of evaluating how well automatic methods are able to reproduce this annotation produced by GENCODE. A second goal of EGASP was to assess the completeness of the GENCODE annotation and, in this regard, EGASP was designed such that, in a follow-up step, a number of computational gene predictions not included in GENCODE were tested experimentally.

In what follows, we first describe the organization and structure of the EGASP experiment. We then present the results of the evaluation of the submitted predictions against the GENCODE annotation, and finally we present the results of the experimental verification of the novel predictions.

Table 1

The 44 selected sequences within the ENCODE region

Sequence set	Manual picks	Random picks Mouse homology			Gene density
		Low	Medium	High	
Training	ENm006	ENr132	ENr231	ENr333	High
			ENr232	ENr334	
	ENm004	-	ENr222	ENr323	Medium
			ENr223	ENr324	
	-	ENr111	-	-	Low
		ENr114			
Test	ENm002	ENr131	ENr233	ENr331	High
	ENm005	ENr133		ENr332	
	ENm007				
	ENm008				
	ENm009				
	ENm010				
	ENm011				
	ENm001	ENr121	ENr221	ENr321	Medium
	ENm003	ENr122		ENr322	
	ENm012	ENr123			
	ENm013				
	ENm014				
	-	ENr112	ENr211	ENr311	Low
		ENr113	ENr212	ENr312	
			ENr213	ENr313	

ENCODE sequences were assigned to either the training or the test set based on annotation data availability (see the section 'The EGASP experiment'). For the performance evaluation, only the test set sequences were used. The numeric code for the randomly picked sequence names correspond to the non-exonic conservation with the mouse genome, the density of previously identified genes, and the sequence number, respectively; numbers vary from 1 (low), to 3 (high). Manually selected sequences range in size from 500 kbp to 2 Mbp, while random regions are 500 kbp. The selection and stratification criteria for all the sequences is described at the ENCODE project web site [34].

The EGASP experiment

Data: the benchmark sequence of 44 selected ENCODE regions
Description of the sequence
The 44 ENCODE regions represent 30 Mb (approximately 1%) of the human genome [30]. Approximately half of the sequence corresponds to a set of 14 manually selected regions including well-studied genes and for which a significant amount of prior comparative sequence data was available. The remaining 30 genomic regions were chosen based on a stratified random sampling based on two measures: gene density (from previous annotations) and non-exonic conservation with the mouse genome sequence.

Briefly, each portion of the human genome sequence was classified as high, medium, or low if it fell in the top 20%, the middle 30%, or the bottom 50%, respectively, of the above two measures. Several 500 kb sequences were chosen from each of the nine classifications created by this stratification procedure.

Table 1 lists the 44 selected sequences within the ENCODE region and classifies them based on random/manual selection, previously known gene density and non-exonic conservation to the mouse genome. It also describes the size differences between the sequences. Information about the criteria used to select the regions and their characteristics can be found on the ENCODE website [34]. The sequences of the ENCODE regions (as well as multiple functional annotations) can be downloaded from the UCSC ENCODE browser [35].

We defined the sequences used for the EGASP experimental evaluation by taking advantage of the prior work of the HAVANA team [32], which had previously comprehensively annotated and released annotation for several human chromosomes [36-42]. Updated annotation for the 13 ENCODE regions on these chromosomes was released in January 2005 as a 'training' set for the EGASP experiment. The manual annotation of the other 31 ENCODE regions was held back from release until after the automated gene predictions had been received. The 31 EGASP test regions represent a total of 21.6 million base-pairs (bp) of sequence. Further information is available at the GENCODE website [31].

The reference gene set: the GENCODE annotations
The ENCODE regions had been subjected to an exhaustive annotation strategy prior to EGASP by the HAVANA team. In short, the annotators initially build coding transcripts manually based on alignments of known mRNA, EST and protein sequences to the human genome. The initial gene map delineated in this way was then experimentally refined through reverse transcription (RT)-PCR and rapid amplification of cDNA ends (RACE), which essentially confirmed the existence of the mRNA sequences of the hypothesized genes. Finally, the initial annotation was refined by the annotators based on these experimental results. While the initial annotation by the HAVANA team is augmented by some experimentally verified *ab initio* and dual-genome gene predictions without *a priori* transcript sequence support, these constitute a marginal fraction of the entire GENCODE annotation set. The strategy is described in detail elsewhere in this issue [33]. We used this final annotation as the reference set for EGASP, and refer to it as the GENCODE annotation.

The protein coding GENCODE annotation for all 44 ENCODE regions consists of 2,471 total transcripts representing 434 unique protein coding gene loci. There are 1,097 coding transcripts that code for 993 unique proteins. The annotation identifies 5.7 total transcripts per locus, with an average of 2.52 coding transcripts. Of the 434 coding loci, 393 contain multi-exon transcripts. In line with earlier estimates [43], 86% of the multi-exon loci exhibit alternative splicing in either the coding or non-coding transcripts. Sixty percent of multi-exon loci have alternative coding transcripts. See [33] in this issue for additional details.

Incomplete annotation

The GENCODE annotation includes incomplete genes and transcripts. These are caused both by the truncation of some features at the end of the ENCODE regions and by transcript annotations that may be incomplete due to lack of evidence. In the rare case that an exon crossed an ENCODE region boundary, the exon was truncated at the ENCODE region boundary in both the annotations and the predictions to ensure that the nucleotide level evaluation statistics were computed correctly (see Materials and methods).

EGASP: a community experiment

To determine an automatic method's ability to reproduce the GENCODE annotation, we organized EGASP in the following way: In January 2005, the GENCODE annotation for 13 of the 44 ENCODE regions (the 'training regions' defined above) was publicly released. With the release of this annotation, EGASP was officially announced: gene and other DNA feature prediction groups world-wide were asked to submit genome annotations on the remaining 31 ENCODE regions, for which the GENCODE annotations would not be released until the deadline for submission expired. Participating groups had access to the annotation of the 13 training regions, as well as to the sequences and all additional publicly available data for all 44 ENCODE regions. No other pre-defined and pre-selected auxiliary data, such as cDNA databases, EST sequences or other genome alignments, were given to the submitters. However, many of the 31 test regions had been previously and extensively annotated by other groups. For example, *ENm001*, the greater cystic fibrosis transmembrane receptor (CFTR) region, has been extensively studied [44].

Participants were asked to submit their genome annotations on the 31 ENCODE test regions, using whatever methods and data were available to them. To be able to better compare different DNA feature prediction methods, we predefined the following prediction categories and asked the submitters to indicate in which category they were submitting: methods using any type of available information; single-genome *ab initio* methods; EST-, mRNA-, and protein-based methods; dual- or multiple-genome based methods; methods predicting unusual genes (non-canonical splicing, short intron-less genes, and so on); and exon-only predictions.

Finally, we allowed an extra category (category 7) for methods predicting other annotation features, including pseudogenes and promoters. Bajic *et al.* [45] have conducted a compre-

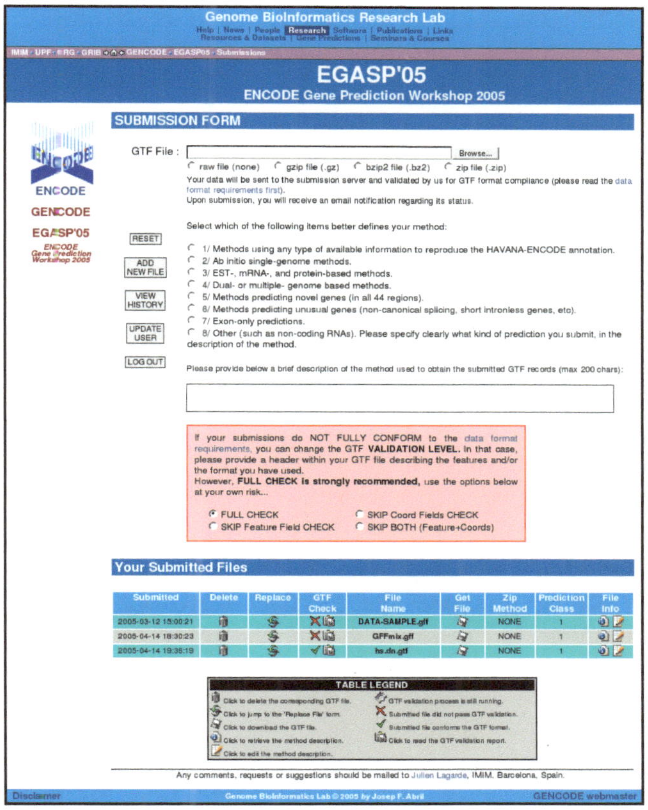

Figure 1
A screenshot of the EGASP submission server [47]. The server was user-authenticated in order to keep the submitted predictions in private before the EGASP workshop. Initially, there were eight suggested submission categories. However, after the workshop, category 5 was not used at all and removed. Promoter and pseudogene predictions from category 8 were then kept as a new category 7, which is not analyzed in this paper (see [45] instead).

hensive evaluation of the promoter predictions and see Zheng and Gerstein [46] for a paper on pseudogenes.

A web server (Figure 1) [47] was set up to collect all the submissions and each group was able to submit predictions for more than one category. The submitted predictions, as well as the GENCODE annotations for the test sequence set, were kept confidential until the submission deadline on 15 April 2005. The format for submissions was the Gene Transfer Format (GTF) [48]. An advisory committee (Table 2) was formed to oversee the submission and evaluation processes and provide advice for the evaluation.

By the submission deadline on 15 April 2005, 18 groups had submitted 30 prediction sets (Table 3). All the submitted predictions together with the annotations are available through the GencodeDB Genome Browser (Figure 2) [49], as well as through the UCSC Genome Browser ('EGASP' tracks). They can also be downloaded from the ftp server as plain text GTF files [50].

Table 2

EGASP organizing and advisory committees

Organizers	Advisory board
Jennifer Ashurst (Wellcome Trust Sanger Institute)	Michael Ashburner (Cambridge University)
Ewan Birney (European Bionformatics Institute)	Vladimir B Bajic (Institute for Infocomm Research)
Peter Good (National Human Genome Research Institute)	Tom Gingeras (Affymetrix, Inc.)
Roderic Guigó (Institut Municipal d'Investigació Mèdica)	Suzanna Lewis (Berkeley)
Tim Hubbard (Wellcome Trust Sanger Institute)	Martin Reese (Omicia, Inc.)

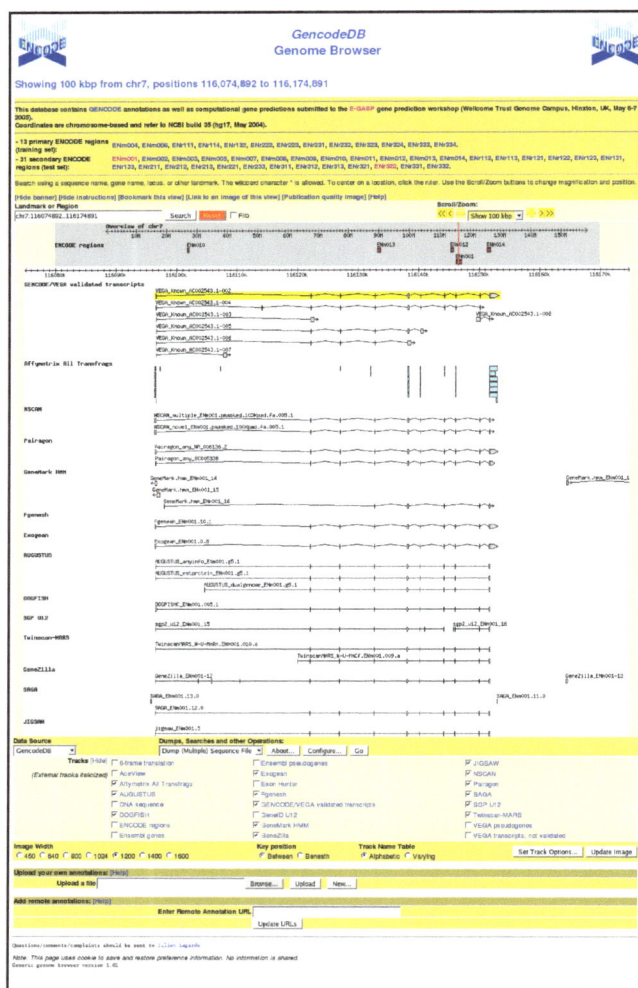

Figure 2

The GencodeDB Genome Browser. A screenshot of the GencodeDB Genome Browser [49], displaying the annotation features on 100 Kbp from the ENm001 region (chr7: 116,074,892-116,174,891). The annotations along with the predicted genes by each submitted method were made publicly available together with further experimental evidence, such as TARs/transfrags.

Predictions were compared with the reference set GENCODE annotations and assessed by members of the

advisory and organizing committees (Table 2), all selected as independent experts in this field. The results of this assessment were presented at a workshop that took place at the Wellcome Trust Genome Campus in Hinxton, UK, on 6 and 7 May 2005. The advisory and organizing committees met on 4 May for a pre-evaluation of the predictions, and to determine a number of summary statistics. Each of the submitting groups was invited to present their methods and submissions at the workshop with a focus on what went right and what went wrong. In total, 16 groups were represented at the workshop. The final prediction evaluation results from the workshop are discussed in the next section.

Results
The evaluation of the predictions against the annotation
The protocol to evaluate the predictions
The main goal of the EGASP experiment was to evaluate the ability of automatic methods of genome annotation to reproduce the manual and experimental annotation of the ENCODE regions described above. By this standard, a perfect prediction strategy would produce annotation completely consistent with the GENCODE annotation.

For the purposes of evaluating the submitted predictions, we considered only the results for the 31 test ENCODE regions, which were the 'blinded' regions for which no GENCODE annotations were available during the submission phase. Potential biases introduced by this restriction will be addressed below. The statistics reported are computed globally for the test region, which means that the total number of prediction successes and failures for all 31 regions are compared directly to the total number of annotated exons, transcripts and genes for all 31 regions.

We evaluated each set of submitted predictions at four distinct levels: nucleotide accuracy, exon accuracy, transcript accuracy, and gene accuracy. At the earlier GASP1 workshop, transcript accuracy levels were not assessed due to the limited transcript information and the lower levels of alternatively spliced transcripts in *Drosophila melanogaster* [27]. For this study we also made a distinction between the statistics calculated for the coding portions of the mRNA

Table 3

Summary of programs used to determine predictions submitted for each EGASP category

Submission category	Program	Affiliation	Reference
1 (AUGUSTUS-any)	AUGUSTUS	Georg-August-Universität, Göttingen	[58]
2 (AUGUSTUS-abinit)			
3 (AUGUSTUS-EST)			
4 (AUGUSTUS-dual)			
1	FGENESH++	Softberry Inc.	[56]
1	JIGSAW	The Institute for Genomic Research (TIGR)	[59]
1 (PAIRAGON-any)	PAIRAGON and NSCAN_EST	Washington University, Saint Louis (WUSTL)	[57]
3 (PAIRAGON+NSCAN_EST)			
2	GENEMARK.hmm	Georgia Institute of Technology	[60]
2	GENEZILLA	TIGR	[81]
3	ACEVIEW	National Center for Biotechnology Information (NCBI)	[52]
3	ENSEMBL	The Wellcome Trust Sanger Institute (WTSI) and European Bioinformatics Institute (EBI)	[64]
3	EXOGEAN	Ecole Normale Superieure, Paris	[62]
3	EXONHUNTER	University of Waterloo	[63]
4	ACESCAN*	Salk Institute	[82]
4	DOGFISH-C	WTSI	[67]
4	NSCAN	WUSTL	[57]
4	SAGA	University of California at Berkeley	[66]
4	MARS	WUSTL - EBI	[65]
5	GENEID-U12	Institut Municipal d'Investigació	–
5	SGP2-U12	Mèdica, Barcelona	
6	ASPIC†	Università degli Studi di Milano	[83]
6 (AUGUSTUS-exon)	AUGUSTUS	Georg-August-Universität, Göttingen	[58]
6	CSTMINER‡	Università degli Studi di Milano	[84]
6	DOGFISH-C-E§	WTSI	[67]
6	SPIDA	EBI	[85]
6	UNCOVER§	Duke University	[86]
1	CCDSGene	UCSC tracks [7]	[55]
1	KNOWNGene		[54]
1	REFSEQ (REFGene)		[4]
2	GENEID		[19]
2	GENSCAN		[18]
3	ACEMBLY		[52]
3	ECGene		[53]
3	ENSEMBL (ENSGene)		[6]
3	MGCGene		[5]
4	SGP2		[9]
4	TWINSCAN		[12,13]
-	CODING 20050607	GENCODE annotation	[33]
-	GENES 20050607		

A complete listing of the number of features for each sequence obtained by each method is available at the Supplementary material web page [51]. *The ACESCAN group submitted results only for the training set and, therefore, has not been evaluated. †ASPIC only provided results for the training regions and, therefore, has not been evaluated. Moreover, ASPIC submitted only intron annotations and should be considered in category 6. ‡CSTMINER predicts coding regions but does not provide strand information. §DOGFISH-C-E and UNCOVER predict only novel exons; this makes it difficult to compare these methods with the others in the same category.

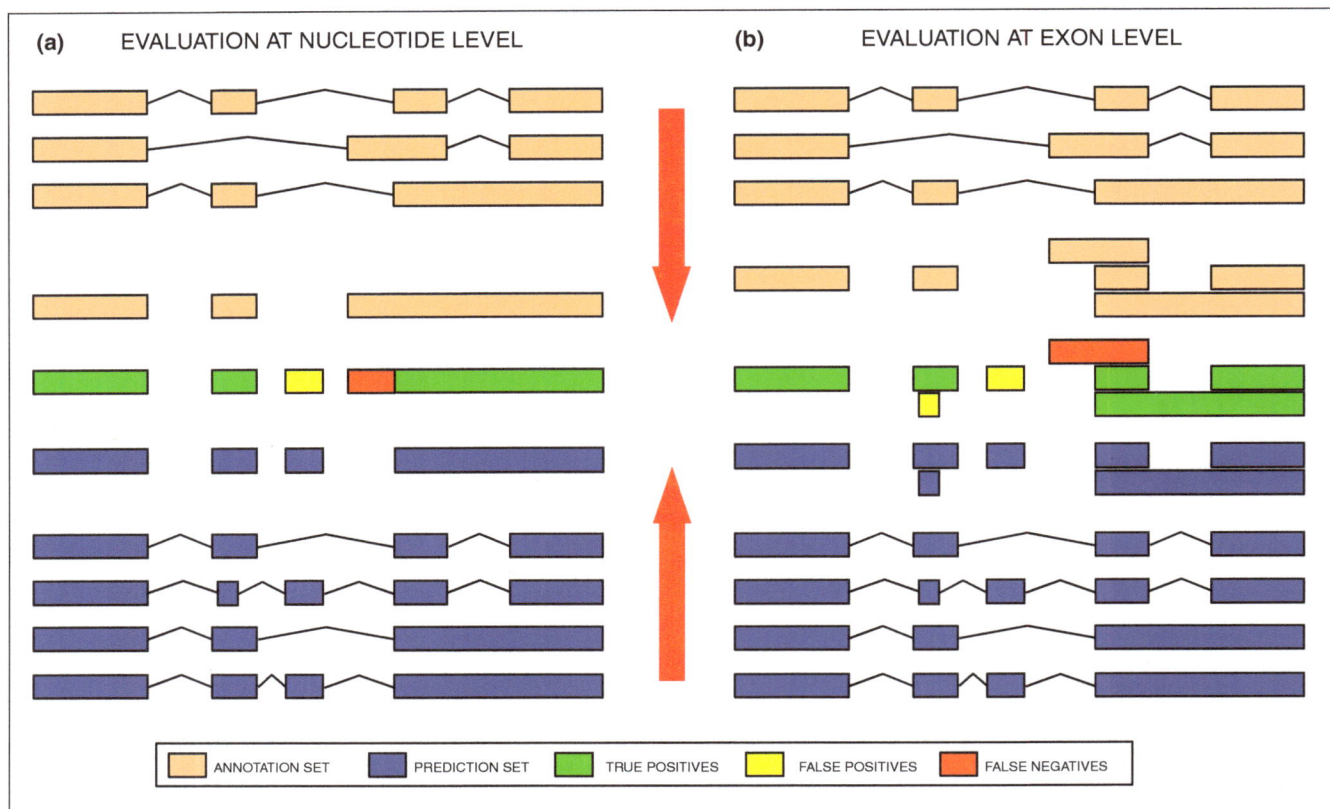

Figure 3
Gene Feature Projection for evaluation. The process of projecting genic features into unique nucleotide and exon coordinates in order to compute the accuracy values (see text for details).

transcripts (coding sequence (CDS) evaluations) and the mRNA transcripts as a whole (mRNA evaluations).

For each of the four levels, we calculated the sensitivity and specificity of the predictions as defined below. In some cases, we have also computed other standard measures previously used in the gene finding literature (see [22-27]). Many additional measures of accuracy have been computed on the EGASP predictions, and they are available through the Supplementary Material web page [51].

Non-EGASP entries
To compare the EGASP results to existing community standards, we also evaluated the performance of 11 gene annotation tracks published in the UCSC Browser [7] just before the start of the EGASP workshop. These tracks included two single genome *ab initio* prediction methods (GENSCAN [18] and GENEID [19]) and two dual-genome prediction methods (TWINSCAN [12,13] and SGP2 [9]). We also considered four methods we classified as using expressed sequence (ENSGENE [6], ACEMBLY [52], MGCGENES [5], and ECGENE [53]) and three we classified as using any information (UCSC 'KNOWN' genes [54], REFSEQ genes [4], and CCDSGENES [55]).

Measures used for evaluating predictions: definitions
Nucleotide level accuracy is a comparison of the annotated nucleotides with the predicted nucleotides. Individual nucleotides appearing in more than one transcript in either the annotation or the predictions are considered only once for the nucleotide level statistics (Figure 3a). Nucleotide predictions must be on the same strand as the annotations to be counted as correct. At the nucleotide level, sensitivity (Sn) is the proportion of annotated nucleotides (as being coding or part of an mRNA molecule) that is correctly predicted, and specificity (Sp) the proportion of predicted nucleotides (as being coding or part of an mRNA molecule) that is so annotated. As a summary measure, we have computed either the simple average of these two measures, or the correlation coefficient between the annotated and the predicted nucleotides (see [22-27]).

The exon level accuracy is calculated with the requirement that an exon in the prediction must have identical start and end coordinates as an exon in the annotation to be counted correct. Only the unique exons in each set are considered (see Figure 3b for a graphical example of how unique exons are collected from both the annotation and prediction sets; also see [22-27] for more details on these definitions). At the exon level, sensitivity is computed as the proportion of annotated exons correctly predicted, and specificity as the proportion of predicted exons that is annotated. As a summary measure, we have computed the average of these two measures. In addition, we have computed 'missing exons' (MEs), the proportion of annotated exons totally missed by the predictions (that is, there is no overlap by a predicted exon by at least 1 bp), and 'wrong exons' (WEs), the proportion of predicted exons not overlapping annotated exons by at least 1 bp. A subset of predicted exons falling in regions annotated as intergenic have been tested experimentally (see the section 'The experimental test of unannotated predictions' below for details). Nucleotide and exon level accuracy are calculated for the CDS evaluation and for the mRNA evaluation. Comparison of the results of these evaluation strategies highlights the differences for those programs that attempt to predict untranslated regions (UTRs) of genes.

The transcript and gene level accuracy measures are more stringent. We consider a transcript accurately predicted for the CDS evaluation if the beginning and end of translation are correctly annotated and each of the 5' and 3' splice sites for the coding exons are correct. Similarly, for the mRNA evaluation, a transcript is counted correct if all of the exons from the start of transcription to the end of transcription are correctly predicted. Thus, at the transcript level, sensitivity is the proportion of annotated transcripts that is correctly predicted, and specificity is the proportion of predicted transcripts that is correct. A gene is counted correct if at least one transcript in the locus is correct as defined above, and sensitivity and specificity are defined accordingly. Using these definitions, transcript accuracy is the most stringent measure for both the CDS evaluation and for the mRNA evaluation (Figure 4).

The accuracy of the prediction methods must be considered in the context of the annotation, which contains a significant fraction of incomplete transcripts. In the case of an incomplete transcript, we made the distinction that if a prediction is completely consistent with the annotation, it will be counted correct. For example, if the annotation contains an incomplete transcript with three exons and a prediction method includes a transcript with these exons plus an additional exon, we consider the prediction to be completely consistent with the annotation and count it as a correct prediction. For the CDS evaluation, if the annotation contains a complete coding transcript, it must be predicted correctly and no additional exons are allowed (Figure 4).

Global results and trends

The evaluation statistics discussed above for the CDS evaluation are provided in Tables 4 and 5 and for the mRNA evaluation in Table 6, which only lists methods that predict full mRNA transcripts. Figures 5-8 display the results for the CDS evaluation at the nucleotide, exon, transcript and gene

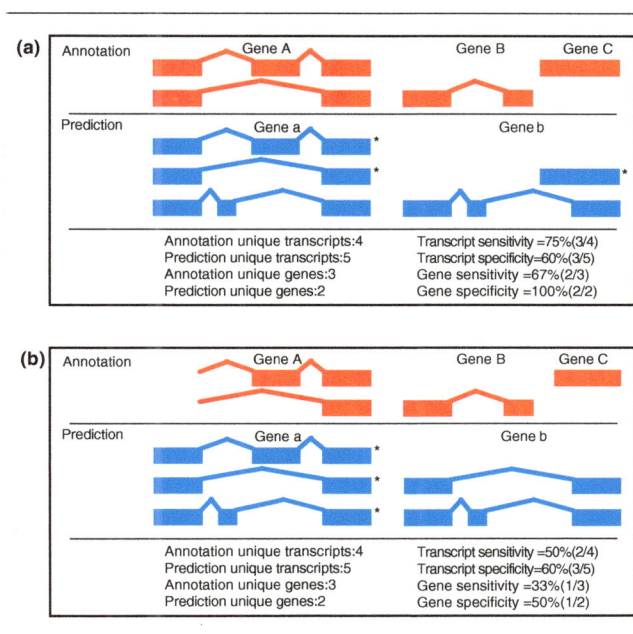

Figure 4
Gene transcript evaluation. Computing sensitivity and specificity at transcript level: **(a)** complete transcript annotation; **(b)** incomplete transcript annotation. Transcripts marked with an asterisk are considered 'consistent with the annotation' and will be scored as correct.

levels. Values are given for programs in categories 1 to 4 (see previous section and Table 3), which constitute the bulk of the submitted predictions. The accuracies of the programs in other categories are often not strictly comparable and, therefore, not shown in these figures. They are, however, given in the Supplementary material [51]. The top panel in Figures 5-8 is a dotplot of sensitivity versus specificity, where each dot represents the performance of one program. The bottom panel includes a boxplot for each program displaying the average of sensitivity and specificity (that is, (Sn + Sp)/2) for the given program on each of 27 test sequences (see Materials and methods). Four test sequences (ENr112, ENr113, ENr311, ENr313) were removed from the original set of 31 because they did not contain any annotated protein coding genes and, therefore, sensitivity and specificity could not be computed for them. The dotplot intends to capture the global balance between sensitivity and specificity for each program, while the boxplots provide the dispersion of the accuracy of each program predictions across test sequences. At similar average accuracies, programs providing more consistent predictions across sequences may be preferable since their behavior can be better anticipated.

No annotation strategy produced perfect predictions, but several clear trends emerged from the evaluations and are summarized here.

Table 4

CDS assessment: summary of accuracy measures for CDS features at the nucleotide and exon levels

	Nucleotide			Exon			
	NSn	NSp	N CC	ESn	ESp	ME	WE
Category 1							
AUGUSTUS-any	94.42%	82.43%	0.88	74.67%	76.76%	8.25%	16.29%
FGENESH++	91.09%	76.89%	0.83	75.18%	69.31%	9.73%	24.64%
JIGSAW	94.56%	92.19%	0.93	80.61%	89.33%	6.22%	7.78%
PAIRAGON-any	87.77%	92.78%	0.90	76.85%	88.91%	11.18%	6.82%
Category 2							
AUGUSTUS-abinit	78.65%	75.29%	0.76	52.39%	62.93%	29.09%	24.82%
GENEMARK.hmm-A	78.43%	37.97%	0.53	50.58%	29.01%	27.86%	63.27%
GENEMARK.hmm-B	76.09%	62.94%	0.69	48.15%	47.25%	31.77%	40.68%
GENEZILLA	87.56%	50.93%	0.66	62.08%	50.25%	19.14%	41.93%
Category 3							
ACEVIEW	90.94%	79.14%	0.84	85.75%	56.98%	4.38%	16.69%
AUGUSTUS-EST	92.62%	83.45%	0.88	74.10%	77.40%	9.01%	15.61%
ENSEMBL	90.18%	92.02%	0.91	77.53%	82.65%	9.99%	9.22%
EXOGEAN	84.18%	94.33%	0.89	79.34%	83.45%	9.88%	5.06%
EXONHUNTER	90.46%	59.67%	0.73	64.44%	41.77%	14.29%	50.94%
PAIRAGON+NSCAN_EST	87.56%	92.77%	0.90	76.63%	88.95%	11.51%	6.85%
Category 4							
AUGUSTUS-dual	88.86%	80.15%	0.84	63.06%	69.14%	16.82%	19.60%
DOGFISH	64.81%	88.24%	0.74	53.11%	77.34%	32.67%	11.70%
MARS	84.25%	74.13%	0.78	65.56%	61.65%	20.26%	26.10%
NSCAN	85.38%	89.02%	0.87	67.66%	82.05%	17.11%	10.93%
SAGA	52.54%	81.39%	0.65	38.82%	50.73%	40.48%	27.85%
UCSC Tracks							
ACEMBLY	96.43%	58.47%	0.74	84.66%	38.32%	2.71%	28.55%
CCDSgene	56.87%	99.52%	0.75	51.95%	97.75%	40.38%	0.27%
ECgene	96.36%	47.30%	0.66	86.22%	35.08%	2.64%	45.92%
ENSgene	91.39%	91.92%	0.92	77.71%	82.39%	9.80%	9.21%
GENEID	76.77%	76.48%	0.76	53.84%	61.08%	27.86%	27.26%
GENSCAN	84.17%	60.60%	0.71	58.65%	46.37%	19.50%	42.91%
KNOWNgene	89.10%	93.61%	0.91	78.11%	82.28%	10.27%	4.30%
MGCgene	44.06%	97.56%	0.65	42.95%	93.61%	49.28%	2.68%
REFgene	85.34%	98.50%	0.92	73.23%	94.67%	15.38%	1.22%
SGPgene	82.81%	82.20%	0.82	60.56%	65.16%	19.36%	22.85%
TWINSCAN	78.16%	84.59%	0.81	58.43%	73.11%	24.64%	16.30%

CC, correlation coefficient.

Table 5

CDS assessment at the transcript and gene levels

	Transcript		Gene		
	TSn	TSp	GSn	GSp	Ratio CDS/UTR
Category 1					
AUGUSTUS-any	22.65%	35.59%	47.97%	35.59%	100.00%
FGENESH++	36.21%	41.61%	69.93%	42.09%	78.25%
JIGSAW	34.05%	65.95%	72.64%	65.95%	100.00%
PAIRAGON-any	39.29%	60.34%	69.59%	61.32%	62.92%
Category 2					
AUGUSTUS-abinit	11.09%	17.22%	24.32%	17.22%	100.00%
GENEMARK.hmm-A	6.93%	3.24%	15.20%	3.24%	100.00%
GENEMARK.hmm-B	7.70%	7.91%	16.89%	7.91%	100.00%
GENEZILLA	9.09%	8.84%	19.59%	8.84%	100.00%
Category 3					
ACEVIEW	44.68%	19.31%	63.51%	48.65%	49.15%
AUGUSTUS-EST	22.50%	37.01%	47.64%	37.01%	100.00%
ENSEMBL	39.75%	54.64%	71.62%	67.32%	65.77%
EXOGEAN	42.53%	52.44%	63.18%	80.82%	59.60%
EXONHUNTER	10.48%	6.33%	21.96%	6.33%	100.00%
PAIRAGON+NSCAN_EST	39.29%	60.64%	69.59%	61.71%	62.89%
Category 4					
AUGUSTUS-dual	12.33%	18.64%	26.01%	18.64%	100.00%
DOGFISH	5.08%	14.61%	10.81%	14.61%	100.00%
MARS	15.87%	15.11%	33.45%	24.94%	100.00%
NSCAN	16.95%	36.71%	35.47%	36.71%	79.80%
SAGA	2.16%	3.44%	4.39%	3.44%	100.00%
UCSC Tracks					
ACEMBLY	33.90%	7.96%	54.39%	21.24%	48.56%
CCDSgene	28.97%	85.58%	55.41%	89.39%	100.00%
ECgene	56.86%	8.84%	79.05%	12.42%	46.11%
ENSgene	40.52%	54.09%	73.99%	68.30%	65.62%
GENEID	4.78%	8.78%	10.47%	8.78%	100.00%
GENSCAN	7.40%	10.13%	15.54%	10.13%	100.00%
KNOWNgene	43.45%	46.93%	77.03%	72.79%	60.03%
MGCgene	23.73%	78.24%	49.32%	82.56%	63.43%
REFgene	41.91%	75.21%	77.03%	82.76%	61.82%
SGPgene	8.17%	12.59%	17.57%	12.59%	100.00%
TWINSCAN	10.63%	20.25%	22.30%	20.25%	100.00%

The ratio CDS/UTR was obtained by summing up all the coding exons' lengths and dividing by the sum of all the exons' lengths. The ratio CDS/UTR for the annotations is 36.78%.

Table 6

mRNA assessment: summary of accuracy measures of mRNA features at the nucleotide and exon levels

	Nucleotide			Exon			
	NSn	NSp	N CC	ESn	ESp	ME	WE
Category 1							
FGENESH++	48.87%	81.16%	0.62	35.84%	58.41%	19.20%	22.84%
PAIRAGON-any	56.31%	89.36%	0.70	41.23%	74.93%	15.83%	7.95%
Category 3							
ACEVIEW	88.08%	79.47%	0.83	64.16%	61.18%	3.60%	10.41%
ENSEMBL	61.61%	95.26%	0.76	41.61%	73.41%	12.84%	7.09%
EXOGEAN	60.58%	94.73%	0.75	48.87%	76.29%	10.38%	4.16%
PAIRAGON+NSCAN_EST	56.22%	89.35%	0.70	41.11%	74.98%	16.06%	7.98%
Category 4							
NSCAN	39.55%	78.69%	0.55	32.41%	65.25%	26.10%	14.69%
UCSC Tracks							
ACEMBLY	91.94%	53.98%	0.70	65.51%	44.28%	2.15%	18.26%
ECgene	93.00%	38.68%	0.59	58.17%	34.83%	1.81%	34.31%
ENSgene	62.43%	95.27%	0.77	41.71%	72.65%	12.62%	7.10%
KNOWNgene	65.74%	91.82%	0.77	43.84%	74.57%	13.58%	2.74%
MGCgene	29.17%	96.73%	0.53	21.21%	74.10%	47.16%	2.22%
REFgene	57.51%	97.07%	0.74	38.35%	83.51%	19.28%	0.91%

Only programs that submitted 5' or 3' UTR exon annotations besides the CDS parts of exons are shown. CC, correlation coefficient.

The prediction methods that used expressed sequence information (category 3) and those that used any information (category 1 prediction methods often used expressed sequence information) were generally the most accurate for all measures.

The three best category 4 dual-genome methods (NSCAN, MARS, and AUGUSTUS-dual) were more accurate than the category 2 single genome *ab initio* prediction methods.

At the nucleotide level, JIGSAW and ENSEMBL both achieved greater than 90% for both sensitivity and specificity for the CDS evaluation, while several other methods scored greater than 80% for both sensitivity and specificity on the same measure, including the NSCAN and AUGUSTUS dual-genome methods (Figure 5). For the mRNA evaluation, ACEVIEW reached 88% sensitivity at 79% specificity, while ENSEMBL and EXOGEAN were more specific with 95% and 94%, respectively, but at much lower sensitivities of 61% and 60%, respectively.

At the exon level, the most accurate predictor of coding exons was JIGSAW with greater than 80% sensitivity while maintaining nearly 90% specificity. ACEVIEW was the most

sensitive prediction method for all exons (coding and non-coding) with greater than 85% (CDS) and 64% (mRNA) exon sensitivity while still being reasonably specific (Figure 6).

At the transcript level, no prediction method correctly identified greater than 45% of the coding transcripts exactly (see sensitivity in Figure 7).

At the gene level, using the measure of averaged sensitivity and specificity, the most accurate gene level predictions in the CDS evaluation were produced by EXOGEAN followed by JIGSAW and ENSEMBL. JIGSAW and ENSEMBL were the only two methods with greater than 70% gene level sensitivity. Of the two, JIGSAW was slightly more sensitive, while ENSEMBL was slightly more specific. EXOGEAN's specificity was higher than 80%, which is more than 13% higher than any other program (Figure 8; Table 5).

Relatively few prediction methods are able to predict multiple transcripts per gene locus. These include four expressed sequence methods from category 3 (PAIRAGON+NSCAN_EST, EXOGEAN, ACEVIEW, and ENSEMBL), FGENESH++ and PAIRAGON-any from category 1, and MARS from category 4.

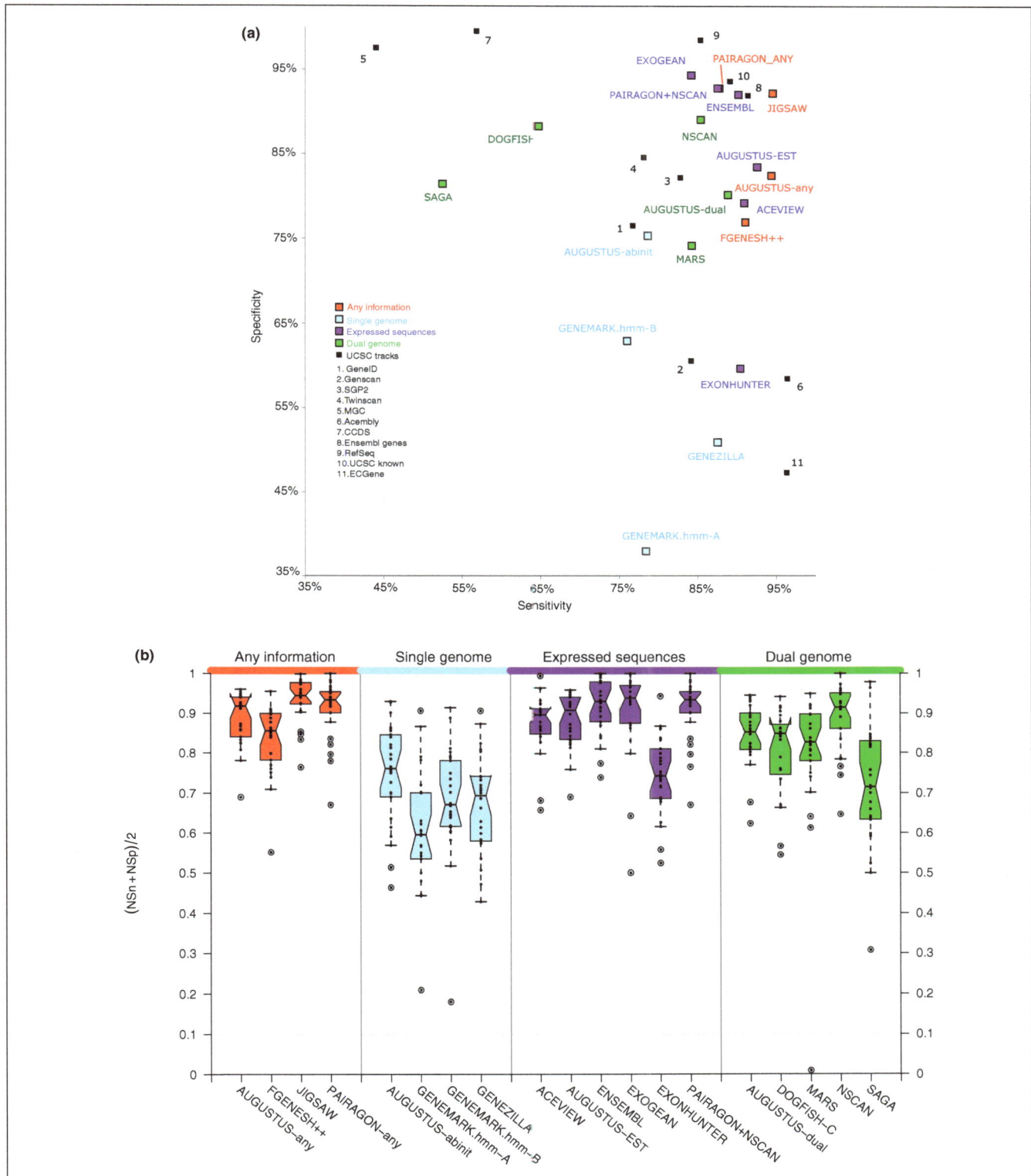

Figure 5
Gene Prediction Accuracy at the nucleotide level: Sensitivity versus specificity. Top panel: dotplot for sensitivity versus specificity at the nucleotide level for CDS evaluation. Each dot represents the overall value for each program on the 31 test sequences. Bottom panel: boxplots of the average sensitivity and specificity ((Sn + Sp)/2) for each program. Each dot corresponds to the average in each of the test sequences for which a GENCODE annotation existed (27 out of 31 sequences).

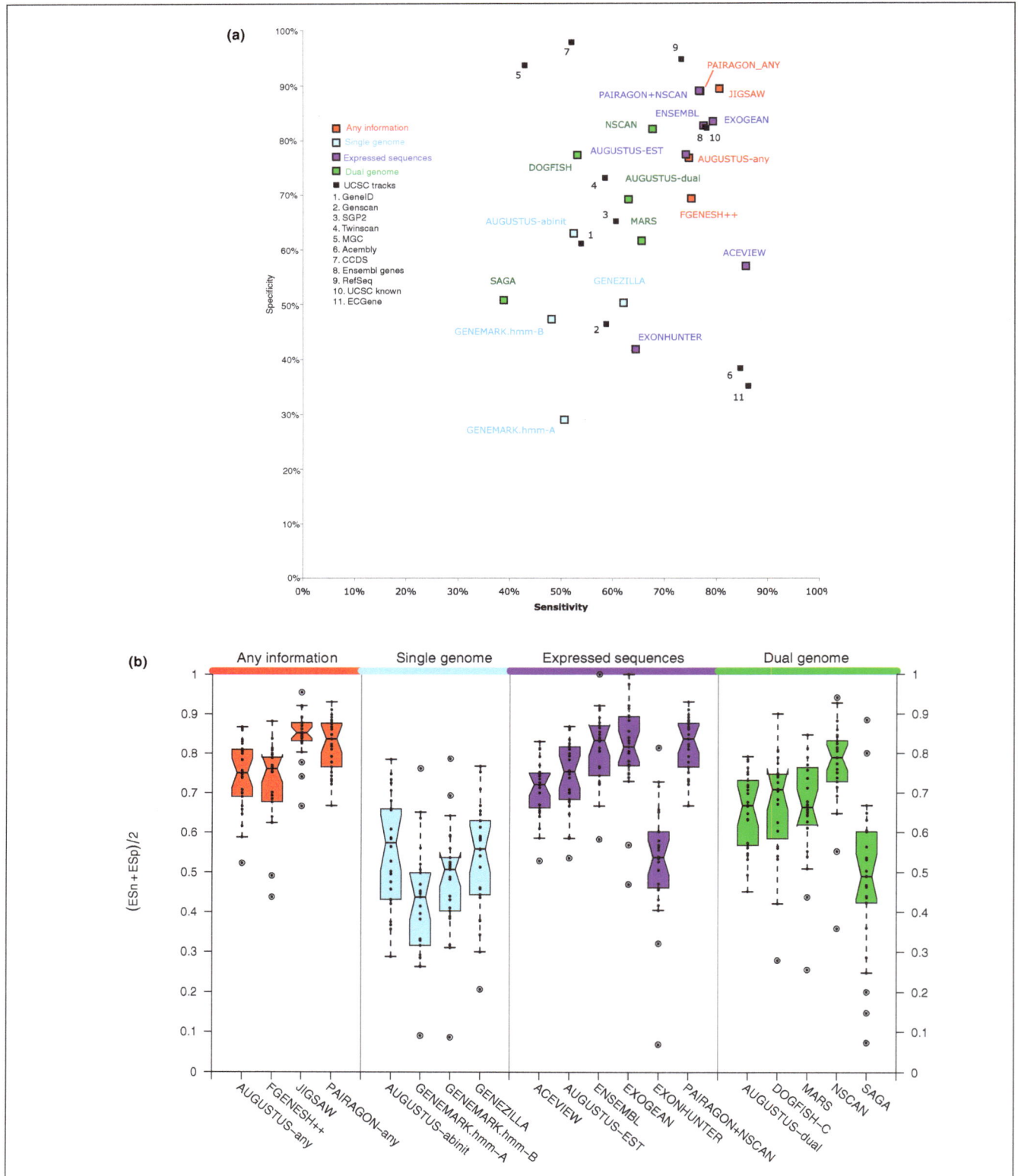

Figure 6
Gene Prediction Accuracy at the exon level: Sensitivity versus specificity. Top panel: dotplot for sensitivity versus specificity at the exon level for CDS evaluation. Each dot represents the overall value for each program on the 31 test sequences. Bottom panel: boxplots of the average sensitivity and specificity for each program. Each dot corresponds to the average in each of the test sequences for which GENCODE annotation existed.

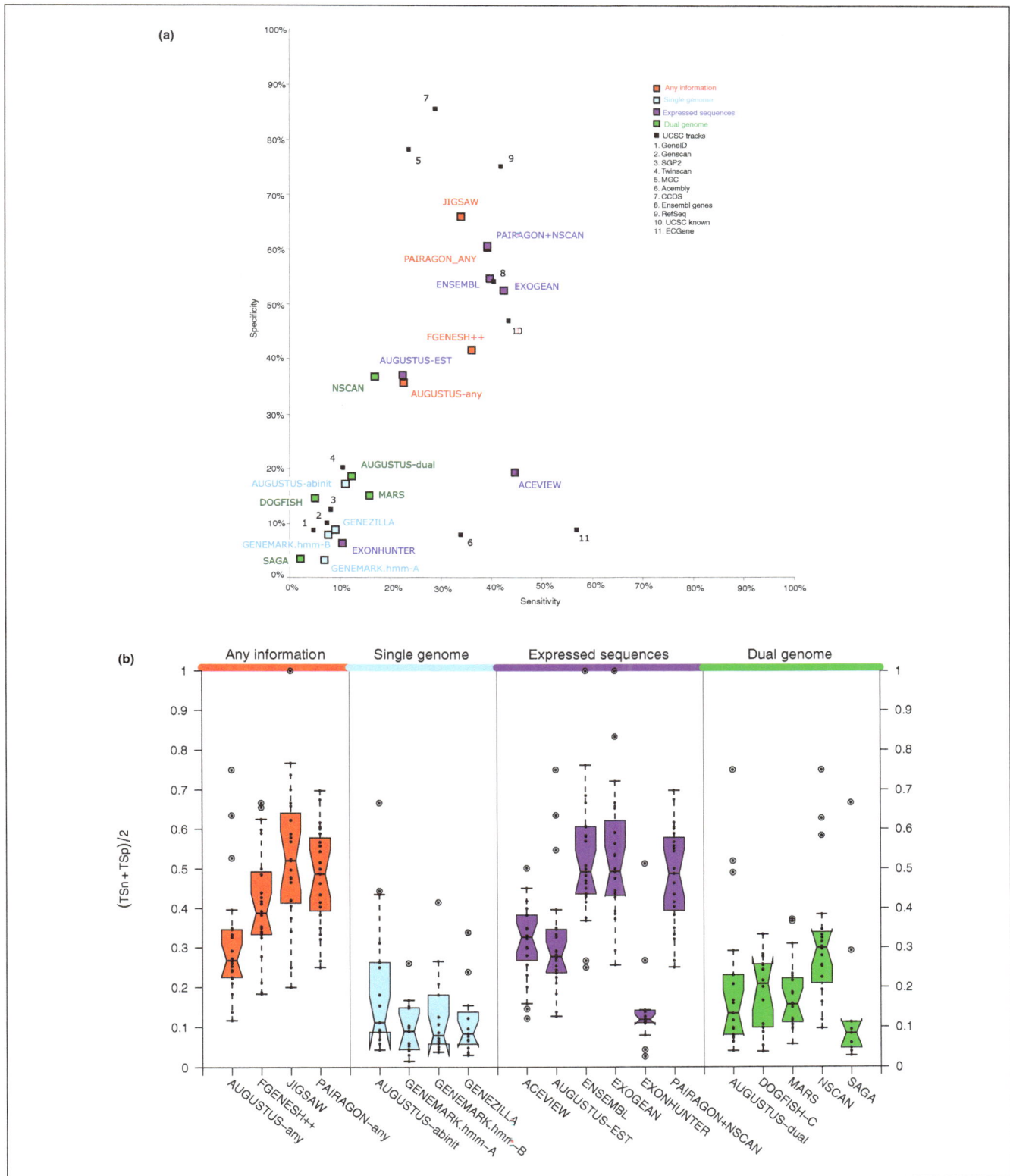

Figure 7
Gene Prediction Accuracy at the transcript level: Sensitivity versus specificity Top panel: dotplot for sensitivity versus specificity at the transcript level for CDS evaluation. Each dot represents the overall value for each program on the 31 test sequences. Bottom panel: boxplots of the average sensitivity and specificity for each program. Each dot corresponds to the average in each of the test sequences for which GENCODE annotation existed.

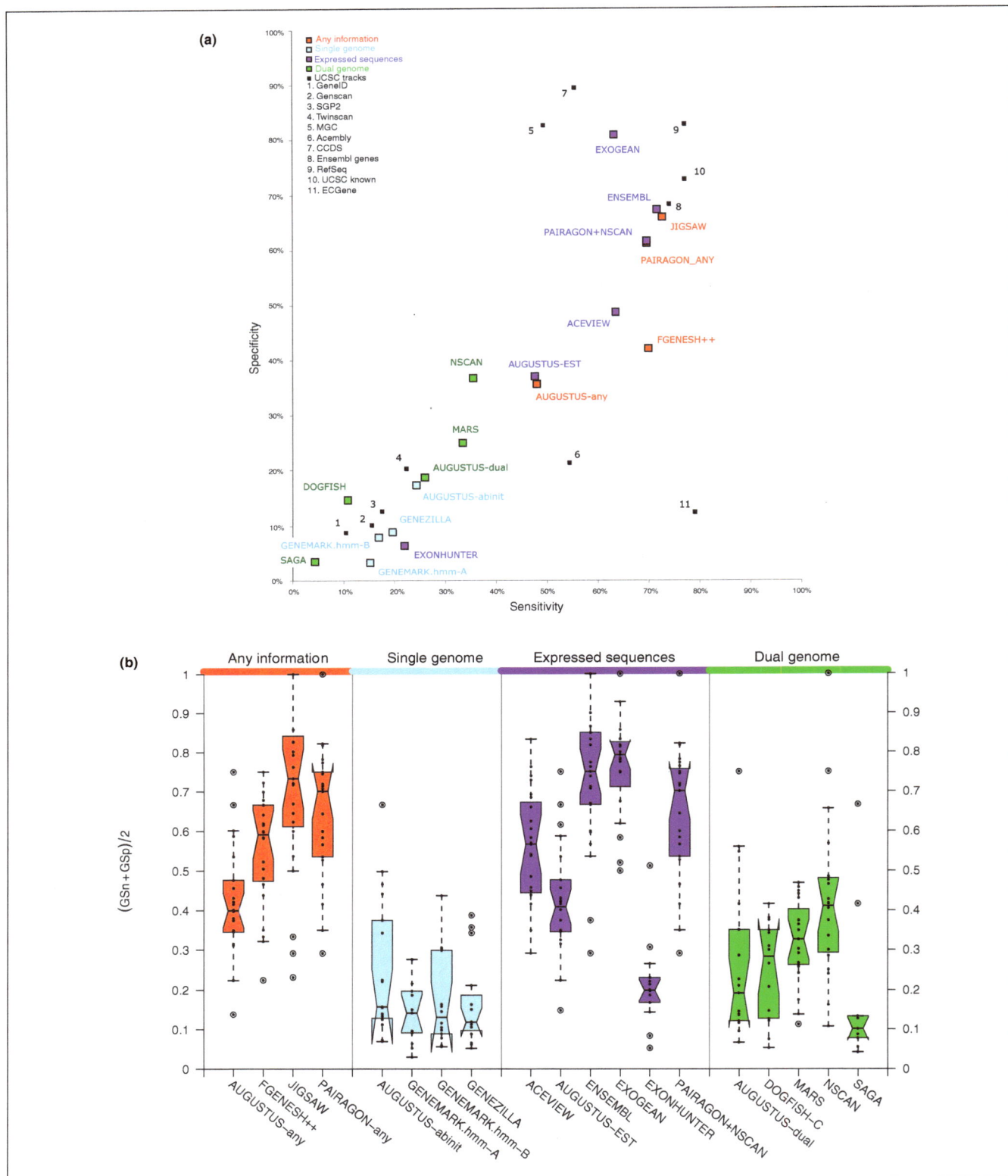

Figure 8
Gene Prediction Accuracy at the gene level: Sensitivity versus specificity. Top panel: dotplot for sensitivity versus specificity at the gene level for CDS evaluation. Each dot represents the overall value for each program on the 31 test sequences. Bottom panel: boxplots of the average sensitivity and specificity for each program. Each dot corresponds to the average in each of the test sequences for which GENCODE annotation existed.

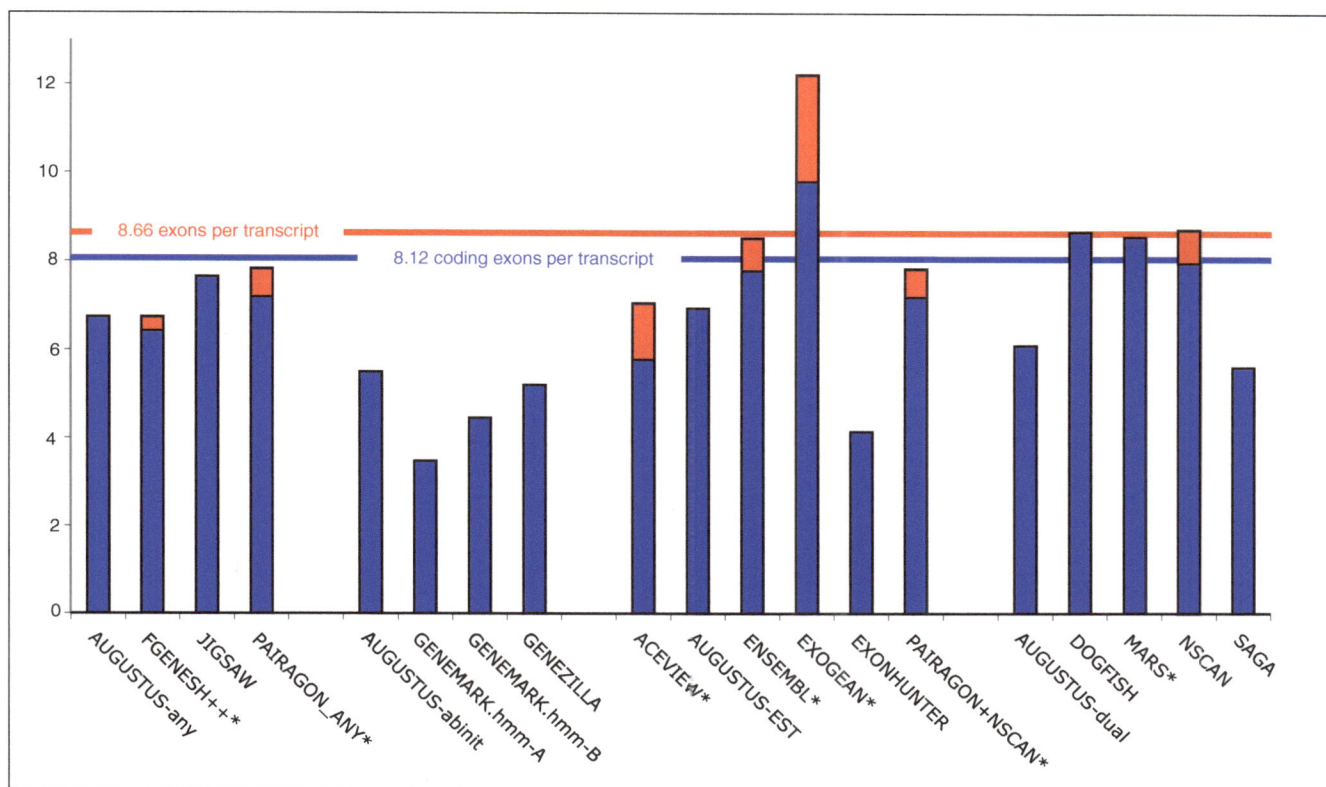

Figure 9
Exon counts per gene transcript. A comparison of the number of exons per transcript and coding exons per transcript in the GENCODE annotation of the 31 test regions and in the predictions. Blue bars show the average number of coding exons per coding transcript for each of the programs in categories 1, 2, 3, and 4; the blue line shows this for the GENCODE annotation. The number of all exons per transcript in the GENCODE annotation is shown with a red line. Those programs that predict non-coding exons are noted with red bars. Programs marked with an asterisk predict multiple transcripts per gene locus.

Most of the methods predict genes that, on average, have fewer coding exons per gene than the GENCODE annotation (Figure 9). The only exceptions to this observation are EXOGEAN, DOGFISH, and MARS, which all predict more coding exons than the annotation.

Prediction tracks from the UCSC browser were generally clustered near the EGASP entries for similar categories. At the transcript level, the BLAT aligned REFSEQ mRNAs ('REFgene') were both more sensitive than all of the prediction methods except EXOGEAN and ACEVIEW, and approximately 8% more specific than the best EGASP entries. The MGC transcripts ('MGCGene') and the CCDS transcripts ('CCDSgene') were 10% and 18% more specific at the transcript level, but had significantly lower sensitivity than the best EGASP method due to the incomplete nature of these sets at the time of the workshop.

In general, the accuracy of the programs varied substantially across test sequences, but some programs appear to behave more consistently than others (as is reflected in the boxplots in Figures 5-8).

Programs performed in general better in the training than in the test sequences, the two exceptions being ACEVIEW from category 3 and FGENESH++ from category 1 (Figure 10).

No overall trend was observed when comparing performance between manually placed ENCODE regions and the random ones (Figure 11). Even though, programs in category 4 performed consistently better in the random picks.

Programs performed clearly better in medium or high gene dense regions than in regions poor in genes (Figure 12). Only the category 6 method SPIDA had higher accuracy in regions of low gene density.

The accuracy of the programs was also related to the level of conservation of the genomic sequence in the mouse genome, with programs performing generally better in the test

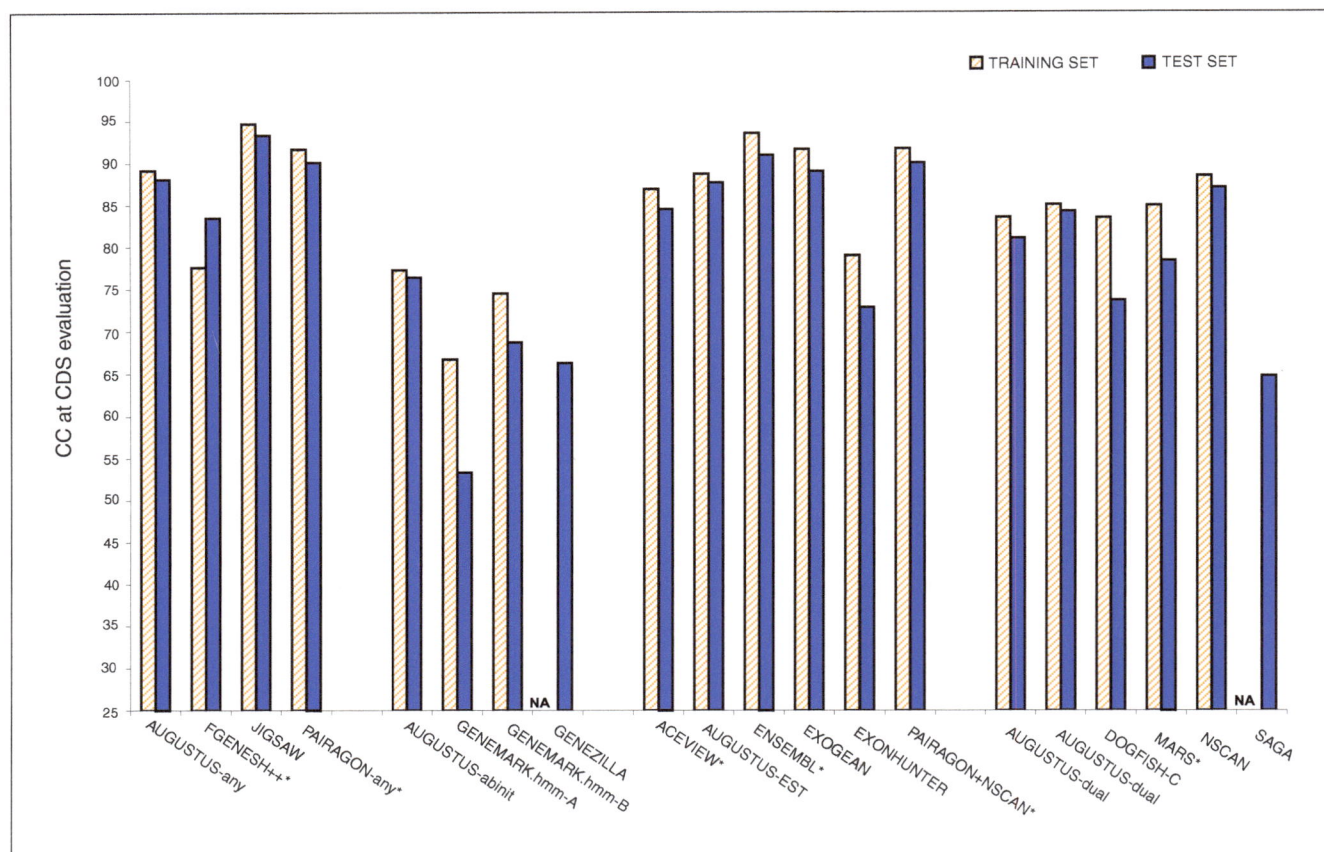

Figure 10
Correlation Coefficient Accuracy for Training and Test Sequences. The correlation coefficient (CC) at the nucleotide level for CDS evaluation for sequences EN_TRN13 and EN_PRD31 for training and test set sequences. NA, not available; because the submitters did not send their results for the training set.

sequences showing stronger conservation in the mouse genome, but the trend was not as strong as with gene density (Figure 13).

Results by category for the CDS evaluation
Category 1: methods using any type of available information
Four prediction methods were considered in EGASP category 1. Of these the FGENESH++ pipeline [56], the PAIRAGON-any pipeline [57], and AUGUSTUS-any [58] are conceptually similar. Each of these approaches uses information from both expressed sequences and from *ab initio* or *de novo* gene prediction strategies.

FGENESH++ and PAIRAGON-any consist of an alignment step followed by *de novo* prediction in the regions where there are not alignments. The sensitivity of these two methods is similar for all levels of the evaluation, but PAIRAGON-any is significantly more specific. AUGUSTUS-any uses both the 'hints' discovered in its expressed sequence (category 3) strategy and those discovered in its dual-genome (category 4) strategy.

Both the AUGUSTUS and the PAIRAGON groups submitted predictions in categories 1 and 3, allowing us to judge the value of the additional information that each of the programs used in producing the category 1 predictions. Neither program shows a significant increase in predictive performance in this category over their respective category 3 predictions (see below). For AUGUSTUS-any, this suggests that its models get very little additional information from the inclusion of the dual-genome prediction information. For PAIRAGON-any, the category 1 prediction set included only two transcripts not included in the category 3 prediction set (PAIRAGON+NSCAN).

JIGSAW [59] is unlike the other three methods. It uses a statistical combination of several sources of evidence to create the best consensus prediction. Considering all the evaluation measures, JIGSAW is the most accurate category 1 prediction method, although both PAIRAGON-any and FGENESH++ are more sensitive than JIGSAW at the transcript level. FGENESH++ and PAIRAGON-any predict multiple transcripts per gene locus.

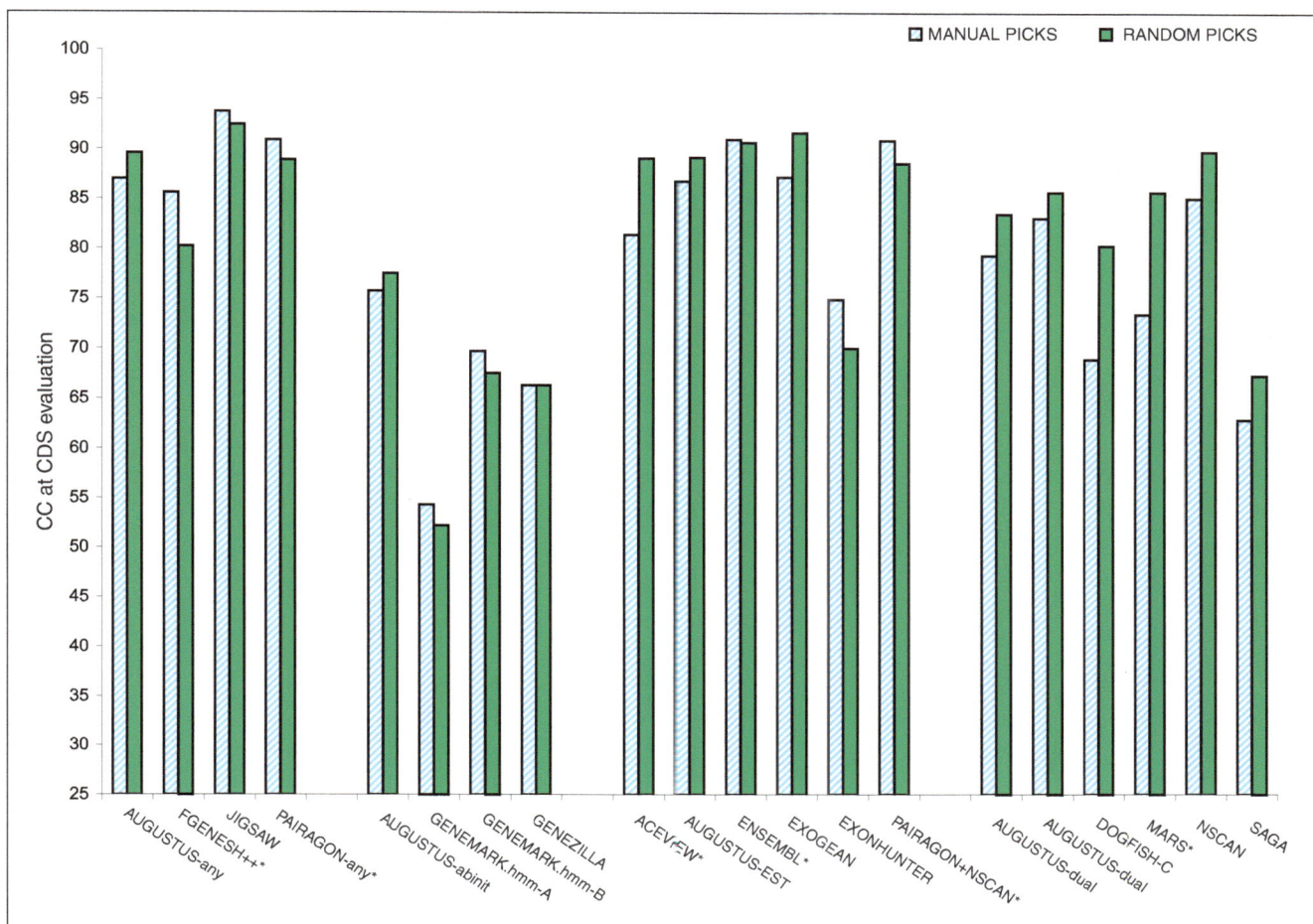

Figure 11
Correlation Coefficient Accuracy for manually and randomly selected Sequences. The correlation coefficient (CC) at the nucleotide level for CDS evaluation for EN_MNLp12 and EN_RNDp19 for manually and randomly selected sequences within the test set.

Category 2: single-genome ab initio methods

Three *ab initio* prediction methods use only the information found in the human genome sequence. All three methods only predict coding transcripts and are thus only considered by the CDS evaluation. Of the three, GENEZILLA is the most sensitive at the nucleotide and exon levels, while AUGUSTUS-abinit is the most specific. AUGUSTUS-abinit is consistently better than the other two at finding the start and end of translation and is thus both more sensitive and more specific at both the transcript and the gene level.

There are two variants of the predictions made by (the human genome version) of the GENEMARK.hmm program [60]. Data marked GENEMARK.hmm-A were produced and submitted prior to the deadline and inadvertently used unmasked genomic sequence (communication at the workshop by M Borodovsky). This is also the case for the GENEZILLA predictions in the single genome category, which were also created using unmasked sequence. There-

fore, we caution the direct comparison of GENEMARK.hmm-A and the GENEZILLA results to the results of the other programs, which in general used masked genomic sequence. It is well known that gene finding programs do worse on unmasked sequences due to the high 'protein-coding-like' content of repetitive elements, resulting in an increase of the number of false positive predictions [61]. Data marked GENEMARK.hmm-B were produced by the same human genome version of the GENEMARK.hmm algorithm run on the masked sequence (communication by M Borodovsky), although this was a post-deadline submission. It is clearly seen that the specificity values for GENEMARK.hmm are higher when run on masked sequence due to the significant decrease in the false positive rate.

Category 3: EST-, mRNA-, and protein-based methods
More submissions were received for category 3 than for any of the other categories and the type of expressed sequence information (EST, mRNA, protein sequence) varied among

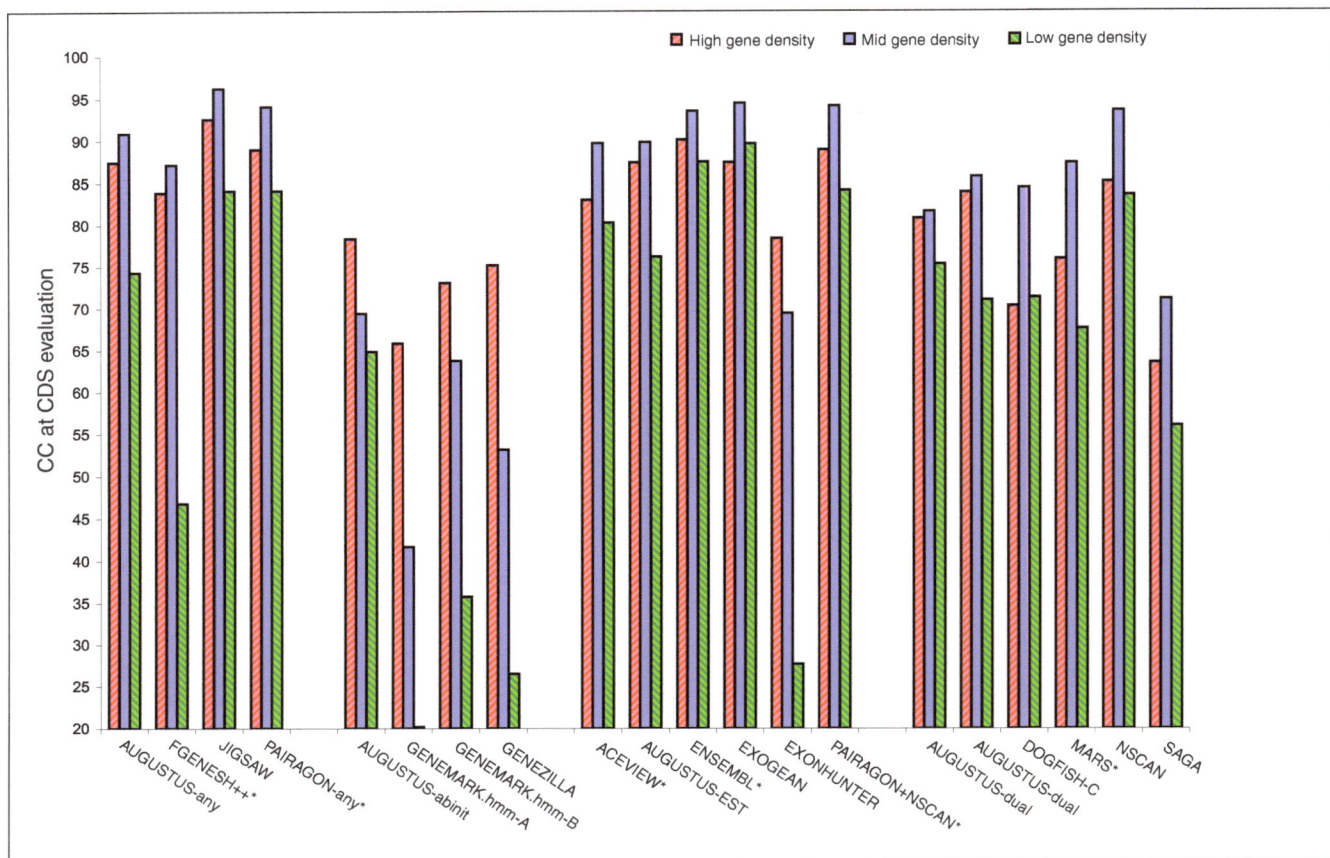

Figure 12
Correlation Coefficient Accuracy in relation to gene density. The correlation coefficient (CC) at the nucleotide level for sequences EN_PGH12, EN_PGM11 and EN_PGL8 for high, mid and low gene density sequence sets within the test set.

the methods, as did the strategy for incorporating the information. As such, it is not surprising that the methods have various strengths and weaknesses depending on the details of the method. For example, ACEVIEW [52] has the highest transcript sensitivity and predicts an average of 4.05 coding transcripts per gene locus. This is nearly twice as many transcripts per gene compared to EXOGEAN [62], which is nearly as sensitive (44.7% and 42.5%, respectively) and predicts only 2.34 coding transcripts per gene locus. ACEVIEW also has the highest coding exon sensitivity, but its high sensitivity comes at a cost of a relatively low specificity.

For the CDS evaluation at the nucleotide level, AUGUSTUS-EST [58] is the most sensitive program and EXOGEAN is the most specific. There is little distinction at the nucleotide level among most of the category 3 programs with the exception of EXONHUNTER [63], which seems to get less information from expressed sequences and scores significantly lower than the other programs.

At the coding exon level, the best programs (EXOGEAN, PAIRAGON+NSCAN_EST, and ENSEMBL) predict more

than 75% of the exons correctly, while maintaining specificity greater than 80%. Of these three, EXOGEAN is the most sensitive, and PAIRAGON+NSCAN_EST is the most specific. A similar story exists at the transcript level, where each of these 3 programs predicts more than 39% of the coding transcripts correctly, with specificity greater than 50%. Again EXOGEAN is the most sensitive (42.5% compared with 39.3% for PAIRAGON+NSCAN_EST and 39.8% for ENSEMBL) and PAIRAGON+NSCAN_EST is the most specific.

At the gene level, ENSEMBL [64] is more sensitive than PAIRAGON+NSCAN_EST (71.6% versus 69.6%) and more specific. EXOGEAN is the most specific program at the gene level at a specificity of 80.8% with a sensitivity of 63.2%.

Category 4: dual- or multiple-genome based methods
Six groups submitted gene structure predictions that were assigned to the dual-genome category. ACESCAN, however, submitted predictions only on the 13 training regions and was, therefore, not evaluated.

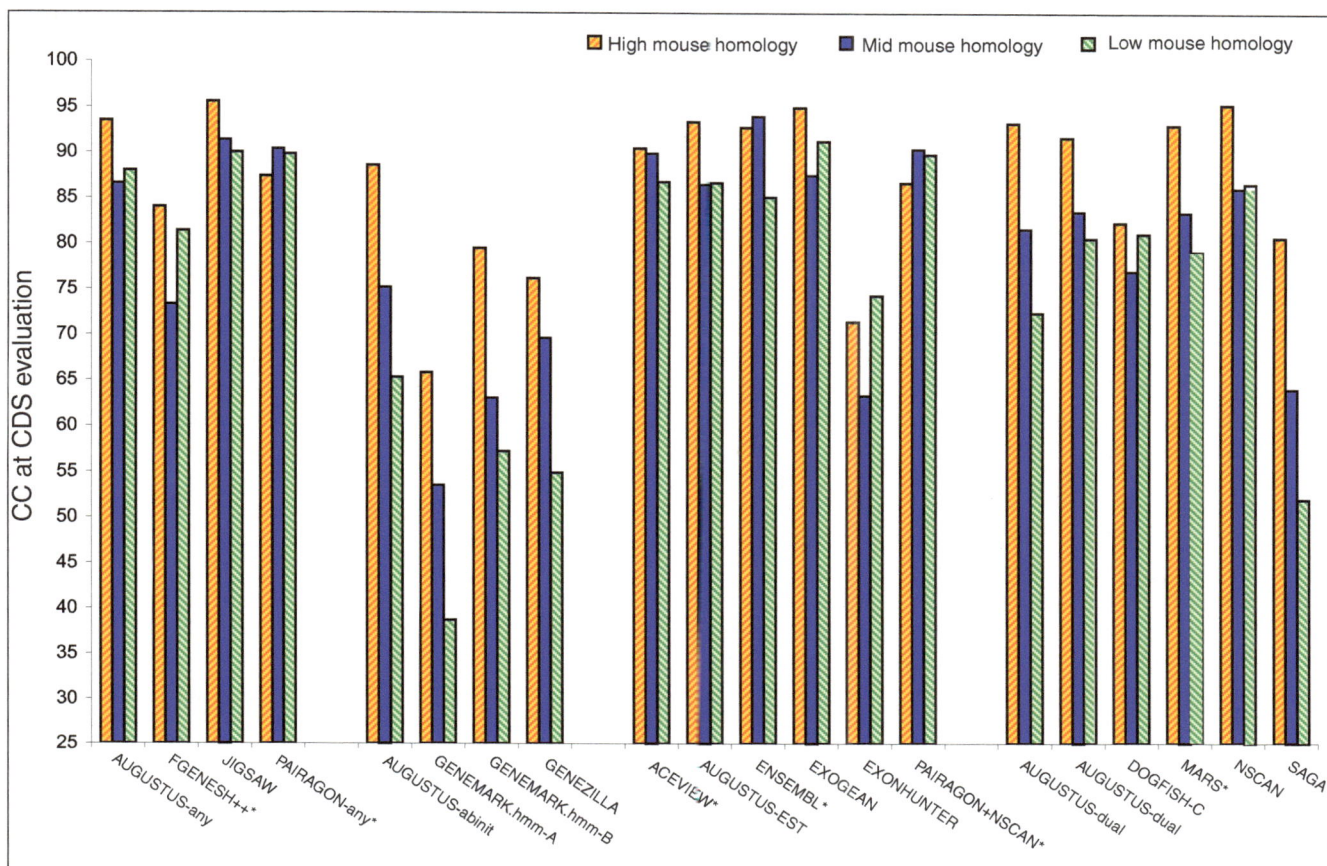

Figure 13
Correlation Coefficient Accuracy in relation to sequence conservation. The correlation coefficient (CC) at the nucleotide level for sequences EN_PMH7, EN_PMM5 and EN_PML7 for high, mid and low conservation with mouse sequences only for the randomly selected sequences in the test set.

Of the dual-genome prediction programs, NSCAN [57] is generally the most sensitive and the most specific for all evaluation levels. The only exception is at the nucleotide level, where AUGUSTUS-dual [58] is more sensitive (88.9% versus 85.4%) at a cost of being less specific than NSCAN (80.2% versus 89.0%). All of the dual-genome predictors except MARS [65] are limited to predicting one transcript per gene locus.

NSCAN is one of the most conservative of the dual-genome gene predictors, which partly explains its high transcript and gene specificity. It predicts approximately 90 fewer genes than SAGA [66], approximately 110 fewer than MARS, and almost 130 fewer than AUGUSTUS-dual. Only DOGFISH [67], which predicts 219 genes, is more conservative.

Other predictions
Two programs submitted predictions on the test regions for category 5 (methods predicting unusual genes, non-canonical splicing, short intronless genes, and so on). Both GENEID-U12 and SGP2-U12 (T. Alioto, unpublished) are optimized to find genes that contain U12 introns (see Patel and Steitz [68] for an in-depth review on U12 splicing).

Six programs submitted predictions that were included in category 6 (exon only predictions). ASPIC predicted only introns for the training regions, CSTMINER predicted coding regions, but did not provide strand information or splice site boundaries, DOGFISH-C-E and UNCOVER predicted only novel exons, and AUGUSTUS-exon and SPIDA predicted exons but they did not attempt to link them into transcript structures.

The programs in categories 5 and 6 have very specialized and diverse goals and cannot easily be compared to each other or to the predictions in other categories. Their accuracy values, however, have been computed when possible, and they are provided in the Supplementary material.

Results for the mRNA evaluation
In the computational gene finding literature, gene predictions have traditionally been evaluated using coding transcripts only. That is, only the exonic structure of the coding fraction of the gene or transcript is taken into account both in the prediction and in the annotation. One reason for this has been the difficulty of experimentally

determining 'full length' cDNAs, which represent a full mRNA transcript. While it is difficult to accurately clone and sequence the 3' UTRs of cDNA clones, it is even harder to obtain and sequence the 5' UTRs of a gene transcript. Besides the limitation of existing experimental data, very little signal information exists in the sequence of 5' and 3' UTRs of genes that can be statistically modeled. Therefore, most of the computational gene finders have historically made no attempt to predict UTRs, and instead predicted genes from the start codon to the stop codon.

Apparently, encouraged by the announcement to explicitly try to "replicate the GENCODE annotations", which included many full mRNA transcript annotations in the training set, several programs submitted predictions of the entire exonic structure of the mRNA molecules. FGENESH++, PAIRAGON+NSCAN_EST, ACEVIEW, ENSEMBL, EXOGEAN, and NSCAN programs all submitted full transcript predictions, including coding and untranslated (UTR) exons. We have compared these predictions with the annotated exonic structure of the mRNA transcripts within the GENCODE annotation. Accuracy results for the mRNA evaluation of these programs are given in Table 6.

In general, programs performed worse when predicting the exonic structure of the entire transcript than when predicting only the coding exons. This is consistent with the fact that the UTR sequences are less constrained than regions coding for amino acid sequences. Note, however, that the 3' and 5' end of the genes are particularly difficult to delineate experimentally. Therefore, a metric that emphasizes prediction of exact exon boundaries will lead to an underestimation of the accuracy of the predictions. Evaluation of the predictions at the intron level, instead of exon level, could partially address this limitation. In any case, given these limitations, ACEVIEW exhibits the highest accuracy of mRNA evaluations and has similar accuracy, at least at the nucleotide level, when considering either the entire mRNA or the CDS. In contrast to other programs, ACEVIEW is more specific in the entire mRNA than on the CDS. It also has the highest sensitivity, although ENSEMBL, EXOGEAN and PAIRAGON+NSCAN_EST are more specific.

Interpreting the results
The 44 ENCODE regions represent 30 Mb (approximately 1%) of the human genome. The 31 EGASP test regions include 21.6 Mb and represent an even smaller fraction of the human genome. Although this is the largest region ever used for benchmarking automatic genome annotation, it is not a random selection of the human genome, and, therefore, results obtained in them should only be extrapolated to the whole genome with appropriate caution. The stratification of the ENCODE regions into 'manually' versus 'randomly' selected and according to gene density and conservation with mouse (Table 1) allows for an investigation into how these factors affect the accuracy of gene predictions.

Figure 14a displays the accuracy of each program (average sensitivity and specificity at the nucleotide level) in the form of boxplots for each individual sequence in which genes were annotated (27 of the 31 EGASP regions) and for the collections of sequences discussed next (random versus manual, training versus test, low, medium and high gene density, and low, medium and high conservation with the mouse genome). The accuracy of the programs varied substantially across sequences (with a median value raging from values below 0.7 (ENm011) to above 0.95 (ENr332). Figure 14b shows similar results but on the exon level. In what follows, we describe potential biases in the evaluation results that can be explained by the characteristics of the raw sequences, instead of the behavior of the prediction methods.

Training versus testing regions
We compared the predictive accuracy of each of the programs on the set of 13 training regions to their performance on the 31 test regions. Most of the gene prediction programs were more accurate on the training set compared to the test set (Figure 10). This can be partially explained by the training set being enriched in gene dense regions (see the section Gene rich versus gene poor regions below; Table 1). Indeed, 11 of the 13 training regions (85%) had a high or medium gene density, compared with 23 out of the 31 test regions (74%).

Random versus manual regions
Within the test set, we compared the performance of each of the gene prediction programs on the set of 12 manually placed ENCODE regions to their performance on the set of 19 ENCODE regions chosen randomly (Figure 11). Some programs performed better in the manual regions, while others did on the random ones, but no overall trend could be observed. Only programs in category 4 (dual- or multiple-genome predictors) performed consistently better in the random than in the manual picks. One possible explanation for this might be that the GENCODE annotation is more exhaustive in the regions selected manually. These regions contain genes of interest, and some of them have been extensively investigated. Therefore, the coverage by cDNAs - on which the GENCODE annotation is based - is likely to be higher in the manual than in the randomly chosen regions, which might explain the difference in class performance.

Gene rich versus gene poor regions
We compared the performance of the prediction programs based on the stratification of high (12 sequences), medium (11 sequences) and low (8 sequences) gene dense regions (see the data description in the section Description of the sequence above). In general, all programs performed better in regions with medium or high gene density than in regions with low gene density (Figure 12). This reflects the low specificity resulting from a higher rate of false positive predictions. Interestingly, single genome *ab initio* gene finders (category 2) performed the best in very gene rich

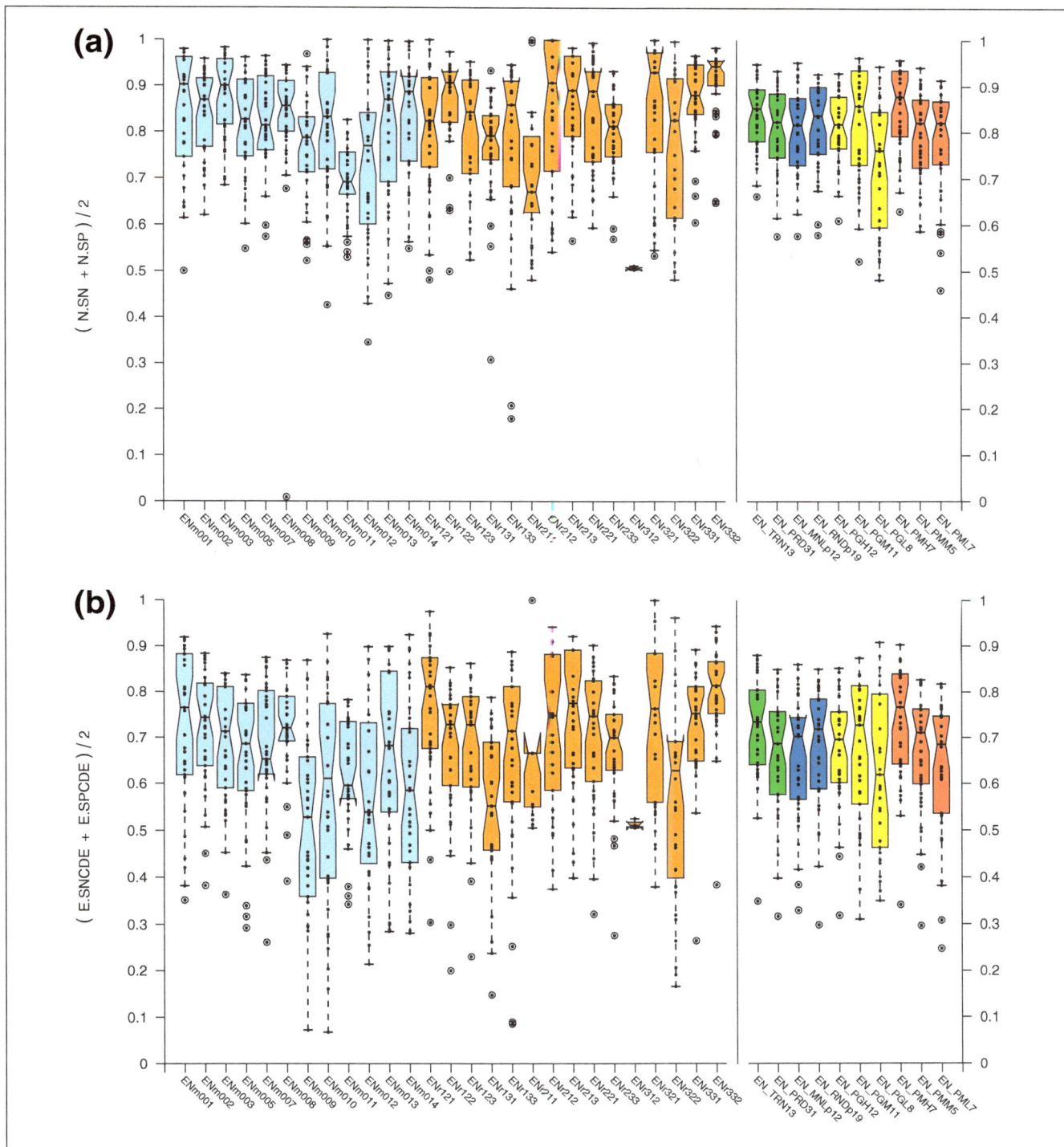

Figure 14

Gene Prediction Accuracy for each ENCODE sequence at the nucleotide and exon levels. Boxplots showing the average sensitivity and specificity at the **(a)** nucleotide level and **(b)** exon level for CDS evaluation of each program on every sequence of the test set. Sequences are displayed across the x-axes. Manual picks are shown in in light blue; random picks are shown in orange. Boxplots corresponding to the overall average sensitivity and specificity at the nucleotide level for CDS evaluation in different subsets of the ENCODE sequences are shown at the right of the graph. EN_TRN13, the set of 13 training regions, and EN_PRD31, the set of 31 test regions, are shown in green. EN_MNLp12, the 12 manual picks in the test set, and EN_RNDp19, the 19 random picks in the test set are shown in dark blue. EN_PGH12/EN_PGM11/EN_PGL8, the subsets of 12 high, 11 medium and 8 low gene dense sequences from the set of test sequences, are shown in yellow. EN_PMH7/EN_PMM5/EN_PML7, the subsets of seven regions with high sequence conservation with mouse, five regions with medium conservation, and seven regions with low conservation from random picks in the test set, are shown in red.

regions, while the programs in all other categories performed the best in regions with an intermediate density of genes. This is possibly due to the training of the programs tuned to balance over- and under-prediction.

High versus low conservation with mouse regions
We also compared the performance of the prediction programs in the randomly selected sequences with high (seven sequences), medium (five sequences) and low (seven sequences) conservation with mouse (see the data description in the section Description of the sequence above). Programs performed generally better in sequences showing higher conservation with mouse, but the trend was not as strong as with gene density (Figure 13). As expected, dual genome predictors performed better in sequences with high mouse homology, but the trend was also observed for single genome predictors. It is possible to speculate that genes conserved across species are also likely to exhibit more typical sequence characteristics, in terms of codon bias and splice site signals, whereas fast evolving genes may have undergone changes toward unusual sequence biases. Single genome predictors are likely to perform better on genes exhibiting typical features.

The experimental test of unannotated predictions
The second major goal of EGASP was to assess the completeness of the GENCODE annotation. This annotation is based on available evidence, and we cannot immediately rule out the possibility that it misses a fraction of protein coding genes. Indeed, many predictions submitted for EGASP did not match any GENCODE annotated exon. Table 7 lists the total number of unique coding and non-coding exons predicted by each program, as well as the number of unannotated exons (that is, predicted exons not overlapped by a GENCODE annotated exon by 1 bp or more). Also listed is the number of unannotated exons predicted in intergenic regions. The unannotated are termed 'wrong exons' in the evaluation section above. It is unclear what fraction of these unannotated exons belong to annotated genes or are exons of novel, unannotated genes. We have carried out an initial investigation by comparing the unannotated exons with transcript data obtained from the hybridization of polyA+ cytosolic RNA onto Affymetrix high-density genome tiling microarrays covering the ENCODE regions. Details of the technology and applications have recently been published [69,70]. Briefly, positive hybridization probes are combined into discrete sites of transcription, which are usually known as TARs (transcriptionally active regions) or transfrags (transcribed fragments). It is important to note that identification of TARs/transfrags is based on selecting a threshold of detection that is derived from estimates of 5% false positive detection using a bacterial sequence spike in controls [71]. Raising or lowering the false positive rates can alter these thresholds. More or fewer detected regions of transcription will likely follow from these changes. TAR/transfrag maps corresponding to different cell lines

and conditions have been downloaded from the ENCODE specific UCSC browser [35]. Table 7 lists the number of predicted unannotated exons that overlap TARs/transfrags by at least 1 bp. Overall, 44.7% of the EGASP predicted exons overlap TARs/transfrags. Importantly, while 71% of the annotated exons overlap TARs/transfrags, only 13% of the unannotated exons do. This difference suggests that many of the predicted, but not annotated, exons are false positives.

Support by TARs/transfrags, however, indicates only that the predicted exon appears to be transcribed and, possibly, processed into an RNA sequence. It does not allow us to infer that the predicted exons assemble into the predicted transcript structure, or that the transcript is a protein-coding RNA. Therefore, to better assess the likelihood of the predicted exonic structures, we selected a subset of the predicted but unannotated exon pairs to be tested experimentally by RT-PCR. We focused our verification efforts on the subset of 8,634 intergenic exons (Table 7), since these predictions could correspond to yet undetected, novel genes. We ranked the predictions based on the predictive specificity at the exon level for the given programs and then selected the top 200 ranking exons. We next identified all predicted introns (exon pairs) radiating from this set of exons. That is, we paired each of these exons with its immediate upstream and downstream neighbors within the same predicted transcript. Selection of those pairs not overlapping any GENCODE annotation resulted in 238 unique non-inclusive exon pairs (pairs in which one of the exons was included in an exon from another selected exon pair were discarded). Of these pairs, 221 could be tested by RT-PCR (see Materials and methods) in 24 tissues. All data files can be accessed through the Supplementary material web page. Of the assayed exon pairs, only seven (3.2%) produced a positive result, all with perfectly predicted exon boundaries. Of the seven validated exon pairs, three were intergenic, presumably representing new transcribed loci, while four extended existing gene annotations. Every positive case was expressed in only a single tissue out of the 24 tested. This result is comparable to that obtained for novel human genes identified using the chicken genome as reference (expressed on average in 3.3 tissues) [72] or for the recently described chimeric transcripts (expressed on average in 2.5 tissues) [73]. This result is significantly below the 7 to 8 average positive tissues out of 12 tested found for known mammalian genes [14,74], suggesting that the majority of yet unannotated genes have a restricted pattern of expression. This also suggests an explanation for why these transcripts have eluded identification by experimental means until now.

The number of exon pairs (introns) tested per program, and the number of positive verifications are given in Table 8 (see also the Supplementary material web page for information about the positive cases [51]). There appear to be differences in the success rate by program, but the numbers are too small to draw significant conclusions. Interestingly, the

Table 7

Exons predicted by the programs not overlapping GENCODE annotated exons, and supported by transfrag evidence from genome tiling microarrays

	Total number of unique exons	Number of exons overlapping TARs/transfrags (%)	Number of non-annotated exons	Number of non-annotated exons overlapping TARs/transfrags (%)	Number of intergenic exons	Number of intergenic exons overlapping TARs/transfrags (%)
Category 1						
AUGUSTUS-any	4,160	2,718 (65.3%)	484	74 (15.3%)	281	38 (13.5%)
FGENESH++	4,784	2,766 (57.8%)	1,071	146 (13.6%)	885	123 (13.9%)
JIGSAW	3,935	2,673 (67.9%)	206	34 (16.5%)	130	19 (14.6%)
PAIRAGON-any	4,414	3,080 (69.8%)	284	84 (29.6%)	221	68 (30.8%)
Category 2						
AUGUSTUS-abinit	3,699	2,336 (63.2%)	776	175 (22.6%)	482	111 (23.0%)
GENEMARK.hmm	6,897	2,552 (37.0%)	3,796	319 (8.4%)	2,826	244 (8.6%)
GENEZILLA	3,415	1,535 (44.9%)	1,361	110 (8.1%)	970	69 (7.1%)
Category 3						
ACEVIEW	8,410	5,756 (68.4%)	539	118 (21.9%)	449	106 (23.6%)
AUGUSTUS-EST	4,073	2,700 (66.3%)	439	73 (16.6%)	253	37 (14.6%)
ENSEMBL	4,505	3,094 (68.7%)	251	27 (10.8%)	187	17 (9.1%)
EXOGEAN	5,014	3,480 (69.4%)	183	21 (11.5%)	100	15 (15.0%)
EXONHUNTER	6,376	2,843 (44.6%)	2,782	257 (9.2%)	2,055	187 (9.1%)
PAIRAGON+NSCAN_EST	4,404	3,073 (69.8%)	284	84 (29.6%)	221	68 (30.8%)
Category 4						
AUGUSTUS-dual	4,024	2,588 (64.3%)	629	114 (18.1%)	364	65 (17.9%)
DOGFISH	3,194	2,290 (71.7%)	267	115 (43.1%)	225	99 (44.0%)
MARS	4,623	2,801 (60.6%)	948	161 (17.0%)	528	89 (16.9%)
NSCAN	3,996	2,686 (67.2%)	500	133 (26.6%)	342	92 (26.9%)
SAGA	2,115	1,147 (54.2%)	564	36 (6.4%)	433	26 (6.0%)
All unique exons (18 progs)	26,818	12,001 (44.7%)	12,023	1,563 (13.0%)	8,634	1,163 (13.5%)

positive predictions tended to be classified high in our ranking based on the specificity of the programs: 3 out of the 7 positive predictions ranked among the top 50 ranked predictions, and 6 ranked among the 100 top ranking predictions. This suggests that combining multiple sources of evidence helps to identify the computational predictions that correspond to 'bona fide' genes. Consistent with these observations, TARs/transfrags overlap with about 20% of the exons classified among the 200 top ranking ones, but only with 13% of the intergenic predicted exons overall. Two of the positive predictions had already been included in later releases of the GENCODE annotation, but were unknown at the time of experimental verification. Two are extending 'putative' GENCODE loci, and three could correspond to novel genes - one being antisense to an annotated GENCODE locus. In the GENCODE annotation, the sequence of the

RT-PCR products is passed back into the GENCODE pipeline, where it is used as another source of transcript sequence evidence. Future versions of GENCODE will incorporate the validated computational predictions.

Since TAR/transfrag support has not been used to prioritize predicted exons for experimental verification, it is possible to investigate whether the results of the RT-PCR experiments are consistent with the TAR/transfrag data, and whether these data can be used to prioritize verification experiments. For example, one would expect the likelihood of RT-PCR success to be higher when the two predicted exons to be tested are both supported by TARs/transfrags from the same cell line and condition, suggesting that the two exons are connected into the same RNA sequence. In only seven of the 221 exon pairs tested were the two exons

Table 8

Number of exon pairs (introns) tested per program, and the number of positive verifications

	Number of tested exon pairs	Number of positive RT-PCR exon pairs and % over tested	Number of tested exon pairs suported by TARs/transfrags and % over tested pairs	Number of positive RT-PCR pairs supported by TARs/transfrags and % over supported pairs	% of positive RT-PCR pairs supported by TARs/transfrags over positive
Category 1					
AUGUSTUS-any	31	2 (6.5%)	3 (9.7%)	1 (33.3%)	50.0%
FGENESH++	119	3 (2.5%)	3 (2.5%)	1 (33.3%)	33.3%
JIGSAW	19	2 (10.5%)	3 (15.8%)	1 (33.3%)	50.0%
PAIRAGON-any	1	0 (0.0%)	0 (0.0%)	0 (0.0%)	0.0%
Category 2					
AUGUSTUS-abinit	29	2 (6.9%)	2 (6.9%)	1 (50.0%)	50.0%
GENEMARK.hmm	99	1 (1.0%)	3 (3.0%)	1 (33.3%)	100.0%
GENEZILLA	34	2 (5.9%)	1 (2.9%)	0 (0.0%)	0.0%
Category 3					
ACEVIEW	13	4 (30.8%)	2 (15.4%)	2 (100.0%)	50.0%
AUGUSTUS-EST	31	2 (6.5%)	3 (9.7%)	1 (33.3%)	50.0%
ENSEMBL	10	1 (10.0%)	2 (20.0%)	1 (50.0%)	100.0%
EXOGEAN	18	1 (5.6%)	3 (16.7%)	1 (33.3%)	100.0%
EXONHUNTER	23	1 (4.3%)	2 (8.7%)	0 (0.0%)	0.0%
PAIRAGON+NSCAN_EST	1	0 (0.0%)	0 (0.0%)	0 (0.0%)	0.0%
Category 4					
AUGUSTUS-dual	26	1 (3.8%)	3 (11.5%)	1 (33.3%)	100.0%
DOGFISH	11	2 (18.2%)	1 (9.1%)	1 (100.0%)	50.0%
MARS	47	7 (14.9%)	5 (10.6%)	3 (60.0%)	42.9%
NSCAN	26	0 (0.0%)	2 (7.7%)	0 (0.0%)	0.0%
SAGA	9	0 (0.0%)	1 (11.1%)	0 (0.0%)	0.0%
All unique exon pairs	238	7 (2.9%)	7 (2.9%)	3 (42.9%)	42.9%

The percentage of success has been computed in the table on the 238 selected exon pairs. For technical reasons, only 221 of them could be tested by RT-PCR. In the text the percentages are given with respect to this number.

supported by TARs/transfrags from the same cell line. Interestingly, three of these cases were positive by RT-PCR. This is a success rate of 43%, compared with 4 successful RT-PCRs out of 214 exons not having consistent transfrag support (less than 2% success rate). While the numbers are too small for significant conclusions, the trend is quite striking: consistent transfrag support of computational predictions is strongly indicative of RT-PCR success. Conversely, the reasons why exon pairs fail RT-PCR verification when supported by consistent transcription evidence from the same cell line and condition are multiple. Depending on the primers chosen, for instance, wrong prediction of the exon boundaries, even by a small offset,

may lead to failed RT-PCR amplification. Moreover, TAR/transfrag maps have been obtained from cell lines different from the tissues used for RT-PCR. Given the extremely restricted expression pattern that these novel transcripts appear to show, transcripts expressed in one given cell line may not be expressed in any of the 24 tissues analyzed. In this regard, it is interesting to note that the four negative RT-PCR exon pairs cluster into a single locus, and even share some sequence (see the Supplementary material web page), and, therefore, may represent the same transcript, while the three other transfrag-supported positive RT-PCR exon pairs correspond to three distinct loci mapped to three different ENCODE regions.

Discussion

The unfolding of the instructions encoded in the DNA sequence is initiated by the transcription of DNA to RNA, and the subsequent processing of the primary transcript to functional RNA sequences. According to the central dogma, most of these processed RNAs correspond to mRNAs that are eventually translated to proteins. Despite the fact that the identification of the protein-coding mRNAs (or genes) is essential for our understanding of how the genome sequence translates into biological phenomena, uncertainty still remains with respect to the set of human genes. The lack of an accurate and complete gene catalogue undermines the impact of the genome sequence on human biology and bio-medical research. Experimental determination of expressed mRNA sequences and computational mapping of this sequence onto the sequence of the genome constitutes the most reliable approach to identify the exonic structure, and chromosomal location, of protein-coding genes. However, this approach has limitations. First, it is unclear what fraction of low and specifically expressed transcripts can be effectively sequenced, and high throughput mRNA sequencing often leads to only partial sequences. Second, computational mapping of mRNA to genomic sequences is not trivial, and it is complicated by fragmentary mRNA sequences, sequencing errors, sequence polymorphism, and the highly repetitive nature of the human genome. Moreover, the high pseudogene content of the human genome, and the presence of small exons, leads to uncertain or incorrect mapping of exon boundaries. Therefore, substantial manual intervention is required to delineate an accurate protein coding gene map from the available mRNA sequence data.

We organized EGASP as a community experiment with the goal of assessing the ability of computational methods to automatically reproduce the accurate protein-coding gene map produced by a team of expert human curators. Such a map [33], subsequently verified experimentally, has been obtained for only 1% of the human genome selected by the ENCODE project [30]. Scaling the map to the entire human genome will require substantial additional resources, and it will enormously benefit from improved computational strategies for gene finding. With its focus on this 1% of the human genome, EGASP has indeed demonstrated progress in the performance of newly developed computational gene finding pipelines, with accuracies of about 80% at the coding exon level for both sensitivity and specificity, and of nearly 90% at the coding nucleotide level (Table 4). However, the success of these metrics is significantly tempered by the relatively low numbers of coding transcripts that are predicted correctly. Programs relying on mRNA and protein sequences were the most accurate in reproducing the manually curated annotation. This is not unexpected, and, to some extent, circular, since the manually curated annotation relies on mRNA and protein sequences as well. Notably, however, programs based on sequence comparisons across two or

more genomes - which do not use information from known mRNA or protein sequences - also exhibited impressive accuracy at the nucleotide and exon levels (Table 6). Dual genome prediction programs, however, were significantly less accurate at finding complete genes than the expressed sequence based methods. Finally, with few exceptions, all of the methods struggled to predict correctly the non-coding exons of transcripts. Indeed, UTRs are often predicted as mere extensions of first and terminal exons, if predicted at all. Thus, while the computational methods are quite reliable in predicting the protein coding components of transcripts, they have difficulties in linking them into transcript structures. Indeed, the most accurate programs were only able to correctly predict about 40% of the annotated transcripts, meaning the correct prediction of all of the exons constituting a transcript (Table 5). The results of coding gene predictions were more encouraging. For up to 80% of human genes the exact structure of the coding part, including all the splice junctions and start/stop codons, could be predicted correctly in at least one transcript.

Contributing to the difficulty is the unexpected complexity of the protein coding loci in higher eukaryotic genomes. Indeed, as revealed in the GENCODE annotation, most protein coding loci appear to encode a mixture of coding and non-coding transcripts, sharing part of their sequence. Additional transcriptional activity, including chimeric, overlapping and antisense transcripts, transcripts within introns, and other transcriptional phenomena, appear to be less exceptional than had been previously suspected. Thus, the model of a eukaryotic gene currently implicit in most computational methods is too simple to capture this complexity, leading to relatively poor prediction performance.

The second goal of EGASP was to assess the completeness of the manual/computational/experimental GENCODE annotation. This annotation is based on available evidence, and thus may miss some protein coding genes and exons. Indeed, in EGASP, computational methods predict many exons and transcripts that are not included in the GENCODE annotation (Table 7), a trend accentuated in *ab initio* and comparative gene finders, which do not rely on available evidence from transcript sequences. While we were not able to confirm experimentally the bulk of these predictions and they are likely to be false positives, some might be real.

To assess what fraction of the predicted exons unannotated in GENCODE could correspond to novel genes, we prioritized - based on the reliability of the programs predicting them - a subset of intergenic predicted exon pairs, and attempted to experimentally verify them by RT-PCR in 24 human tissues. Only 3.2% of these pairs tested positive, a result consistent with most of the computational predictions outside of GENCODE being false positives. All verified cases tested positive in only one tissue among the 24 tested, emphasizing the extremely restricted expression patterns of these novel, unannotated exons. Since many more tissues and cell lines exist, it cannot be ruled out that some other predictions could also be positive in other tissues. Support for a larger fraction of predictions corresponding to real exons comes from the observation that 13% of these predictions overlap sites of transcription (or TARs/transfrags) as detected by genome tiling experiments. Interestingly, the success rate of RT-PCR was much higher (at least 40%) for those few tested exon pairs that both overlapped TARs/transfrags and were detected in the same cell line and condition. Thus, consistent TAR/transfrag support is strongly indicative of an underlying transcript, including exons predicted to be connected. In total, about 100 unannotated predicted exons in EGASP are consistently supported by TARs/transfrags, and are, therefore, likely to belong to transcribed RNAs. In summary, a non-negligible fraction of unannotated exons predicted in EGASP have some evidence of transcription (not necessarily associated with protein coding), but only a small fraction of the predicted structures connecting exons could be verified experimentally here.

In this regard, the EGASP experiment seems to indicate that the GENCODE annotation of protein coding genes is quite complete, although it is still unclear what fraction of all the alternative transcript diversity of gene loci is captured by GENCODE. EGASP was also useful in helping to identify the software tools that can contribute to reduce the amount of human intervention required to delineate the GENCODE annotation. Programs accelerating and improving the mapping of cDNA sequences (partial or complete) into the genome sequence could be particularly useful towards that end.

Overall, we believe that the EGASP project has given a fair assessment of the state-of-the-art of gene prediction in human DNA. This will allow biologists to interpret better the annotations presented to them in public genome databases such as GenBank, the UCSC browser, ENSEMBL and others. It has also clearly shown that we are still far from being able to computationally predict human gene structures with total accuracy from the DNA sequence alone. Furthermore, while we believe the experiment has shown that only very few protein-coding human genes seem to missing from the annotations, the exact protein sequences are annotated for roughly over 50% of the sequences. Getting a complete protein sequence correct is also made difficult by the existence of many splice forms, mis-assembled cDNAs and additional contamination in cDNA/EST sequences in the public databases. Each can lead to various spurious protein sequence annotations. Unfortunately, there are very few processes in place to remove erroneous sequences and annotations from the public databases, so it will still take some time to get a better picture of exact gene structures. It has to be noted that the human genome and its annotation for protein coding genes are still works in progress.

Another class of genes, non-protein coding transcripts, which were not generally considered by EGASP, are thought to be especially difficult to predict. These genes, such as those that encode miRNAs and snoRNAs, were not addressed in this experiment; nevertheless, they seem to play a very important role in physiological processes such as development and disease.

One of the most difficult problems in gene prediction accuracy assessment is the definition of a reference set against which to evaluate. Ultimately, this reference set should be 'unknown' to the prediction teams. In EGASP, the delayed publication of the GENCODE annotations partially achieved this goal, although a significant amount of the annotation information was known from previously submitted cDNA and EST sequences to public databases such as ENSEMBL or Genbank. This is slightly different to GASP1 [27], where novel cDNA sequences had been withheld before the experiment. Additionally, it may be optimal if each group used the same auxiliary data for their predictions. One suggestion would be to 'freeze' databases of auxiliary data and allow only the inclusion in the predictions of these frozen databases, so that progress in these assessment experiments can be measured independently of growing experimental data.

Furthermore, while our assessments have started to evaluate gene annotations on the transcript level, better and additional evaluation methods for evaluating UTRs are needed. One suggestion would be to evaluate the transcript performance at the intron level (similar to the exon evaluation above). This measure would exclude the beginning and end of a gene, two coordinates that are considered the most difficult to obtain experimentally, but would include non-coding introns that are determined by their splice sites.

One of the major benefits of this kind of experiment is that it allows prediction teams to measure their programs and methods against each other, to learn from their failures, and, as a community, to identify the open and difficult questions in this area of research.

Materials and methods
Submitted predictions
Files submitted to the EGASP server were validated to conform to the GTF specifications [33] and the use of standard annotation features such as 'exon', 'CDS', 'stop codon' and 'start codon'. Submissions not conforming to this format were rejected, although the participants were allowed to fix prediction files accordingly and to resubmit to the server (Figure 1) [47]. Submissions were clipped to the ENCODE region sequence boundaries. The clipping criteria were the following: (*feature_start* < 1 and *feature_end* >= 1 then *feature_start* == 1) and (*feature_end* > *sequence_end*

and *feature_start* <= *sequence_end* , then *feature_end* ==
sequence_end); while those records where (*feature_end* < 1
or *feature_start* > *sequence_end*) were removed from the
GENCODE annotation and the submitted predictions before
performing the evaluations.

Evaluations

We used different programs to obtain accuracy values at
nucleotide, exon, gene/transcript and clustered transcripts.
These programs included software developed by the authors
at IMIM and at the EBI. We also used the Eval package [75].
We confirmed the results obtained with the evaluation
programs by comparison. The programs can be downloaded,
along with a small description on how to use them, from the
Supplementary material web page [51].

When comparing annotations against the predictions for
each individual sequence in the test sets for the boxplots,
sequences that contained no feature annotations either in
the annotations or in the predictions were excluded from the
analysis for the boxplots. This did not happen when the
numbers were computed globally for the sequences. We
considered the following sequence sets build up by
concatenating (without overlap) the coordinates of different
sequence annotation and prediction sets: EN_TRN13, all
training set sequences (13 sequences); EN_PRD31, all
evaluation set sequences (31 sequences); EN_MNLp12,
evaluation set sequences, manual picks (12 sequences);
EN_RNDp19, evaluation set sequences, random picks (19
sequences); EN_PGH12/EN_PGM11/EN_PGL8, all the
sequences of the test set were collected into three sequence
sets based on their gene density into three sequence sets, for
high, medium and low densities (12, 11 and 8 sequences,
respectively); EN_PMH7/EN_PMM5/ENPML7, in this case,
the random sequences from the evaluation set were
considered, depending on their sequence conservation with
mouse, into three sequences, for high, medium and low
conservation (7, 5 and 7 sequences, respectively). See the
Supplementary materials web page for the complete set of
results on all sequences and sequence sets [51].

The box-and-whisker plots [76] (simply 'boxplots') describe
graphically how the data being analyzed are distributed. The
horizontal line within the box shows the median value of the
data set, while the top and the bottom of the box correspond to
the third and first quartiles, respectively; therefore, the box
represents the interquartile range (IQR). The whiskers
represent the range of the data and show a maximum and a
minimum, which are based on 1.5 times the length of the IQR.
The notches centered on the median correspond to the 5%
interval of confidence for this median (median ± 1.57 · IQR/√n,
as defined in R [77]).

RT-PCR

Primers mapping in the two predicted exons spanning the
exon junction to be tested were designed using Primer3 [78]

with the following parameters: 18 ≤ primer size ≤ 27, optimal
size = 20, 57°C ≤ primer Tm ≤ 63°C, optimal Tm = 60°C,
20% ≤ primer GC percentage ≤ 80%. Similar amounts of 24
human cDNAs (brain, heart, kidney, spleen, liver, colon,
small intestine, muscle, lung, stomach, testis, placenta, skin,
peripheral blood lymphocytes, bone marrow, fetal brain,
fetal liver, fetal kidney, fetal heart, fetal lung, thymus,
pancreas, mammary glands, prostate, final dilution 1,000·)
were mixed with JumpStart REDTaq ReadyMix (Sigma-
Aldrich, St. Louis, MO, USA) and 4 ng/µl primers (Sigma-
Genosys, Cambridge, U.K.) with a BioMek 2000 robot
(Beckman, Fullerton, CA, USA) as described and modified
[14,79,80]. The 10 first cycles of PCR amplification were
performed with a touchdown annealing temperature
decreasing from 60°C to 50°C; the annealing temperature of
the next 30 cycles was 50°C. Amplimers were separated on
'Ready to Run' precast gels (Amersham Pharmacia,
Sunnyvale, CA, USA) and sequenced. We tested 221 exon
pairs out of the 238 exon pairs with an exon ranked in the
top 200. The remaining 17 exon pairs were not
experimentally evaluated because either the targeted
amplimer was too small (8 cases) or one of the exons was too
short to allow us to design a primer (9 cases).

Acknowledgements
We would like to thank Peter Good from the National Human Genome
Research Institute (NHGRI) for motivating EGASP and, together with
Elise Feingold and Mark Guyer, also from NHGRI, for their continuous
support. We thank all participants for their enthusiasm, commitment and
efforts to meet all the EGASP deadlines. We also acknowledge Evan
Keibler for providing support for the Eval software package. We are par-
ticularly thankful to the National Human Genome Research Institute for
inspiring and providing funding for EGASP. We also thank the Wellcome
Trust Conference staff for their assistance during the workshop. RG is
supported by grants from the NHGRI ENCODE Project, the European
Biosapiens Project, and from the Spanish Ministry of Education and
Science. PF is supported by EMBL. AR acknowledges the Swiss National
Science Foundation for financial support. MR is partially supported by the
NHGRI.

This article has been published as part of *Genome Biology* Volume 7,
Supplement 1, 2006: EGASP '05. The full contents of the supplement are
available online at http://genomebiology.com/supplements/7/S1.

References
1. Venter JC, Adams MD, Myers EW, Li PW, Mural RJ, Sutton GG,
   Smith HO, Yandell M, Evans CA, Holt RA, *et al.*: **The sequence of
   the human genome.** *Science* 2001, **291:**1304-1351.
2. Lander ES, Linton LM, Birren B, Nusbaum C, Zody MC, Baldwin J,
   Devon K, Dewar K, Doyle M, FitzHugh W, *et al.*: **Initial sequencing and analysis of the human genome.** *Nature* 2001, **409:**860-
   921.
3. International Human Genome Sequencing Consortium: **Finishing
   the euchromatic sequence of the human genome.** *Nature*
   2004, **431:**931-945.
4. Pruitt KD, Tatusova T, Maglott DR: **NCBI Reference Sequence
   (RefSeq): a curated non-redundant sequence database of
   genomes, transcripts and proteins.** *Nucleic Acids Res* 2005, **33
   (Database issue):**D501-504.
5. Gerhard DS, Wagner L, Feingold EA, Shenmen CM, Grouse LH,
   Schuler G, Klein SL, Old S, Rasooly R, Good P, *et al.*: **The status,
   quality, and expansion of the NIH full-length cDNA project:
   the Mammalian Gene Collection (MGC).** *Genome Res* 2004,
   14:2121-2127.

6. Hubbard T, Andrews D, Caccamo M, Cameron G, Chen Y, Clamp M, Clarke L, Coates G, Cox T, Cunningham F, *et al.*: **Ensembl 2005.** *Nucleic Acids Res* 2005, **33(Database issue):**D447-453.

7. Karolchik D, Baertsch R, Diekhans M, Furey TS, Hinrichs A, Lu YT, Roskin KM, Schwartz M, Sugnet CW, Thomas DJ, *et al.*: **The UCSC Genome Browser Database.** *Nucleic Acids Res* 2003, **31:**51-54.

8. Maglott D, Ostell J, Pruitt KD, Tatusova T: **Entrez Gene: gene-centered information at NCBI.** *Nucleic Acids Res* 2005, **33(Database issue):**D54-58.

9. Parra G, Agarwal P, Abril JF, Wiehe T, Fickett JW, Guigo R: **Comparative gene prediction in human and mouse.** *Genome Res* 2003, **13:**108-117.

10. Cawley S, Pachter L, Alexandersson M: **SLAM web server for comparative gene finding and alignment.** *Nucleic Acids Res* 2003, **31:**3507-3509.

11. Alexandersson M, Cawley S, Pachter L: **SLAM: cross-species gene finding and alignment with a generalized pair hidden Markov model.** *Genome Res* 2003, **13:**496-502.

12. Korf I, Flicek P, Duan D, Brent MR: **Integrating genomic homology into gene structure prediction.** *Bioinformatics* 2001, **17 (Suppl 1):**S140-148.

13. Flicek P, Keibler E, Hu P, Korf I, Brent MR: **Leveraging the mouse genome for gene prediction in human: from whole-genome shotgun reads to a global synteny map.** *Genome Res* 2003, **13:** 46-54.

14. Guigo R, Dermitzakis ET, Agarwal P, Ponting CP, Parra G, Reymond A, Abril JF, Keibler E, Lyle R, Ucla C, *et al.*: **Comparison of mouse and human genomes followed by experimental verification yields an estimated 1,019 additional genes.** *Proc Natl Acad Sci USA* 2003, **100:**1140-1145.

15. Wu JQ, Shteynberg D, Arumugam M, Gibbs RA, Brent MR: **Identification of rat genes by TWINSCAN gene prediction, RT-PCR, and direct sequencing.** *Genome Res* 2004, **14:**665-671.

16. Eyras E, Reymond A, Castelo R, Bye JM, Camara F, Flicek P, Huckle EJ, Parra G, Shteynberg DD, Wyss C, *et al.*: **Gene finding in the chicken genome.** *BMC Bioinformatics* 2005, **6:**131.

17. Tenney AE, Brown RH, Vaske C, Lodge JK, Doering TL, Brent MR: **Gene prediction and verification in a compact genome with numerous small introns.** *Genome Res* 2004, **14:**2330-2335.

18. Burge C, Karlin S: **Prediction of complete gene structures in human genomic DNA.** *J Mol Biol* 1997, **268:**78-94.

19. Parra G, Blanco E, Guigo R: **GeneID in Drosophila.** *Genome Res* 2000, **10:**511-515.

20. Jaillon O, Aury JM, Brunet F, Petit JL, Stange-Thomann N, Mauceli E, Bouneau L, Fischer C, Ozouf-Costaz C, Bernot A, *et al.*: **Genome duplication in the teleost fish Tetraodon nigroviridis reveals the early vertebrate proto-karyotype.** *Nature* 2004, **431:**946-957.

21. Brent MR, Guigo R: **Recent advances in gene structure prediction.** *Curr Opin Struct Biol* 2004, **14:**264-272.

22. Burset M, Guigo R: **Evaluation of gene structure prediction programs.** *Genomics* 1996, **34:**353-367.

23. Bajic VB: **Comparing the success of different prediction software in sequence analysis: a review.** *Brief Bioinform* 2000, **1:**214-228.

24. Baldi P, Brunak S, Chauvin Y, Andersen CA, Nielsen H: **Assessing the accuracy of prediction algorithms for classification: an overview.** *Bioinformatics* 2000, **16:**412-424.

25. Guigo R, Agarwal P, Abril JF, Burset M, Fickett JW: **An assessment of gene prediction accuracy in large DNA sequences.** *Genome Res* 2000, **10:**1631-1642.

26. Rogic S, Mackworth AK, Ouellette FB: **Evaluation of gene-finding programs on mammalian sequences.** *Genome Res* 2001, **11:** 817-832.

27. Reese MG, Hartzell G, Harris NL, Ohler U, Abril JF, Lewis SE: **Genome annotation assessment in Drosophila melanogaster.** *Genome Res* 2000, **10:**483-501.

28. Dunbrack RL Jr, Gerloff DL, Bower M, Chen X, Lichtarge O, Cohen FE: **Meeting review: the Second meeting on the Critical Assessment of Techniques for Protein Structure Prediction (CASP2), Asilomar, California, December 13-16, 1996.** *Fold Des* 1997, **2:**R27-42.

29. Ashburner M, Misra S, Roote J, Lewis SE, Blazej R, Davis T, Doyle C, Galle R, George R, Harris N, *et al.*: **An exploration of the sequence of a 2.9-Mb region of the genome of Drosophila melanogaster: the Adh region.** *Genetics* 1999, **153:**179-219.

30. ENCODE Project Consortium: **The ENCODE (ENCyclopedia Of DNA Elements) Project.** *Science* 2004, **306:**636-640.

31. **The GENCODE Project** [http://genome.imim.es/gencode/]

32. **The HAVANA Team** [http://www.sanger.ac.uk/HGP/havana/]

33. Harrow J, Denoeud F, Frankish A, Reymond A, Chen C-K, Chrast J, Lagarde J, Gilbert JGR, Storey R, Swarbreck D, *et al.*: **GENCODE: Producing a reference annotation for ENCODE.** *Genome Biology* 2006, 7 (Suppl 1) :S4.

34. **ENCODE Project Target Selection Process and Target Regions** [http://genome.gov/10506161]

35. **ENCODE Project at UCSC Genome Browser** [http://genome.cse.ucsc.edu/ENCODE/]

36. Mungall AJ, Palmer SA, Sims SK, Edwards CA, Ashurst JL, Wilming L, Jones MC, Horton R, Hunt SE, Scott CE, *et al.*: **The DNA sequence and analysis of human chromosome 6.** *Nature* 2003, **425:**805-811.

37. Humphray SJ, Oliver K, Hunt AR, Plumb RW, Loveland JE, Howe KL, Andrews TD, Searle S, Hunt SE, Scott CE, *et al.*: **DNA sequence and analysis of human chromosome 9.** *Nature* 2004, **429:**369-374.

38. Deloukas P, Earthrowl ME, Grafham DV, Rubenfield M, French L, Steward CA, Sims SK, Jones MC, Searle S, Scott C, *et al.*: **The DNA sequence and comparative analysis of human chromosome 10.** *Nature* 2004, **429:**375-381.

39. Dunham A, Matthews LH, Burton J, Ashurst JL, Howe KL, Ashcroft KJ, Beare DM, Burford DC, Hunt SE, Griffiths-Jones S, *et al.*: **The DNA sequence and analysis of human chromosome 13.** *Nature* 2004, **428:**522-528.

40. Deloukas P, Matthews LH, Ashurst J, Burton J, Gilbert JG, Jones M, Stavrides G, Almeida JP, Babbage AK, Bagguley CL, *et al.*: **The DNA sequence and comparative analysis of human chromosome 20.** *Nature* 2001, **414:**865-871.

41. Collins JE, Goward ME, Cole CG, Smink LJ, Huckle EJ, Knowles S, Bye JM, Beare DM, Dunham I: **Reevaluating human gene annotation: a second-generation analysis of chromosome 22.** *Genome Res* 2003, **13:**27-36.

42. Ross MT, Grafham DV, Coffey AJ, Scherer S, McLay K, Muzny D, Platzer M, Howell GR, Burrows C, Bird CP, *et al.*: **The DNA sequence of the human X chromosome.** *Nature* 2005, **434:** 325-337.

43. Kan Z, Rouchka EC, Gish WR, States DJ: **Gene structure prediction and alternative splicing analysis using genomically aligned ESTs.** *Genome Res* 2001, **11:**889-900.

44. Ellsworth RE, Jamison DC, Touchman JW, Chissoe SL, Braden Maduro VV, Bouffard GG, Dietrich NL, Beckstrom-Sternberg SM, Iyer LM, Weintraub LA, *et al.*: **Comparative genomic sequence analysis of the human and mouse cystic fibrosis transmembrane conductance regulator genes.** *Proc Natl Acad Sci USA* 2000, **97:**1172-1177.

45. Bajic VB, Brent MR, Brown RH, Frankish, A, Harrow, J, Ohler U, Solovyev VV, Tan SL: **Performance assessment of promoter predictions on ENCODE regions in the EGASP experiment.** *Genome Biology* 2006, **7(Suppl1):**S3.

46. Zheng D, Gerstein M: **A computational approach for identifying pseudogenes in the ENCODE regions.** *Genome Biol* 2006, **7 (Suppl 1):**S13.

47. **The EGASP Submission Server for Predictions** [http://genome.imim.es/cgi-bin/EGASP2005/submission]

48. **Gene Transfer Format Specifications** [http://genes.cs.wustl.edu/ GTF2.html]

49. **GencodeDB Genome Browser** [http://genome.imim.es/cgi-bin/ gbrowse/encode]

50. **EGASP ftp Server** [ftp://genome.imim.es/pub/projects/gencode/ data/egasp05/submitted_predictions/]

51. **Supplementary Material** [http://genome.imim.es/datasets/ egasp2005/]

52. **AceView** [http://www.ncbi.nlm.nih.gov/IEB/Research/Acembly/]

53. Kim N, Shin S, Lee S: **ECgene: genome-based EST clustering and gene modeling for alternative splicing.** *Genome Res* 2005, **15:**566-576.

54. **The UCSC Known Genes Track** [http://genome.ucsc.edu/ cgi-bin/hgTrackUi?hgsid=63708116&c=chr7&g=knownGene]

55. **The Consensus CDS (CCDS) Project** [http://www.ncbi.nlm.nih. gov/CCDS/]

56. Solovyev V, Kosarev P, Seledsov I, Vorobyev D: **Automatic annotation of eukaryotic genes, pseudogenes and promoters.** *Genome Biology* 2006, **7(Suppl 1):**S10.

57. Arumugam M, Wei C, Brown RH, Brent MR: **Pairagon+N-SCAN_EST: a model-based gene annotation pipeline.** *Genome Biology* 2006, **7(Suppl 1):**S5.

58. Stanke M, Tzvetkova A, Morgenstern B: **AUGUSTUS at EGASP: using EST, protein and genomic alignments for improved gene prediction in the human genome.** *Genome Biology* 2006,**7**(Suppl 1):S11.
59. Allen JE, Majoros WH, Pertea M, Salzberg SL: **JIGSAW, GeneZilla and GlimmerHMM: puzzling out the features of human genes in the ENCODE regions.** *Genome Biology* 2006, **7**(Suppl 1):S9.
60. Besemer J, Borodovsky M: **GeneMark: web software for gene finding in prokaryotes, eukaryotes and viruses.** *Nucleic Acids Res* 2005, **33**(Web Server issue):W451-454.
61. Bedell JA, Korf I, Gish W: **MaskerAid: a performance enhancement to RepeatMasker.** *Bioinformatics* 2000, **16**:1040-1041.
62. Djebali S, Delaplace F, Roest Crollius H: **Exogean: a framework for annotating protein-coding genes in eukaryotic genomic DNA.** *Genome Biology* 2006, **7** (Suppl 1): S7.
63. Brejova B, Brown DG, Li M, Vinar T: **ExonHunter: a comprehensive approach to gene finding.** *Bioinformatics* 2005, **21**(Suppl 1):i57-i65.
64. Curwen V, Eyras E, Andrews TD, Clarke L, Mongin E, Searle SM, Clamp M: **The Ensembl automatic gene annotation system.** *Genome Res* 2004, **14**:942-950.
65. Flicek P, Brent MR: **Using several pair-wise informant sequences for de novo prediction of alternatively spliced transcripts.** *Genome Biology* 2006, **7** (Suppl 1) :S8.
66. Chatterji S, Pachter L: **Large multiple organism gene finding by collapsed Gibbs sampling.** *J Comput Biol.* 2005 Jul-Aug; 12(6): 599-608.
67. Carter D, Durbin R: **Vertebrate gene finding from multiple-species alignments using a two-level strategy.** *Genome Biology* 2006, **7**(Suppl 1):S6.
68. Patel AA, Steitz JA: **Splicing double: insights from the second spliceosome.** *Nat Rev Mol Cell Biol* 2003, **4**:960-970.
69. Kapranov P, Cawley SE, Drenkow J, Bekiranov S, Strausberg RL, Fodor SP, Gingeras TR: **Large-scale transcriptional activity in chromosomes 21 and 22.** *Science* 2002, **296**:916-919.
70. Cheng J, Kapranov P, Drenkow J, Dike S, Brubaker S, Patel S, Long J, Stern D, Tammana H, Helt G, *et al.*: **Transcriptional maps of 10 human chromosomes at 5-nucleotide resolution.** *Science* 2005, **308**:1149-1154.
71. Kampa D, Cheng J, Kapranov P, Yamanaka M, Brubaker S, Cawley S, Drenkow J, Piccolboni A, Bekiranov S, Helt G, *et al.*: **Novel RNAs identified from an in-depth analysis of the transcriptome of human chromosomes 21 and 22.** *Genome Res* 2004, **14**:331-342.
72. Castelo R, Reymond A, Wyss C, Camara F, Parra G, Antonarakis SE, Guigo R, Eyras E: **Comparative gene finding in chicken indicates that we are closing in on the set of multi-exonic widely expressed human genes.** *Nucleic Acids Res* 2005, **33**:1935-1939.
73. Parra G, Reymond A, Dabbouseh N, Dermitzakis ET, Castelo R, Thomson TM, Antonarakis SE, Guigo R: **Tandem chimerism as a means to increase protein complexity in the human genome.** *Genome Res* 2006, **16**:37-44.
74. Reymond A, Marigo V, Yaylaoglu MB, Leoni A, Ucla C, Scamuffa N, Caccioppoli C, Dermitzakis ET, Lyle R, Banfi S, *et al.*: **Human chromosome 21 gene expression atlas in the mouse.** *Nature* 2002, **420**:582-586.
75. Keibler E, Brent MR: **Eval: A software package for analysis of genome annotations.** *BMC Bioinformatics* 2003, **4**:50.
76. Tukey JW: *Exploratory Data Analysis.* Reading, MA: Addison-Wesley; 1977.
77. R Development Core Team: *R: A Language and Environment for Statistical Computing.* Vienna, Austria: R Foundation for Statistical Computing; 2005.
78. **Primer3** [http://frodo.wi.mit.edu/cgi-bin/primer3/primer3_www.cgi]
79. Reymond A, Friedli M, Henrichsen CN, Chapot F, Deutsch S, Ucla C, Rossier C, Lyle R, Guipponi M, Antonarakis SE: **From PREDs and open reading frames to cDNA isolation: revisiting the human chromosome 21 transcription map.** *Genomics* 2001, **78**:46-54.
80. Reymond A, Camargo AA, Deutsch S, Stevenson BJ, Parmigiani RB, Ucla C, Bettoni F, Rossier C, Lyle R, Guipponi M, *et al.*: **Nineteen additional unpredicted transcripts from human chromosome 21.** *Genomics* 2002, **79**:824-832.
81. Majoros WH, Pertea M, Salzberg SL: **TigrScan and GlimmerHMM: two open source ab initio eukaryotic gene-finders.** *Bioinformatics* 2004, **20**:2878-2879.
82. Yeo GW, Van Nostrand E, Holste D, Poggio T, Burge CB: **Identification and analysis of alternative splicing events conserved in human and mouse.** *Proc Natl Acad Sci USA* 2005, **102**:2850-2855.
83. Bonizzoni P, Rizzi R, Pesole G: **ASPIC: a novel method to predict the exon-intron structure of a gene that is optimally compatible to a set of transcript sequences.** *BMC Bioinformatics* 2005, **6**:244.
84. Castrignano T, Canali A, Grillo G, Liuni S, Mignone F, Pesole G: **CSTminer: a web tool for the identification of coding and noncoding conserved sequence tags through cross-species genome comparison.** *Nucleic Acids Res* 2004, **32**(Web Server issue):W624-627.
85. Keefe D: **SPIDA: Substitution Periodicity Index and Domain Analysis.** ftp://ftp.ebi.ac.uk/pub/databases/ensembl/encode/egasp05
86. Ohler U, Shomron N, Burge CB: **Recognition of unknown conserved alternatively spliced exons.** *PLoS Comput Biol* 2005, **1**:113-122.

Review

Performance assessment of promoter predictions on ENCODE regions in the EGASP experiment

Vladimir B Bajic[1], Michael R Brent[2], Randall H Brown[2], Adam Frankish[3], Jennifer Harrow[4], Uwe Ohler[5], Victor V Solovyev[6] and Sin Lam Tan[7]

Addresses: [1]South African National Bioinformatics Institute (SANBI), University of the Western Cape, Bellville 7535, South Africa. [2]Laboratory for Computational Genomics and Department of Computer Science, Washington University in St Louis, USA. [3]Human and Vertebrate Analysis and Annotation Group, Wellcome Trust Sanger Institute, Wellcome Trust Genome Campus, Hinxton, Cambridge CB10 1SA, UK. [4]Wellcome Trust Sanger Institute, Wellcome Trust Genome Campus, Hinxton, Cambridgeshire CB10 1HH, UK. [5]Institute for Genome Sciences and Policy, Science Dr, Duke University, Durham, NC 27708, USA. [6]Royal Holloway, University of London, London, UK. [7]Knowledge Extraction Lab, Institute for Infocomm Research, Heng Mui Keng Terrace, Singapore 119613.

Correspondence: VB Bajic. Email: vlad@sanbi.ac.za

Published: 7 August 2006

Genome Biology 2006, 7(Suppl 1):S3

The electronic version of this article is the complete one and can be found online at http://genomebiology.com/2006/7/S1/S3

Abstract

Background: This study analyzes the predictions of a number of promoter predictors on the ENCODE regions of the human genome as part of the ENCODE Genome Annotation Assessment Project (EGASP). The systems analyzed operate on various principles and we assessed the effectiveness of different conceptual strategies used to correlate produced promoter predictions with the manually annotated 5' gene ends.

Results: The predictions were assessed relative to the manual HAVANA annotation of the 5' gene ends. These 5' gene ends were used as the estimated reference transcription start sites. With the maximum allowed distance for predictions of 1,000 nucleotides from the reference transcription start sites, the sensitivity of predictors was in the range 32% to 56%, while the positive predictive value was in the range 79% to 93%. The average distance mismatch of predictions from the reference transcription start sites was in the range 259 to 305 nucleotides. At the same time, using transcription start site estimates from DBTSS and H-Invitational databases as promoter predictions, we obtained a sensitivity of 58%, a positive predictive value of 92%, and an average distance from the annotated transcription start sites of 117 nucleotides. In this experiment, the best performing promoter predictors were those that combined promoter prediction with gene prediction. The main reason for this is the reduced promoter search space that resulted in smaller numbers of false positive predictions.

Conclusions: The main finding, now supported by comprehensive data, is that the accuracy of human promoter predictors for high-throughput annotation purposes can be significantly improved if promoter prediction is combined with gene prediction. Based on the lessons learned in this experiment, we propose a framework for the preparation of the next similar promoter prediction assessment.

Background
Complexity of the target

Accurate determination of transcription start sites (TSSs) is one of the most difficult problems in genomics. The reference genomic location from which a transcript will be generated has remained elusive for many years, mainly due to our insufficient understanding of the transcription initiation process. The transcript promoter region surrounds the TSS and serves as the docking DNA segment that binds the preinitiation complex and various transcription factors that jointly create the biochemical conditions to initiate transcription [1,2]. Consequently, the analysis of promoter regions for binding sites of transcription factors can reveal many crucial aspects of how, where and when the transcript will be generated.

The naive concept of a gene having one TSS was abandoned long ago. Current data suggest that TSSs can be found scattered across the gene loci, generally more concentrated at the 5' end, but also more downstream, sometimes in exons, introns, and interestingly in the 3' untranslated regon (UTR) [3]. Moreover, one gene region may frequently have several promoters, and within one promoter several alternative TSS locations close to each other could be found. To make this complex picture even more complicated, promoter regions are frequently shared or overlap each other, such as in sense/antisense genes and in bidirectionally promoted genes [4]. All these considerably complicate the development of strategies for attacking the problem of promoter prediction. To avoid confusion, in this report, by 'promoter prediction' we mean the prediction of the TSS locations and not the prediction of a region surrounding a TSS.

Potential use of accurate TSS locations

Promoters are among the key genomic control regions for transcriptional regulation of every gene [1,2,5]. Thus, accurate TSS location makes determination of promoters more accurate, which allows for more accurate analysis of transcriptional regulatory elements necessary for any subsequent transcriptional regulatory network analyses. Furthermore, even when there are no expressed sequence data (expressed sequence tag (EST), cDNA, mRNA or different tags such as CAGE (cap-analysis of gene expression), SAGE (serial analysis of gene expression) and so on), the computational prediction of promoters and TSSs can allow for gene discovery.

Historical perspective

Realizing the importance of predicting promoters accurately, different experimental and computational methods have been developed. The large number of gene loci in eukaryotic genes inevitably calls for high-throughput large-scale technologies for determining TSS locations. Among the most efficient ones are those based on oligo-capping [6] and CAP-trapping [7]. Another group of methods is based on the use of multiple aligned ESTs and cDNA/mRNA fragments, and

an assessment of TSS location as groups of identical 5' ends or the most 5' located end within the same locus. The third group of methods is based on the assessment of the binding location of DNA-associated RNA polymerase from ChIP-chip experiments [8]. However, the TSS location cannot be determined precisely from these experiments. In summary, none of the mentioned methods is sufficiently accurate or complete; this makes it difficult to obtain a proper reference dataset - one with high coverage and accuracy - to use for evaluation of promoter predictions.

An alternative to experimental methods are computational ones, but they generally are imperfect due to our insufficient understanding of the transcription initiation process. Several reviews have been published aiming at presenting the most crucial aspects and principles used in the construction of promoter prediction systems, as well as in the assessment of performance of promoter predictors [9-14]. Solutions proposed [15-31] were based on different concepts and exhibited various degrees of performance. PromoterInspector [28] was the first study to present computational predictions with an acceptable level of false positives (FPs) with human data, after the first genome scale evaluation as part of GASP had earlier shown promising results for *Drosophila* [32]. Encouraged by this, several efficient methods were later proposed [15-20,23-27,29-31]. The performance of many of these solutions have been extensively evaluated in [12].

Two strategies for designing promoter predictors

There is a lot of evidence that in mammalian genomes transcription initiates at various and unusual positions, such as intergenic regions far from currently known genes, 3' UTRs of known protein-coding genes, coding exons, and introns [3,4,33]. One gene may overlap another and promoters of such genes could fall anywhere on the body of the other gene [4]. The destiny of transcripts that are initiated is decided at various levels in the post-transcriptional processing, and many such transcripts are later degraded. However, it is difficult to estimate what proportion of all transcripts that the cell generates is functional. It is also difficult to determine which TSSs generate non-functional transcripts and whether they always generate such transcripts. For a long time biologists focused on protein-coding genes and this is one of the reasons that today most of the data we have relate to that transcript group. However, non-coding transcripts have recently been recognized as important for regulation of gene expression. A significant proportion of transcripts also cannot be accurately classified as being in either the coding or the non-coding group. For all these reasons it would be valuable to make the inventory of all TSSs in one genome and to investigate their functional properties.

For some purposes, a comprehensive list of potential TSSs may be most useful, even if the list contains FPs and TSSs of non-functional transcripts; for other applications, a list containing fewer FPs and non-functional TSSs may be better, even if it systematically omits interesting TSSs whose functions are less common or less well understood. Given our current state of knowledge, we must choose; predicting all and only functional TSSs is not currently feasible. Thus, TSS prediction programs have been designed around two strategies: use only the local genomic context (that is, model some aspects of the biological transcription initiation process or look at distinguishing characteristics of the region that immediately surrounds the TSS); or also take into account possible gene presence to restrict the search to regions that are most likely to contain promoters. The latter approach may use any of the available methods of gene prediction, including *de novo* prediction and prediction based on aligning ESTs, cDNA sequences, and/or proteins. It is also possible to utilize the annotation of genes if it is available. Using evidence about the presence of nearby genes may considerably enhance the performance of systems that work by analyzing the local promoter context. In general, on the genome scale, such a combination will reduce sensitivity to some extent, but it will significantly reduce the total number of predictions and will increase specificity.

On the other hand, to understand biological mechanisms of regulatory regions and to cover broad spectra of such regions, we probably should not use necessarily gene identification as a part of a strategy for pinpointing TSSs. The gene finding models introduce many implicit assumptions that reduce coverage of various types of TSSs that could be of interest. Also, linking promoter predictors to gene finders does not directly model the way in which transcription is initiated in the cell. A comprehensive solution is most likely to come from modeling the information cells use to determine where to initiate transcription, including the local promoter sequence and its epigenetic state [34].

Goals of this assessment

The ENCODE Genome Annotation Assessment Project (EGASP) is explained in detail in the main EGASP report [35]. The main goal of the project has been to assess the accuracy of prediction of protein coding genes, as well as the completeness of current human genome annotations of the ENCODE regions [36] covering approximately 1% of the human genome sequence. The reference gene set against which all predictions were assessed was created by manual annotation of the ENCODE regions by the HAVANA group [37] at the Sanger Institute, within the GENCODE project [38].

In our study, we attempt to make a critical assessment of the promoter prediction field in its current state relative to the HAVANA gene annotation [39] of the ENCODE regions. Thus, we assessed the extent of correlation of promoter predictions with the 5' gene ends of the HAVANA annotation. We argue that using promoter predictors together with gene predictors or as a complement to the manual annotation of genes is a good intermediate step to improve promoter prediction performance because this constrains the search space based on information about the gene. We propose promising strategies for future development of promoter prediction systems on the basis of the current performance assessment.

Results

The method for counting correct and wrong predictions is explained in Materials and methods. We have analyzed predictions on all 44 ENCODE regions (total length 29,998,060 base-pairs (bp)), with the training set consisting of 13 regions of total length 8,538,447 bp and the remaining part as the test set with a length of 21,459,613 bp. The genomic sequences were from the human genome Build hg17. The performance results are summarized in Figures 1 and 2. Figure 1 contains results where true positive (TP) predictions were allowed to be within a maximum distance of 1,000 nucleotides from the reference TSS; Figure 2 contains results where the maximum distance allowed was 250 nucleotides. We present results within three categories: for the test ENCODE regions, for the training ENCODE regions, and for all ENCODE regions. We considered only predictions of promoters for known genes that contained coding sequence (CDS) based on the HAVANA annotation that was submitted for the EGASP workshop. In total, there were 994 unique TSSs, of which 319 were in the ENCODE training set and 675 were within the ENCODE test set. In our analysis, the reference data against which the performance of promoter predictors was evaluated were the estimated TSS locations based on the 5' ends of genes in the HAVANA annotation. It is important to note, however, that HAVANA annotation does not attempt to specifically predict TSSs but rather to best represent the exon structure, CDS and UTRs of a gene and its splice variants.

In arriving at our conclusions, we used various measures of performance, as presented in [11]. The use of these different performance measures ensure that the final conclusions are less influenced by the choice of performance measures. The main reference for discussion is the current performance achieved on the ENCODE test regions. Since the ENCODE training regions have higher GC content (44.69%) than the average of the human genome, the results on the ENCODE training set and comprehensive ENCODE set are less representative.

Figure 1 shows that TSS locations compiled from DBTSS [40] and H-Invitational [41] databases, when used as predicted TSS locations and compared to the reference manual HAVANA annotation, show only 58% sensitivity (Se) and 92% positive predictive value (ppv). N-SCAN [30] has achieved a greater ppv of 93%. However, all promoter predictors had a ppv >79%, which is a considerable improvement over the last assessment [12]. The sensitivity, however,

ALL ENCODE REGIONS

	TP	FP	Number of hits for TP	Unclear	Se	ppv	AE	DIP1	DIP2	CC	ASM	MaxTol
7-80-8 McPromoter	380	48	180	466	0.3823	0.8879	258.6632	0.6278	435.6963	0.5826	5.1818	1,000
7-81-8 McPromoter	339	57	152	518	0.341	0.8561	263.1239	0.6745	490.3552	0.5403	6.5455	1,000
41-108-8 Fprom	482	59	249	225	0.4849	0.8909	216.9295	0.5265	280.6293	0.6573	3.8182	1,000
20_76_4 N-SCAN	559	43	283	115	0.5624	0.9286	240.5313	0.4434	195.5468	0.7226	2.6364	
DBTSS	608	45	352	127	0.6117	0.9311	116.7237	0.3944	206.6641	0.7547	1.090	1,000
DGSF	456	46	197	385	0.4588	0.9084	324.4912	0.5489	344.7403	0.6455	3.9091	1,000
DPF	614	151	242	1,175	0.6177	0.8026	282.4896	0.4302	932.7682	0.7041	3.8182	1,000
FEF	593	120	246	900	0.5966	0.8317	271.2968	0.4371	553.3941	0.7044	3.6364	1,000

TRAINING ENCODE REGIONS

	TP	FP	Number of hits for TP	Unclear	Se	ppv	AE	DIP1	DIP2	CC	ASM	MaxTol
7-80-8 McPromoter	142	17	68	188	0.4451	0.8931	258.6831	0.5651	154.2631	0.6305	5.0909	1,000
7-81-8 McPromoter	123	22	55	209	0.3856	0.8483	266.5447	0.6329	181.0026	0.5719	6.7273	1,000
41-108-8 Fprom	145	20	70	68	0.4545	0.8788	194.9724	0.5588	88.2841	0.632	5	1,000
20_76_4 N-SCAN	199	16	98	37	0.6238	0.9256	225.1859	0.3835	57.9034	0.7599	3.1818	1,000
DBTSS	216	13	124	42	0.6771	0.9432	115.2778	0.3278	58.6827	0.7992	1.9091	1,000
DGSF	180	20	79	151	0.5643	0.9	353.4556	0.4471	111.7661	0.7126	3.6364	1,000
DPF	235	53	86	654	0.7367	0.816	258.0311	0.3213	254.7557	0.7753	3.5455	1,000
FEF	239	36	96	333	0.7492	0.8691	278.6862	0.2829	131.5462	0.8069	2.0909	1,000

TESTING ENCODE REGIONS

	TP	FP	Number of hits for TP	Unclear	Se	ppv	AE	DIP1	DIP2	CC	ASM	MaxTol
7-80-8 McPromoter	238	31	112	278	0.3526	0.8848	258.6513	0.6576	276.843	0.5585	5.0909	1,000
7-81-8 McPromoter	216	35	97	309	0.32	0.8606	261.1759	0.6941	306.12	0.5248	6.1818	1,000
41-108-8 Fprom	337	39	179	157	0.4993	0.8963	226.3769	0.5114	191.764	0.6689	3.2727	1,000
20_76_4 N-SCAN	360	27	185	78	0.5333	0.9302	249.0139	0.4719	136.8374	0.7044	2.2727	1,000
DBTSS	392	32	228	85	0.5807	0.9245	117.5204	0.426	146.9693	0.7327	1.1818	1,000
DGSF	276	26	118	234	0.4089	0.9139	305.6014	0.5973	225.7974	0.6113	3.8182	1,000
DPF	379	98	156	1,121	0.5615	0.7945	297.6552	0.4843	665.8586	0.6679	4.0909	1,000
FEF	354	84	150	567	0.5244	0.8082	266.3079	0.5128	410.7287	0.651	4.3636	1,000
					0.32-0.56	>0.79	226-305					

Figure 1

Prediction results for the distance criterion of 1,000 nucleotides. The light blue row shows the results of comparison of DBTSS+H-Invitational data to the manual HAVANA annotation. We used this as a reference to enable assessment of promoter predictor performance. The highlighted blue fields denote the score for the best performing promoter predictor. MaxTol is the maximum allowed mismatch between the predictions and the reference TSS locations. The programs with names in red officially participated in the EGASP data submission. The results shown are for the MaxTol = 1,000 nucleotides. AE is the average mismatch of predictions relative to the most close TSS location from the HAVANA annotation. It is divided by 1,000 to scale for the graph presentation. DIP1 and DIP2 are two measures representing distance from the ideal predictor as defined in [10]. ASM is the average score measure as defined in [10].

ranged from 32% to 56%. Positional mismatch of the predicted TSS locations relative to the reference ones was, on average, in the range 226 to 305 nucleotides for promoter predictors, while it was 117 nucleotides for DBTSS and H-Invitational TSS predictions. The correlation coefficient (CC; see Materials and methods) ranged from 0.52 to 0.70 for promoter predictors, and was 0.73 for DBTSS and H-Invitational TSS estimates. Figures 3 and 4 are bar graphs of different performance indicators. When the maximum allowed mismatch of the prediction from the reference TSS for counting TP predictions was 1,000 nucleotides, the best predictor, based on 11 measures of prediction success, was N-SCAN, followed by Fprom, Dragon Gene Start Finder (DGSF) [17,18], Dragon Promoter Finder (DPF) [15,16], First Exon Finder (FEF) [19], and McPromoter [23,24].

When this maximum allowed distance was reduced to 250 nucleotides (Figure 2), the obtained sensitivity and positive predictive value were, as expected, lower. With this distance constraint, the DBTSS and H-Invitational prediction set produced a sensitivity of 49%, ppv of 89%, and an average mismatch of predictions to the reference TSS of 41 nucleotides. Promoter predictors achieved a sensitivity in the range 17% to 33%, a ppv in the range 58% to 81%, and an average positional error in the range 77 to 126 nucleotides. Correlation coefficients ranged from 0.35 to 0.51, while for the DBTSS and H-Invitational set it was 0.66. In this case, the best ranked predictors based on a cocktail of 11 measures were Fprom and N-SCAN, followed by DGSF, FEF, McPromoter (the standard system), DPF, and McPromoter (with the post-processing of shadowed predictions).

ALL ENCODE REGIONS

	TP	FP	Number of hits for TP	Unclear	Se	ppv	AE	DIP1	DIP2	CC	ASM	MaxTol
7-80-8 McPromoter	228	73	124	497	0.2294	0.7575	85.5351	0.8079	560.6727	0.4168	5.5455	250
7-81-8 McPromoter	194	83	101	543	0.1952	0.7004	86.799	0.8588	624.3458	0.3697	7.2727	250
41-108-8 Fprom	327	82	188	263	0.329	0.7995	74.8746	0.7003	373.2796	0.5128	3.7273	250
20_76_4 NSCAN	350	78	212	151	0.3521	0.8178	119.9086	0.673	296.8066	0.5366	2.0909	250
DBTSS + H-Inv	509	60	321	143	0.5121	0.8946	40.0884	0.4992	261.5766	0.6768	1	250
DGSF	239	79	128	421	0.2404	0.7516	128.6695	0.7992	501.8674	0.4251	5.3636	250
DPF	349	215	159	1794	0.3511	0.6188	112.4345	0.7526	1631.599	0.4661	6.2727	250
FEF	350	190	159	917	0.3521	0.6481	115.06	0.7373	933.3758	0.4777	4.7273	250

TRAINING ENCODE REGIONS

	TP	FP	Number of hits for TP	Unclear	Se	ppv	AE	DIP1	DIP2	CC	ASM	MaxTol
7-80-8 McPromoter	84	28	45	200	0.2633	0.75	68.8929	0.7779	212.3781	0.4444	5.3636	250
7-81-8 McPromoter	74	32	35	219	0.232	0.6981	84.2838	0.8252	236.0147	0.4024	7.5455	250
41-108-8 Fprom	102	28	51	79	0.3197	0.7846	69.902	0.7135	112.7385	0.5009	4.1818	250
20_76_4 NSCAN	132	30	74	47	0.4138	0.8148	125.2348	0.6148	92.829	0.5806	2.9091	250
DBTSS + H-Inv	179	20	109	50	0.5611	0.8995	37.2123	0.4502	80.5915	0.7104	1	250
DGSF	80	33	45	172	0.2508	0.708	132.6	0.8041	201.0301	0.4214	6.7273	250
DPF	151	72	63	658	0.4734	0.6771	111.4517	0.6177	489.8666	0.5661	4.1818	250
FEF	142	62	65	338	0.4451	0.6961	118.6901	0.6326	294.1788	0.5566	4.0909	250

TESTING ENCODE REGIONS

	TP	FP	Number of hits for TP	Unclear	Se	ppv	AE	DIP1	DIP2	CC	ASM	MaxTol
7-80-8 McPromoter	144	45	79	297	0.2133	0.7619	95.2431	0.8219	346.0235	0.4032	5.6364	250
7-81-8 McPromoter	120	51	66	324	0.1778	0.7018	88.35	0.8746	385.7174	0.3532	7	250
41-108-8 Fprom	225	54	137	184	0.3333	0.8065	77.1289	0.6942	260.3228	0.5185	2.7273	250
20_76_4 NSCAN	218	48	138	104	0.323	0.8195	116.6835	0.7007	203.195	0.5145	2.7273	250
DBTSS + H-Inv	330	40	212	93	0.4889	0.8919	41.6485	0.5224	180.2347	0.6603	1	250
DGSF	159	46	83	249	0.2356	0.7756	126.6918	0.7967	301.1515	0.4274	4.7273	250
DPF	198	143	96	1136	0.2933	0.5806	113.1841	0.8217	1129.876	0.4127	6.7273	250
FEF	208	128	94	579	0.3081	0.619	112.5817	0.7898	632.6296	0.4367	5.4545	250
					0.17-0.33	>0.58	77-126					

Figure 2

Prediction results for the distance criterion of 250 nucleotides. The light blue row shows the results of comparison of DBTSS+H-Invitational data to the manual HAVANA annotation. We used this as a reference to enable assessment of promoter predictor performance. The highlighted blue fields denote the score for the best performing promoter predictor(s). MaxTol is the maximum allowed mismatch between the predictions and the reference TSS locations. AE is the average mismatch of predictions relative to the closest TSS location from the HAVANA annotation. It is divided by 1,000 to scale for the graph presentation. DIP1 and DIP2 are two measures representing distance from the ideal predictor as defined in [10]. ASM is the average score measure as defined in [10]. The programs with names in red officially participated in the EGASP data submission. The results shown are for the MaxTol = 250 nucleotides.

Discussion

We have analyzed four sets of promoter predictions that were submitted as a response to the EGASP call. These include McPromoter (the standard system), McPromoter (with post-processing of shadowed predictions), Fprom and N-SCAN. These submissions received internal EGASP coding 7-80-8, 7-81-8, 41-108-8, 20-76-4, respectively. The internal coding of submissions by the three numbers is explained in [42]. For the control set we used the estimated TSS locations inferred from the DBTSS and H-Invitational databases. These TSS estimates are based on flcDNAs, with those from DBTSS being derived from the oligo-capped full-length cDNAs (flcDNAs), and thus such a control set is expected to largely reflect the real TSS locations. Additionally, we also considered the predictions of three other programs, FEF, DPF and DGSF, as these were found in a recent comparative study [12] on the whole human genome to have reasonably good performance. The best performing programs in study [12] were DGSF and FEF. Thus, it was of interest to see how they would perform in EGASP. For these additional three programs, the predictions were run under the same conditions as in [12]. These collections of predictions formed the basis for the assessment of performance and promoter prediction strategies. It should be noted that all programs included in this study make assessments of the TSS locations.

Based on the results shown in Figures 1 and 2, we conclude that the best performance achieved with the ENCODE data is by programs that combine promoter prediction with gene

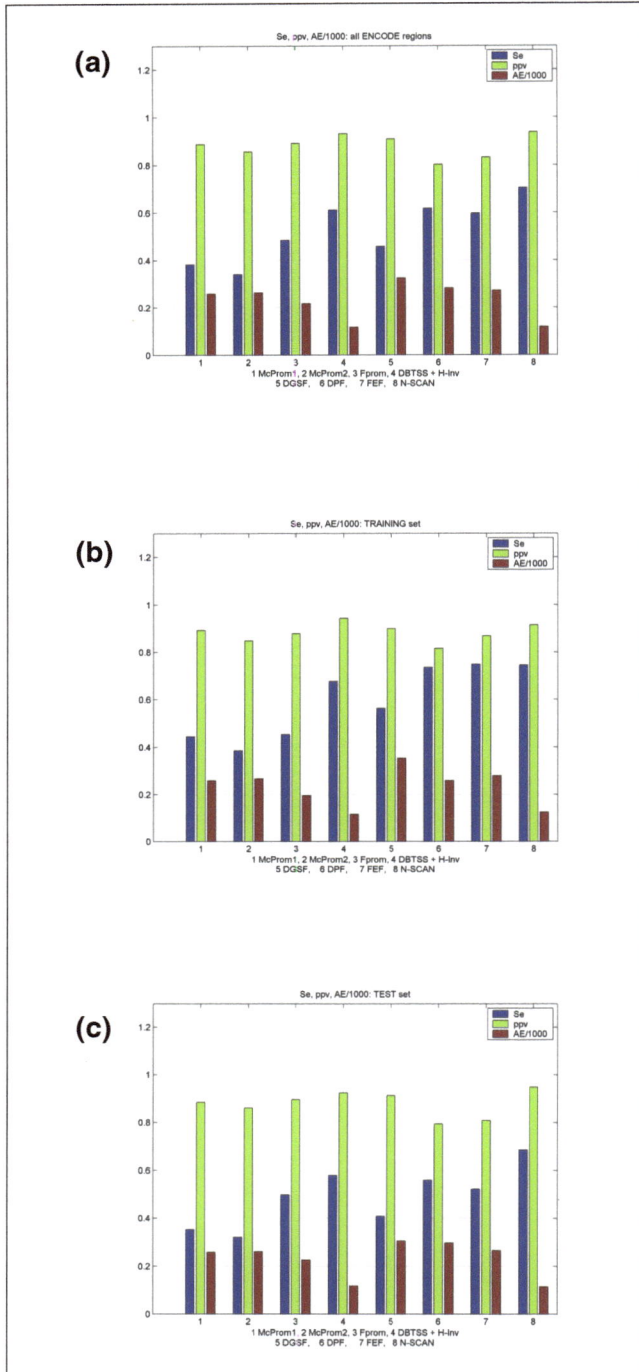

Figure 3
The results for different ENCODE regions. The results presented are for the maximum allowed distance of 1,000 nucleotides between the predicted TSS and the reference one. AE is the average mismatch of predictions relative to the most close TSS location from the HAVANA annotation. It is divided by 1,000 to scale for the graph presentation. Results are presented for: all ENCODE regions; the training set; and the test set. Relation of scores to the predictor performance is as follows: for Se and ppv, the higher the score, the better the performance. The scores for these two measures range from 0 to 1. For AE, the lower the score, the better.

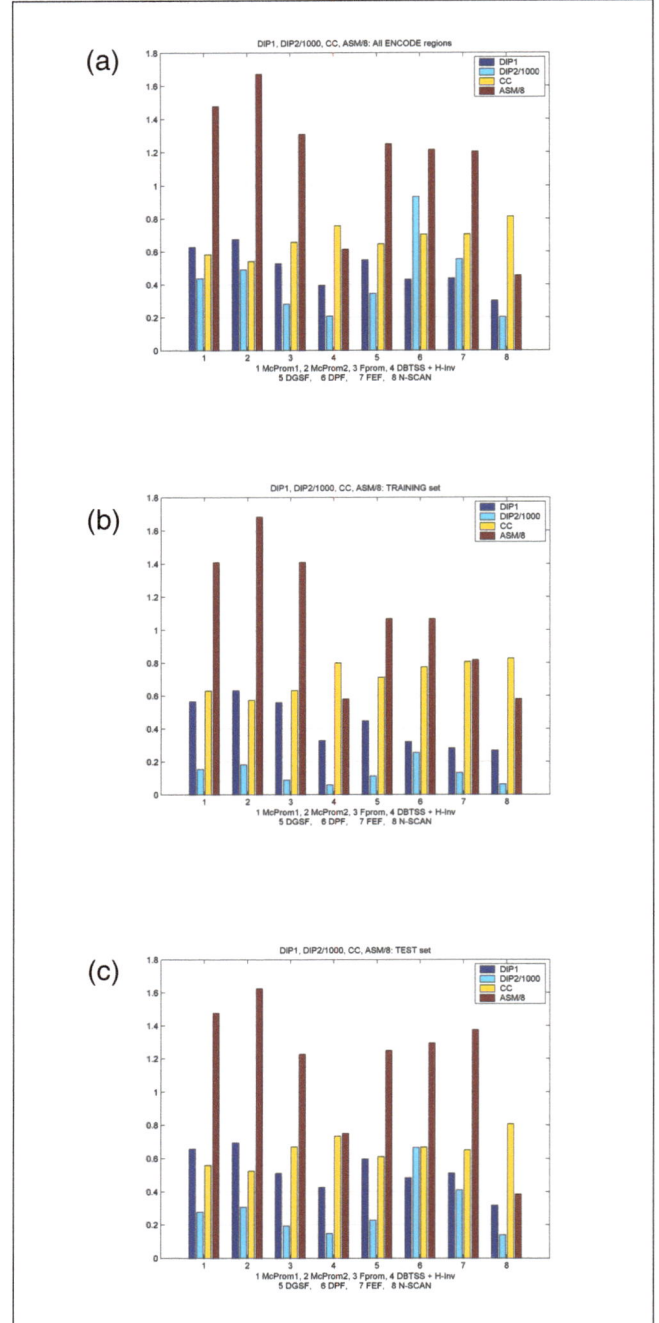

Figure 4
Another set of results for ENCODE regions. The results presented are for the maximum allowed distance of 1,000 nucleotides between the predicted TSS and the reference one. DIP1 and DIP2 are two measures of prediction qualities expressed as distances from the ideal predictor [10]. CC is the Pearson correlation coefficient. ASM is the average score measure as defined in [10]. DIP2 and ASM are scaled down to fit into the graph. Results are presented for all ENCODE regions, for the training set and for the test set. Relation of scores to the predictor performance is as follows: for distances from the ideal predictor (DIP1 and DIP2), as well as for ASM, the lower the score, the better. ASM represents the averaged rank position of the predictor calculated based on the individual measures of success. For CC, the greater the score, the better. CC ranges from -1 to +1.

prediction. This directly reduces the search space for promoters and minimizes the number of FP predictions since promoter searches are localized to the regions close to the estimated 5' end of genes. This also reduces the overall number of predictions. As a consequence, the accuracy of such programs (N-SCAN and Fprom) is somewhat increased compared to other programs. It is obvious that one could use the existing gene annotation to restrict the promoter search space. However, all programs evaluated in this study use *ab initio* predictions and do not rely on gene annotation. Moreover, programs that may rely on gene annotation to enhance promoter prediction would not work efficiently in a situation where such annotation does not exist.

The other three programs (McPromoter, DPF, DGSF) did not utilize gene structure prediction, while FEF used only a partial prediction of gene structure. In particular, McPromoter is a representative example of the aforementioned first group of successful *ab initio* genome-wide predictors, given that its version tuned for human data is essentially unchanged since its publication [24]. FEF uses an internal recognition of the first exon that is part of the overall gene structure, although it does not attempt to predict other parts of the gene structure. Also, DPF and DGSF use rough, simplified models of intron and exon domains in the promoter recognition process. These four programs (FEF, McPromoter, DPF, DGSF) have been tuned to search for promoters when no information except a single DNA sequence is available. This requires much tighter tuning in order to reduce FPs and maximize TPs. Still, their performance is considerable, keeping in mind that many of the FP predictions of these programs could be eliminated if some form of the complete gene structure prediction is used. One should note that the idea that promoter predictions can benefit from gene prediction is not new. One of the early suggestions in this direction was given in [9]. Although a similar idea has been contemplated by others, such as in [19], it has never been supported by comprehensive data. Our report seems to be the first one to provide such evidence on a larger scale.

The comparison analysis [12] focused on programs that do not use additional gene prediction. That study has demonstrated that a strong beneficial effect in accuracy can be achieved for many promoter predictors if masking repeats is used and the promoter search is restricted to non-masked regions. In the current study we reach a similar general conclusion on improved accuracy when restricted search space is used, in the context of combining promoter prediction with gene prediction. Note that N-SCAN has also used masking repeats in the context of their gene prediction.

Finally, we comment on the generally better performance of promoter predictors on the training ENCODE set as opposed to the test ENCODE set. The simple explanation could be that it is a consequence of the increased GC content of the

training ENCODE regions (44.69%) compared to the test ENCODE regions (42.33%). Usually, GC rich isochores represent more dense gene regions than the GC depleted isochors (at least based on current data). We also know [12] that many promoter predictors more efficiently predict GC rich promoters, which complies with the results on the ENCODE regions. However, since the DBTSS and H-Invitational TSS set shows better concordance with the HAVANA annotation data, it is also possible that part of the answer is in a more detailed and accurate annotation of the training set.

The reference TSS locations and TSS estimates from DBTSS and H-Invitational databases

We have used the HAVANA group's manual annotation of the ENCODE regions and considered the annotated 5' ends of transcripts as the reference TSS locations. As an alternative, we also used DBTSS and H-Invitational data-bases as a source of another collection of estimated TSS locations. Since this second collection is based on flcDNAs, of which many were oligo-capped, the TSS estimates based on this dataset should largely correspond or be close to genuine TSS locations. Actually, a recent report [43] indicates that 7% of the TSSs estimated from the oligo-capped flcDNAs of DBTSS mismatch by more than 100 bp those from the Eukaryotic Promoter Database (EPD) [44], while no precise estimates of distance mismatch are given for the remaining 93% of the DBTSS TSSs that fall within 100 bp of the corresponding EPD TSSs.

We then compared the HAVANA annotation and the TSS predictions based on the DBTSS and H-Invitational data-bases. It was somewhat disconcerting to find that sensitivity was only 58% with the DBTSS and H-Invitational data relative to the HAVANA reference set. Moreover, the ppv was only 92%. This estimation was done using the maximum allowed distance mismatch of 1,000 nucleotides between the estimated TSS and HAVANA annotated 5' gene ends. For those DBTSS and H-Invitational TSSs that did satisfy the distance criterion, the average positional error relative to HAVANA based estimates was 117 nucleotides, again a significant difference. Of the DBTSS and H-Invitational TSSs, 42% were more than 1,000 nucleotides apart from the closest HAVANA annotated 5' gene end. Although HAVANA gene structures may be based on the same mRNA evidence as DBTSS and H-invitational TSS predictions, HAVANA annotation may introduce a bias towards the most 5' TSS for some genes as gene structures are extended as far as other mRNAs and ESTs with identical exon structures support them (see Materials and methods). However, HAVANA annotation only uses spliced mRNA and ESTs as evidence to extend gene structures and, as such, would fail to extend the 5' end of a gene upstream where only single exon evidence supported it. Furthermore, mRNAs used by DBTSS and H-invitational to predict TSSs may not be used in HAVANA annotation to support coding genes, or possibly any gene structure, if their predicted CDSs appear questionable in its genomic context. The annotation of coding genes and splice variants supported by human ESTs and non-human mRNAs and ESTs by HAVANA may also result in 5' ends of genes being identified that are not represented in the current DBTSS and H-invitational databases. However, being aware that the experimental support for accurate TSS location is not easy to provide, we believe that this issue requires a separate and in-depth study, particularly when the CAGE data [3] have become available.

Although we used the HAVANA annotation as a reference dataset, we do not treat it as the 'gold standard' for promoter prediction. We are fully aware of the fact that there is no universally accepted genomic scale 'gold standard' for the accurate TSS locations that we could use. Different sets of experimental data bear the bias of the shortcomings of the experimental procedures used in experiments or of the post-processing of these data. One may argue that the TSS estimates based on the DBTSS and H-Invitational database could be more reliable. However, one should not forget that TSS estimates from DBTSS and H-Invitational databases are also not guaranteed to be correct. Thus, blindly assuming that one set is good while the other is not without an in-depth evaluation of the experimental data is not justified. For this reason, we emphasize that the conclusions of our study are based on the constraints and framework defined in EGASP and those of our analysis, and they are valid to that extent.

The differences between the reference set and TSS estimates from the DBTSS and H-Invitational databases may explain the sensitivity results achieved by programs used in this study - for example, the decline in sensitivity for programs such as FEF, DPF, DGSF and McPromoter that were evaluated in [12] where DBTSS data was used as a reference. In any case, the HAVANA annotation currently represents the best gene annotation for the ENCODE regions. We believe that this has resulted in an increased ppv for promoter predictors in this study. Specifically, when we compare the ppv results from [12], we find that FEF, DPF, DGSF and McPromoter all have a much higher ppv on the ENCODE data and the associated HAVANA annotation, likely because of the more accurate annotation of gene loci regions.

Comparison with a previous study on the whole human genome

The direct comparison of the results of this study and the one performed recently on the whole human genome [12] is not possible simply for the reason that the reference data against which assessments are made are different. In [12] we used the whole human genome and the data from DBTSS; in the current study we used HAVANA annotation as the reference and focus only on ENCODE regions that make up about 1% of the whole human genome. In addition, the two datasets are not very similar, as we have already shown.

However, in spite of these differences in the reference dataset, we are still in a position to make some global observations. Compared to the previous whole human genome analysis [12], in this study we used a more stringent distance constraint: the maximum allowed mismatch of the predicted TSS from the reference TSS was 1,000 nucleotides. In [12] as the maximum allowed was 2,000 nucleotides. Because of this, one would expect the decrease in ppv, but we observe the opposite trend for all programs that were evaluated in [12] (FEF, DPF, DGSF and McPromoter). In [12] the reported ppv was in the range 25% to 67%. In the current study, with the stringent distance criterion, the ppv for these programs is in the range 79% to 91%, which is a positive surprise. For N-SCAN and Fprom, which were not included in [12], the ppv is also very high at 94% and 89%, respectively. Sensitivities for FEF, DPF, DGSF and McPromoter were, in [12], in the range 54% to 80% and in this study, as expected, they have been reduced, falling to the range 32% to 56%. However, one should be cautious in drawing conclusions as the DBTSS and H-Invitational TSS set shows only 58% sensitivity and a 92% ppv relative to the HAVANA annotation.

Another positive surprise is the positional accuracy of promoter predictors. Note that for experimental DBTSS and H-Invitational TSSs the positional error is 117 nucleotides. All promoter predictors in the current study achieved an average positional error in the range 226 to 305 nucleotides relative to the HAVANA annotation. This is only two- to three-fold larger than the average positional error of the DBTSS and H-Invitational experimental data.

Future developments

The lessons from EGASP relative to promoter predictions is that it is beneficial to combine the TSS/promoter predictors with gene finding programs irrespective how gene prediction is done. Using such an approach it will be possible to retune promoter predictors and also to partly change their design philosophy since more relaxed conditions will be required due to the restricted search space.

However, this cannot be a final solution as it will inevitably bias the predictions to only those towards the 5' gene end, or, at best, extend predictions to cover the whole body of the gene. The intergenic space will be covered only to the extent provided by the abilities of gene finding programs to detect new genes by *ab initio* methods.

Although most of the promoter predictors today can detect TSSs on the basis of an *ab initio* approach, we need to enhance their predictive ability. The ultimate solution will be to mimic the cellular transcription initiation process through technical implementation in promoter predictors. That is likely to allow efficient detection of a broad range of genuine TSSs in arbitrary genomic sequence irrespective of the support from experimental data or gene predictions. This is a challenging task and requires more sophisticated technical solutions that take advantage of the molecular biology of promoter regulation.

We also observe that the positional accuracy of promoter predictors requires further improvement. A recent review [14] proposed that the next goal in positional accuracy of promoter predictors is a 20 nucleotides mismatch relative to the experimental TSS locations, that is, on the same scale as naturally observed variation in the initiation process. However, this leaves the open issue of a good reference dataset. But, if we intend to achieve this goal, we have to incorporate more of the relevant biological information in the recognition algorithms. Related to this is also the following problem. Due to the massive expressed data (EST/cDNA/mRNA) available, annotation naturally uses such sources of information. Promoter prediction programs that utilize expressed sequences should generate predictions most close to the annotation based reference dataset, as this is more or less how the reference annotation is derived as well. This brings into focus an issue of circularity that will just confirm that promoter predictors that use such strategies comply well with the annotated data.

Scenario for promoter prediction for future experiments

Lessons from the current experiment motivate us to propose a framework for future promoter prediction assessment. It is absolutely necessary to conduct promoter prediction experiments within different categories of conditions that programs utilize, so as to be in a position to compare individual contributions of different types of information used. Two broad scenarios are of interest: one that assesses the genomic context within which the predictions are made, and another that assesses types of data/information used in deriving predictions.

In the first group, it will be helpful to consider separately methods that utilize only the immediate region surrounding a TSS (say [-200,+200]), as opposed to those that use a much broader genomic context. The reason for this is to evaluate the contribution of global and local signals in promoter predictions. The latter methods can include those that make use of gene structure prediction.

The second group could include: *ab initio* predictions based exclusively on the use of genomic sequence from one genome; *ab initio* predictions that use only genomic sequences from multiple genomes; predictions that utilize different support information (that is, known protein mapping, and so on), but not transcript data (that is, mRNA/EST mapping); and predictions that use information from mapping transcript data, as well as any other information. The comparison of programs would make sense only within categories, but not across various categories.

Conclusions

The current study argues in favor of combining promoter predictions with gene structure predictions as an intermediate improvement for promoter prediction accuracy. The long term goal has to be the development of a positionally accurate *ab initio* promoter prediction solution. For the next EGASP or similar project, different categories of promoter predictions should be provided, to enable the comparison of approaches differing on a large scale and the assessment of contributions of different types of information used in solutions. These in return would allow for more efficient promoter prediction programs.

Materials and methods
EGASP participants
We analyzed the following prediction sets provided in response to the EGASP call for submissions: 7-80-8 (McPromoter, the standard system), 7-81-8 (McPromoter with post-processing of shadowed predictions), 41-108-8 (Fprom), 20_76_4 (N-SCAN).

Additional prediction sets
To make the assessment of promoter predictions more complete, we also added four additional set of predictions, the TSSs estimated based on the DBTSS and H-Invitationsl data, which represent a large-scale experimental TSS dataset based on capped flcDNA, and those from FEF, DGSF, and DPF.

McPromoter
McPromoter is an *ab initio* system for predicting transcription start sites and was among the first fully probabilistic approaches to this problem. It uses a sequence of six Markov chain models for different subregions and elements within a core promoter spanning position -250 to +50, such as TATA-box, spacer, and initiator regions. As the core promoter is considerably different for distantly related eukaryotes, we have trained two separate models on vertebrate (mammalian) and invertebrate (fly) sequences. The *Drosophila* system has been under constant development [23], motivated by the identification of additional core promoter elements such as DPE (reviewed in [2]). The mammalian system has essentially remained constant throughout several years, including the data set it is trained on (a set of 565 sequences taken from the EPD) [24]. Small differences result from different strategies for the post-processing of the initial posterior probabilities of the predictor: For instance, submission 7-81-8 addressed the issue of shadow predictions, that is, simultaneous predictions on both strands of a core promoter caused by a strong signal in base composition. Here, we removed a lower scoring prediction if it fell within 1 kb of a higher scoring prediction of the standard system (7-80-8) on the opposite strand. However, as the results clearly show, this simple strategy actually decreased the performance slightly, indicating that a fraction of TPs is accompanied by stronger scoring predictions on the opposite

strand in close proximity. The version of the McPromoter program used is MM:II, with a threshold of +0.005. The program can be found at [45].

Fprom: Softberry Pol-II promoter recognition approach

The task of finding eukaryotic polymerase II promoter involves two internal issues: finding the exact position of TSSs within long upstream regions of eukaryotic genes; and avoiding FP predictions within exon and intron sequences. To resolve the second part of this task some authors of promoter finding software include some recognition procedures of gene coding parts inside promoter prediction programs [15,28]. However, gene finding software such as Genscan [46] or Fgenesh [47] provides a much better accuracy in coding exon-intron identification than any such empirical procedures. We think that the best promoter identification strategy is to predict all gene components in one program. In creating such a program, it has currently been decided to use some intermediate variant, which includes the following steps: compute the gene annotation using a gene prediction pipeline and run promoter prediction on 5'-regions upstream of the annotated coding regions of predicted genes.

For promoter location within the selected regions, we used the Fprom (find promoter) program, which is the development of an algorithm realized earlier in the TSSW/TSSG programs [48]. For each potential TSS position of a given sequence, the Fprom program evaluates its possibility to be a TSS using two linear discriminant functions (for TATA+ and TATA- promoters) with characteristics computed in the [-200,+50] region around the given position. For TATA promoter recognition we consider the following features selected by discriminant analysis on the learning set of known promoters: hexamers in region [-200,-45]; hexamers in region [0:+40]; triplets in region [-200,-45]; triplets in region [0,+40]; TATA box maximal weight in interval [-45,-25]; TATA box average score on interval [-45,-25]; CpG-content; position triplet matrix in the [TSS-50,TSS+30] region; similarities between [-200,-100] and [-100,-1] regions; protein-DNA twist; protein-induced deformability; regulatory motif density in region [-200,-101] in the direct chain; and regulatory motif density in region [-100,-1] in the reverse chain.

If we find a TATA-box (using TATA-box weight matrix) in the positions [-45,-25] of the analyzed region, then we compute the value of LDF for TATA+ promoters, otherwise the value of the linear discriminant function (LDF) for TATA-less promoters. Only one prediction, with the highest LDF score and greater than some threshold, is selected within any 300 bp region. We run Fprom on 5' regions extracted from the predicted genes. For each such region, we selected the closest to the CDS predicted promoter and presented it in our results. The Fprom program can be found at the Softberry's web site at [49] and contains no user adjustable parameters.

N-SCAN

N-SCAN [50,51] is an extension of TWINSCAN [52]. N-SCAN's DNA sequence modeling is identical to TWINSCAN with the addition of states modeling 5' UTR exons and introns [30] and the capability to include conserved noncoding states in intergenic regions. N-SCAN's method of incorporating alignment information is quite different from TWINSCAN's method. TWINSCAN utilizes alignment information from one informant genome through a conservation sequence. A conservation sequence is generated by assigning each target sequence base a match, mismatch/gap or unaligned symbol based on a BLASTN alignment of the two genomes. N-SCAN replaces TWINSCAN's conservation sequence with a multiple genome alignment that represents the evolutionary relationships among the target and multiple informants with a Bayesian network rooted at the target genome along with a richer alphabet representing a more detailed modeling of substitution rates, insertions, and deletions across all informants. N-SCAN does not predict TSSs as isolated features, but rather as the 5' boundary of the first exon in a gene structure.

N-SCAN's human gene predictions employed human genome Build hg17 (May 2004), the corresponding RefSeq mappings, and a whole-genome, 8-way, MULTIZ alignment, which were all downloaded from UCSC [53]. The particular alignment subset chose human (hg17) as the target genome and mouse (mm5), rat (rn3), and chicken (galGal2) as informants, with all gaps in the target removed. Build hg17 was masked for interspersed repeats, but not low-complexity or simple repeats as identified by UCSC. The human sequence was further pseudogene masked (MJ van Baren and MR Brent 2005, submitted). The RefSeq mappings were filtered to remove probable errors; parameters were trained on three-quarters of the filtered RefSeq mappings. The program design and setting is explained in the companion article [54].

First Exon Finder

The main idea implemented in FEF [19] is that promoter prediction should be derived from prediction of the first exon. This is implemented by splitting the first exons into two groups, one that is GC rich and another that is GC poor. Several types of compositional features are used in the recognition process that is implemented as a rule-based solution with several quadratic discriminant functions. In [12], FEF was found to be among the best *ab initio* promoter predictors. It was also found that its performance benefits if combined with masking repeats by RepeatMasker. The recommendations from [12] were implemented with the default FEF parameter setting: a cutoff value for the first-exon *a posteriori* probability of 0.5, a cutoff value for the promoter *a posteriori* probability of 0.4, and a cutoff value of the splice-donor *a posteriori* probability of 0.4. We used the download version of the program. The web-server implementation can be found at [55].

Dragon Promoter Finder

DPF [15,16] uses three types of models for promoter regions, exonic regions and intronic regions. It utilizes position weight matrices of overlapping pentamers in these three regions to derive its predictions. The program uses separation of promoters to GC rich and GC poor groups and uses five different prediction models for different levels of sensitivity. It uses only 200 nucleotides DNA segments to make predictions. In this study, it was used with the default parameters and according to recommendations from [12], which combine predictions with masking repeats by RepeatMasker and uses clustering of its predictions. This means that predictions are clustered if the distance between the neighboring predictions is 1,000 nucleotides or less. Such clusters are represented by the average position of predictions in the cluster. The program version 1.5 was run with the expected sensitivity of 0.65 and according to recommendations from [12]. The program can be found at [56].

Dragon Gene Start Finder

DGSF [17,18] uses predictions of DPF in the region it assesses to be a CpG island. The program is aimed at finding the approximate start of gene loci. It first localizes the CpG island and then identifies the most likely DPF prediction within that region. Version 1.0 of the program was run with its default threshold parameter of 0.994 and according to recommendations from [12]. The program can be found at [57].

Counting predictions and other performance measures

The counting of TP and FP predictions is illustrated in Figure 3. If the maximum allowed distance of the prediction form the closest reference TSS on the same strand is D nt, then, if one or more predictions fall on the region [-D,+D] relative to the reference TSS location and on the same strand where the TSS resides, the TSS is counted as TP. If the reference TSS is missed based on this type of counting, then such a TSS is a false negative (FN). All reference TSS locations that were missed by this counting of TP predictions represent true negatives (TN). Every other prediction that falls on the annotated part of the gene loci in the segment [+D+1,EndOfTheGene] at the same strand where TSS resides counts as a FP. One has to be aware that some real TSSs/promoters could be in the regions [+1001,EndOfTheGene]. The other predictions were not taken for the determination of TPs and FPs. Figure 5 illustrates the counting method.

The measures of performance were those used in [12]. In determining the average distance of predictions, only the minimum distance of one prediction from all reference TSSs was considered. Sensitivity is the proportion of correct predictions of TSSs relative to all experimental TSSs, defined as:

$$Se = TP/(TP + FN)$$

A ppv is the proportion of correct predictions of TSSs out of all counted positive predictions, defined as:

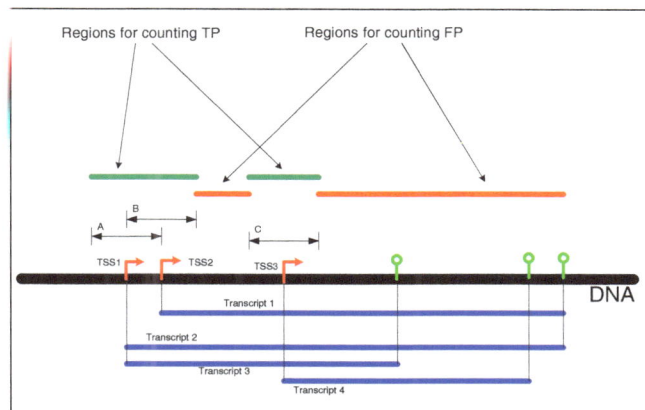

Figure 5

The counting method for TPs and FPs. All hits to the 'orange' segments count as FPs. Only one hit within A, B, or C counts as a TP for a unique position of TSS (for example, three hits within C will count only as one TP). Note that all TSS locations that were mutually different were considered as valid reference TSSs. So, alternative TSSs were considered different TSSs. Each of these had to be predicted. If one prediction falls on the intersection of A and B, then that prediction identifies two TSS locations (one that correspond to TSS related to A, and the other corresponding to TSS related to B). In other words, one prediction correctly identifies all reference TSS locations within the distance criterion.

$$ppv = TP/(TP + FP)$$

The CC is the Pearson correlation coefficient, defined as:

$$CC = (TP \times TN - FP \times FN)/ ((TP + FP)(TP + FN)(TN + FP)(TN + FN))^{1/2}$$

Data

The ENCODE regions mapped at the human genome Build hg17 (May 2005) were used. Out of the HAVANA annotation for ENCODE regions we analyzed only the category of known genes with CDS (category 2). After eliminating the redundant TSS locations, we obtained 994 unique TSSs for all ENCODE regions, 319 unique TSSs in the ENCODE 'training' set (13 regions), and 675 unique TSSs in the ENCODE test set. Note that the region ENr313 does not have any annotation. The length of DNA sequences in these regions is: all regions 29,998,060 bp; 'training' regions 8,538,447 bp; and testing regions 21,459,613 bp.

Reference TSS locations: HAVANA annotated 5' end of gene objects

All HAVANA/GENCODE annotation is based on primary EST, mRNA and protein evidence and structures are only extended as far as the supporting evidence allows. No automated predictions are used to support gene objects. Main gene structures are based on human (and where a novel structure with canonical splicing is supported) non-human mRNA and EST evidence identified in the nucleotide sequence databases and aligned by wuBLASTN [58]. Significant hits are re-aligned to the unmasked genomic sequence using est2genome [59] and proteins aligned by wuBLASTX. All evidence is navigated using the Blixem alignment viewer [60]. The 5' ends of gene structures are extended using only splicing human mRNA and EST evidence that agrees completely with the structure of the gene object that it is used to extend. As such, where mRNAs and ESTs support an identical gene structure but have different length 5' UTRs, they are merged into a single gene structure that is extended as far as the longest supported 5' UTR, that is to the most 5' aligned base of the most 5' EST or mRNA. Where sequence from the 5' end of mRNA and EST evidence is missing from the Est2genome alignment, visual inspection of the dot-plot output from the Dotter tool [60] is used in an attempt to identify any alignment with the genomic sequence upstream of the identified end of homology. Where a very short length of sequence (<15 bases) is missing from the 5' end of the alignment, a dot-plot is unsuitable due to the difficulty in seeing very short alignments at the edge of the display and the AcedB Restriction Analysis tool (essentially a pattern matching tool) [61] is used to try and identify any alignment with the genome. As such, the annotated 5' ends of gene objects are specified according to the best possible alignment of transcriptional evidence to the genome rather than specifically identifying TSSs in the genomic sequence. As new transcript evidence is added to the databases, so novel 5' exons and 5' extensions of existing exons continue to be identified.

TSS estimates from DBTSS and H-Invitational databases

Using DBTSS data (version 4.2, 11 Jan 2005), we obtained 12,763 TSS estimates for hg17, and of these, 286 were mapped to ENCODE regions. These were complemented by H-Invitational TSS data. We used 95% identity and 90% homology in BLAST mapping of H-Invitational data to hg17. This provided us with 20,116 TSS estimates. Within the ENCODE regions we found 325 TSS estimates not overlapping with DBTSS data. In total, the DBTSS and H-Invitational datasets provided 611 experimental TSS locations. These are provided as Additional data files 1 and 2.

Additional data files

The following additional data are available with the online version of this paper. Additional data file 1 lists the DBTSS TSS locations. Additional data file 2 lists the H-Invitational TSS locations.

Acknowledgements

This article has been published as part of *Genome Biology* Volume 7, Supplement 1, 2006: EGASP '05. The full contents of the supplement are available online at http://genomebiology.com/supplements/7/S1.

References

1. Weinzierl ROJ: *Mechanisms of Gene Expression: Structure, Function, and Evolution of the Basal Transcriptional Machinery.* London: Imperial College Press; 1999.
2. Smale ST, Kadonaga JT: **The RNA polymerase II core promoter.** *Annu Rev Biochem* 2003, **72**:449-479.
3. FANTOM Consortium; RIKEN Genome Exploration Research Group and Genome Science Group (Genome Network Project Core Group): **The transcriptional landscape of the mammalian genome.** *Science* 2005, **309**:1559-1563.
4. RIKEN Genome Exploration Research Group, Genome Science Group (Genome Network Project Core Group) and FANTOM Consortium: **Antisense transcription in the mammalian transcriptome.** *Science* 2005, **309**:1564-1566.
5. Pedersen AG, Baldi P, Chauvin Y, Brunak S: **The biology of eukaryotic promoter prediction - a review.** *Computers Chem* 1999, **23**:191-207.
6. Maruyama K, Sugano S: **Oligo-capping: a simple method to replace the cap structure of eukaryotic mRNAs with oligoribonucleotides.** *Gene* 1994, **138**:171-174.
7. Carninci P, Kvam C, Kitamura A, Ohsumi T, Okazaki Y, Itoh M, Kamiya M, Shibata K, Sasaki N, Izawa M, *et al.*: **High-efficiency full-length cDNA cloning by biotinylated CAP trapper.** *Genomics* 1996, **37**:327-336.
8. Kim TH, Barrera LO, Qu C, Van Calcar S, Trinklein ND, Cooper SJ, Luna RM, Glass CK, Rosenfeld MG, Myers RM, Ren B. **Direct isolation and identification of promoters in the human genome.** *Genome Res* 2005, **15**:830-839.
9. Fickett JW, Hatzigeorgiou AG: **Eukaryotic promoter recognition.** *Genome Res* 1997, **7**:861-878.
10. Prestridge DS: **Computer software for eukaryotic promoter analysis.** *Methods Mol Biol* 2000, **130**:265-295.
11. Bajic VB: **Comparing the success of different prediction software in sequence analysis: A review.** *Brief Bioinform* 2000, **1**:214-228.
12. Bajic VB, Tan SL, Suzuki Y, Sugano S: **Promoter prediction analysis on the whole human genome.** *Nat Biotechnol* 2004, **22**:1467-1473.
13. Ohler U, Frith M: **Models for complex eukaryotic regulatory DNA sequences.** In *Information Processing and Living Systems.* Edited by Bajic VB, Tan TW. London, UK: Imperial College Press, 2005, 575-610.
14. Bajic VB, Werner T: **Promoter prediction.** In *Encyclopedia of Genetics, Genomics, Proteomics and Bioinformatics.* Part 4. Bioinformatics, 4.2. Gene Finding and Gene Structure. (Editors: Dunn MJ, Jorde LB, Little PF, Subramaniam S); John Wiley and Sons, Ltd; Hoboken, New Jersey 2005: DOI: 10.1002/047001153X.g402301.
15. Bajic VB, Seah SH, Chong A, Zhang G, Koh JLY, Brusic V: **Dragon Promoter Finder: recognition of vertebrate RNA polymerase II promoters.** *Bioinformatics* 2002, **18**:198-199.
16. Bajic VB, Seah SH, Chong A, Krishnan SP, Koh JL, Brusic V: **Computer model for recognition of functional transcription start sites in RNA polymerase II promoters of vertebrates.** *J Mol Graphics Model* 2003, **21**:323-332.
17. Bajic VB, Seah SH: **Dragon Gene Start Finder identifies approximate locations of the 5' ends of genes.** *Nucleic Acids Res* 2003, **31**:3560-3563.
18. Bajic VB, Seah SH: **Dragon Gene Start Finder: an advanced system for finding approximate locations of the start of gene transcriptional units.** *Genome Res* 2003, **13**:1923-1929.
19. Davuluri RV, Grosse I, Zhang MQ: **Computational identification of promoters and first exons in the human genome.** *Nature Genetics* 2001, **29**:412-417.
20. Down TA, Hubbard TJ: **Computational detection and location of transcription start sites in mammalian genomic DNA.** *Genome Res* 2002, **12**:458-461.
21. Reese MG: **Application of a time-delay neural network to promoter annotation in the *Drosophila melanogaster* genome.** *Computers Chem* 2001, **26**:51-56.
22. Knudsen S: **Promoter2.0: for the recognition of PolII promoter sequences.** *Bioinformatics* 1999, **15**:356-361.
23. Ohler U, Liao GC, Niemann H, Rubin GM: **Computational analysis of core promoters in the Drosophila genome.** *Genome Biol* 2002, **3**:RESEARCH0087.
24. Ohler U, Stemmer G, Harbeck S, Niemann H: **Stochastic segment models of eukaryotic promoter regions.** *Proc Pacific Sym Biocomputing* 2000, **5**:380-391.
25. Ponger L, Mouchiroud D: **CpGProD: identifying CpG islands associated with transcription start sites in large genomic mammalian sequences.** *Bioinformatics* 2002, **18**:631-633.
26. Hannenhalli S, Levy S: **Promoter prediction in the human genome.** *Bioinformatics* 2001, **17(Suppl)**:S90-S96.

27. Ioshikhes IP, Zhang MQ: **Large-scale human promoter mapping using CpG islands.** *Nat Genet* 2000, **26:**61-63.
28. Scherf M, Klingenhoff A, Werner T: **Highly specific localization of promoter regions in large genomic sequences by PromoterInspector: a novel context analysis approach.** *J Mol Biol* 2000, **297:**599-606.
29. Solovyev VV, Shahmuradov IA, Prom H: **Promoters identification using orthologous genomic sequences.** *Nucleic Acids Res* 2003, **31:**3540-3545.
30. Brown RH, Gross SS, Brent MR: **Begin at the beginning: predicting genes with 5' UTRs.** *Genome Res* 2005, **15:**742-747.
31. Liu R, States DJ: **Consensus promoter identification in the human genome utilizing expressed gene markers and gene modeling.** *Genome Res* 2002, **12:**462-469.
32. Reese MG, Hartzell G, Harris NL, Ohler U, Abril JF, Lewis SEL: **Genome annotation assessment in *Drosophila melanogaster.*** *Genome Res* 2000, **10:**483-501.
33. Carninci P, Sandelin A, Lenhard B, Katayama S, Shimokawa K, Ponjavic J, Semple CA, Taylor MS, Engstrom PG, Frith MC, *et al.*: **Genome-wide analysis of mammalian promoter architecture and evolution.** *Nat Genet* 2006, **38:**626-635.
34. Lee CK, Shibata Y, Rao B, Strahl BD, Lieb JD: **Evidence for nucleosome depletion at active regulatory regions genome-wide.** *Nat Genet* 2004, **36:**900-905.
35. Guigó R, Flicek P, Abril JF, Reymond A, Lagarde J, Denoeud F, Antonarakis S, Ashburner M, Bajic VB, Birney E, *et al.*: **EGASP: The human ENCODE Genome Annotation Assessment Project.** *Genome Biol* 2006, **7(Suppl 1):**S2.
36. ENCODE Project Consortium: **The ENCODE (ENCyclopedia Of DNA Elements) Project.** *Science* 2004, **306:**636-640.
37. **The HAVANA Team** [http://www.sanger.ac.uk/HGP/havana/]
38. **The GENCODE Project** [http://genome.imim.es/gencode/]
39. Harrow J, Denoeud F, Frankish A, Reymond A, Chen C-K, Chrast J, Lagarde J, Gilbert JGR, Storey R, Swarbreck D, *et al.*: **GENCODE: producing a reference annotation for ENCODE.** *Genome Biol* 2006, **7(Suppl 1):**S4.
40. Suzuki Y, Yamashita R, Sugano S, Nakai K: **DBTSS, DataBase of Transcriptional Start Sites: progress report 2004.** *Nucleic Acids Res* 2004, **32:**D78-D81.
41. Imanishi T, Itoh T, Suzuki Y, O'Donovan C, Fukuchi S, Koyanagi KO, Barrero RA, Tamura T, Yamaguchi-Kabata Y, Tanino M, *et al.*: **Integrative annotation of 21,037 human genes validated by full-length cDNA clones.** *PLoS Biol* 2004, **2:**e162.
42. **EGASP Submissions** [ftp://genome.imim.es/pub/projects/gencode/data/egasp05/egasp_submissions_20050503/submissions_bysubmitter.pdf]
43. Yamashita R, Suzuki Y, Wakaguri H, Tsuritani K, Nakai K, Sugano S: **DBTSS: DataBase of Human Transcription Start Sites, progress report 2006.** *Nucleic Acids Res* 2006, **34(Database issue):**D86-89.
44. Praz V, Perier R, Bonnard C, Bucher P: **The Eukaryotic Promoter Database, EPD: new entry types and links to gene expression data.** *Nucleic Acids Res* 2002, **30:**322-324.
45. **McPromoter MM:II** [http://genes.mit.edu/McPromoter.html]
46. Burge C, Karlin S: **Prediction of complete gene structures in human genomic DNA.** *J Mol Biol* 1997, **268:**78-94.
47. Salamov AA, Solovyev VV: **Ab initio gene finding in Drosophila genomic DNA.** *Genome Res* 2000, **10:**516-522.
48. Solovyev VV, Salamov AA: **The Gene-Finder computer tools for analysis of human and model organisms genome sequences.** In *Proceedings of the Fifth International Conference on Intelligent Systems for Molecular Biology: 21-25 June; Halkidiki, Greece.* Edited by Rawling C, Clark D, Altman R, Hunter L, Lengauer T, Wodak S. AAAI Press; Menlo Park, CA, USA 1997:294-302.
49. **Fprom** [http://www.softberry.com/berry.phtml?topic=fprom&group=programs&subgroup=promoter]
50. Gross SS, Brent MR: **Using multiple alignments to improve gene prediction.** In *Research in Computational Molecular Biology: Proceedings of the 9th Annual International Conference, RECOMB 2005; Boston.* Edited by Miyano S, Mesirov JP, Kasif S, Istrail S, Pevzner PA, Waterman MS;. Cambridge, MA, Springer; 2005:374-388.
51. Gross SS, Brent MR: **Using multiple alignments to improve gene prediction.** *J Comput Biol* 2006, **13:** 379-393.
52. Korf I, Flicek P, Duan D, Brent MR: **Integrating genomic homology into gene structure prediction.** *Bioinformatics* 2001, **17 (Suppl 1):**S140-S148.
53. **UCSC Browser** [http://genome.ucsc.edu/]
54. Arumugam M, Wei C, Brown RH, Brent MR: **Pairagon+N-SCAN_EST: a model-based gene annotation pipeline.** *Genome Biol* 2006, **7(Suppl 1):**S5.
55. **First Exon Finder** [http://rulai.cshl.edu/tools/FirstEF/]
56. **Dragon Promoter Finder** [http://research.i2r.a-star.edu.sg/promoter/promoter1_5/DPF.htm]
57. **Dragon Gene Start Finder** [http://research.i2r.a-star.edu.sg/promoter/dragonGSF1_0/genestart.htm]
58. **wuBLASTN** [http://blast.wustl.edu]
59. Mott R: **EST_GENOME: a program to align spliced DNA sequences to unspliced genomic DNA.** *Comput Appl Biosci* 1997, **13:**477-478.
60. Sonnhammer EL, Wootton JC: **Integrated graphical analysis of protein sequence features predicted from sequence composition.** *Proteins* 2001, **45:**262-273
61. Durbin R, Griffiths E: **Acedb genome database.** *Genetics, Genomics, Proteomics and Bioinformatics Online. Volume 4 Bioinformatics.* Modern Programming Paradigms in Biology. Edited by Peter Clote. Boston College, Massachusetts, USA: Wiley Interscience; 2005

Research

GENCODE: producing a reference annotation for ENCODE

Jennifer Harrow*[1], France Denoeud[1†], Adam Frankish*[1], Alexandre Reymond[1‡§], Chao-Kung Chen*, Jacqueline Chrast[§], Julien Lagarde[‡], James GR Gilbert*, Roy Storey*, David Swarbreck*, Colette Rossier[‡], Catherine Ucla[‡], Tim Hubbard[†], Stylianos E Antonarakis[‡] and Roderic Guigo[†¶]

Addresses: *Wellcome Trust Sanger Institute, Wellcome Trust Campus, Hinxton, Cambridge CB10 1SA, UK. [†]Grup de Recerca en Informatica Biomedica, Institut Municipal d'Informatica Medica-Universitat Pompeu Fabra, Pg. Maritim de la Barceloneta, 08003 Barcelona, Catalonia, Spain. [‡]Department of Genetic Medicine and Development, University of Geneva Medical School and University Hospitals of Geneva, Geneva, Switzerland. [§]Center for Integrative Genomics, University of Lausanne, Lausanne, Switzerland. [¶]Centre de Regulacio Genomica, Pg. Maritim de la Barceloneta, 08003 Barcelona, Catalonia, Spain. [1]These authors contributed equally to this work.

Correspondence: Jennifer Harrow. Email: jla1@sanger.ac.uk

Published: 7 August 2006

Genome *Biology* 2006, 7(Suppl 1):S4

The electronic version of this article is the complete one and can be found online at http://genomebiology.com/2006/7/S1/S4

Abstract

Background: The GENCODE consortium was formed to identify and map all protein-coding genes within the ENCODE regions. This was achieved by a combination of initial manual annotation by the HAVANA team, experimental validation by the GENCODE consortium and a refinement of the annotation based on these experimental results.

Results: The GENCODE gene features are divided into eight different categories of which only the first two (known and novel coding sequence) are confidently predicted to be protein-coding genes. 5' rapid amplification of cDNA ends (RACE) and RT-PCR were used to experimentally verify the initial annotation. Of the 420 coding loci tested, 229 RACE products have been sequenced. They supported 5' extensions of 30 loci and new splice variants in 50 loci. In addition, 46 loci without evidence for a coding sequence were validated, consisting of 31 novel and 15 putative transcripts. We assessed the comprehensiveness of the GENCODE annotation by attempting to validate all the predicted exon boundaries outside the GENCODE annotation. Out of 1,215 tested in a subset of the ENCODE regions, 14 novel exon pairs were validated, only two of them in intergenic regions.

Conclusions: In total, 487 loci, of which 434 are coding, have been annotated as part of the GENCODE reference set available from the UCSC browser. Comparison of GENCODE annotation with RefSeq and ENSEMBL show only 40% of GENCODE exons are contained within the two sets, which is a reflection of the high number of alternative splice forms with unique exons annotated. Over 50% of coding loci have been experimentally verified by 5' RACE for EGASP and the GENCODE collaboration is continuing to refine its annotation of 1% human genome with the aid of experimental validation.

Background

The complete sequence of the euchromatic region of the human genome provides a new opportunity to establish the complete catalogue of the human genes. Although automated gene prediction has improved greatly over the years and the human gene count is thought to be between 20,000 and 25,000 protein-coding genes [1], defining a gene is not a trivial issue. According to classic genetics, genes are inheritable units responsible for an associated phenotype. Although in some cases this relationship derives from mutation of non-coding DNA or regulatory elements, in most cases it is synonymous with protein-coding genes. However, in the past four years there has been an explosion in the discovery of transcripts with no apparent coding potential (termed non-coding RNAs) and there are indications these could play as important a role in cellular function as proteins [2,3].

In an effort to investigate and understand all the functional elements in the human genome, the ENCODE project (Encyclopedia of DNA Elements) [4] was established. In this pilot stage, the aim of the ENCODE project is to investigate in great depth, computationally and experimentally, 44 regions totaling 30 Mb of sequence representing approximately 1% of the human genome. As part of this project, the GENCODE consortium [5] was formed to identify and map all protein-coding genes within the ENCODE regions. This is achieved by a combination of initial manual annotation by the HAVANA team [6], experimental validation by the GENCODE consortium, and a refinement of the annotation based on these experimental results (Figure 1).

This annotation is used as a reference set by all the ENCODE consortium members. It also represents the standard to which the automated prediction programs were assessed during the ENCODE Genome Annotation Assessment Project (E-GASP) 05 workshop (see [7] in this issue). This report describes how the manual annotation and experimental verification were performed. It also highlights some interesting features in the GENCODE annotation and indicates the weaknesses of the automated predictions compared to the manual annotation.

Results and discussion

Initial classification of loci

The HAVANA group divides gene features into different categories of which only the first two (known and novel coding sequence (CDS)) are confidently predicted to be protein-coding genes. The common factor between all annotated gene structures is that they must be supported by transcriptional evidence, through homology to cDNA, expressed sequence tags (ESTs) and/or protein sequences. The following are the gene types first applied to the human chromosome 20 annotation [8] and later expanded to fully classify the annotation produced for the ENCODE project.

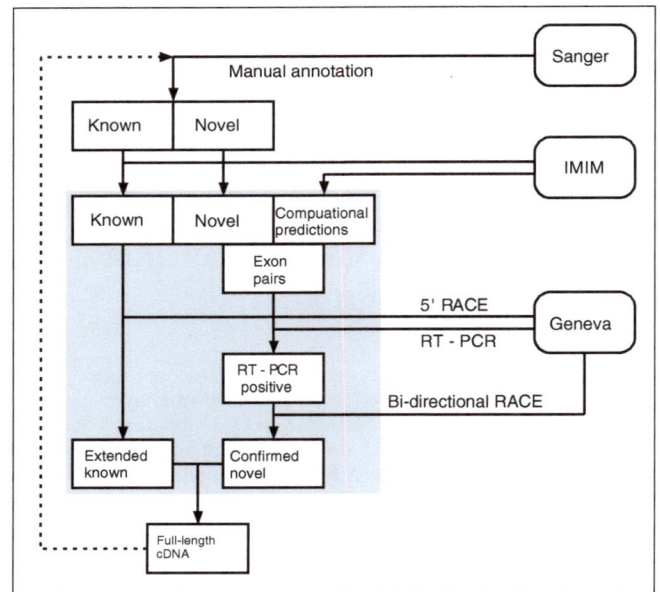

Figure 1
The GENCODE pipeline. This schematic diagram shows the flow of data between the three groups involved in the GENCODE consortium (HAVANA, IMIM and Geneva) to produce an experimentally verified annotation of the ENCODE region.

Known genes

Known genes are identical to human cDNA or protein sequences and identified by a GeneID in Entrez Gene [9].

Novel coding sequence

Novel coding sequences have an open reading frame (ORF) and are identical, or have homology, to cDNAs or proteins but do not fall into the above category; these mRNA sequences are submitted to public databases, but they are not yet represented in Entrez Gene or have not yet received an official gene name from the nomenclature committee [10]. They can also be novel in the sense that they are not yet represented by an mRNA sequence in the species concerned.

Novel transcripts

Novel transcripts are as above but no ORF can be unambiguously assigned; these can be genuine non-coding genes or they may be partial protein-coding genes supported by limited evidence. They should be supported by at least three ESTs from independent sources (not originating from the same clone identifier).

Putative genes

Putative genes are identical, or have homology, to spliced ESTs but lack a significant ORF and polyA features; these are generally short two or three exon genes or gene fragments.

Pseudogenes

Pseudogenes (assumes no expressed evidence) have homology to proteins but generally suffer from a disrupted CDS and an active homologous gene can be found at another locus. This category can be further subdivided into processed or unprocessed pseudogenes. Sometimes these entries have an intact CDS or an open but truncated ORF, in which case there is other evidence used (for example genomic polyA stretches at the 3' end) to classify them as a pseudogene.

Transcribed pseudogenes

Transcribed pseudogenes are not currently given a separate tag within GENCODE and are handled by creating a pseudogene object and an overlapping transcript object with the same locus name.

TEC (To be experimentally confirmed)

To be experimentally confirmed (TEC) is used for non-spliced EST clusters that have polyA features. This category has been specifically created for the ENCODE project to highlight regions that could indicate the presence of novel protein coding genes that require experimental validation, either by 5' rapid amplification of cDNA ends (RACE) or RT-PCR to extend the transcripts or by confirming expression of the putatively encoded peptide with specific antibodies.

Artefact gene

Artefact gene is used to tag mistakes in the public databases (Ensembl/SwissProt/Trembl). Usually, these arise from high-throughput cDNA sequencing projects, which submit automatic annotation sometimes resulting in erroneous CDSs that are, for example, 3' untranslated regions (UTRs).

GENCODE annotation of the ENCODE regions

The first release of the annotation of the 44 ENCODE regions was frozen on 29 April 2005 and was used in the E-GASP workshop. It contained 416 known loci, 26 novel CDS loci, 82 novel transcript loci, 78 putative loci, 104 processed pseudogenes and 66 unprocessed pseudogenes. The current version (release 02) was frozen on 14 October 2005. It contains 411 known loci, 30 novel CDS loci, 81 novel transcript loci, 83 putative loci, 104 processed pseudogenes and 66 unprocessed pseudogenes. The gene content has changed as a result of the experimental validation (see next section). In total, 2.9% of the nucleotides in the ENCODE regions (both strands considered separately) are covered by annotated exons (1.2% by coding and 1.7% by UTRs and non-coding), and 31% are transcribed (covered by annotated exons or introns).

Multiple transcripts are annotated at any locus where supporting evidence is available. Thus, the 487 compiled GENCODE reference loci set (compiled from coding and experimentally verified loci) corresponds to 2,608 transcripts, of which 1,097 are coding. Of the coding loci (known and novel CDS), 78% have alternative splice forms (86% of the multi-exon gene loci), with an average of 5.7 variants per locus. Of the coding variants, approximately 70% have a complete CDS (the remainder are partial); 54% of the coding loci have alternative CDS, indicating that diversity is lower at the protein level than at the transcript level as a substantial proportion of the alternative splice forms affect only the UTRs. The RNPC2 (RNA-binding region (RNP1, RRM) containing 2) gene has 37 variants, which is the highest number in the ENCODE regions, of which only 6 are annotated as coding.

Experimental verification of GENCODE annotation

The initial HAVANA annotation was submitted for experimental verification (Figure 1). First, 5' RACE in 12 different tissues was employed to confirm that annotated coding genes (within both known and novel CDS locus categories) had been extended as far as possible towards the transcriptional start site, to exclude the possibility of additional exons in their 5' UTR and identify a representative full-length transcript for each locus. Of the 420 coding loci tested, 229 RACE products could be sequenced. They supported 5' extensions of 30 loci (extension of the first exon in two-thirds of the cases, new 5' exons in one-third of the cases) and new splice variants (not extending the 5' end) in 50 loci.

Second, RT-PCR in 24 tissues was used for verifying transcript (novel and putative) structures by checking the splice junctions. All 360 splice junctions in the 161 novel and putative transcript loci were tested. Of those tested, 47 loci were validated, consisting of 31 novel and 15 putative transcripts. As expected, the success rate of RT-PCR was higher for the 'novel transcripts' (37%) than for the putative transcripts (19%). Bidirectional RACE was carried out for transcript loci with successfully validated splice junctions. This supported seven loci over their full length but did not extend them.

Third, all annotated non-canonical sites (that is, all introns not conforming to the AG-GT or AG-GC rule) were tested by RT-PCR on 24 tissues. Of the annotated splice sites, 98% are canonical GT-AG and an additional 0.9% are GC-AG. There are 0.2% of AT-AC splice sites, most of them corresponding to canonical U12 introns [11]. Other non-canonical splice sites occur in the remaining 0.9% of the introns. Among 90 non-canonical splice sites tested by RT-PCR in 24 tissues, 78 reactions were negative, 11 provided other canonical junctions (most of them already annotated in other splice forms), and only 1 was confirmed (CT-TG). The very low level of success of the RT-PCRs on non-canonical splice sites in 24 tissues suggests that these events may be artifactual. As a control, we performed RT-PCR on 24 tissues (see Materials and methods) on 96 randomly selected exon pairs from within the GENCODE annotation. After sequencing of the amplimer, the annotated exon pair was confirmed in 84 cases (87%) in at least one tissue. This is essentially the

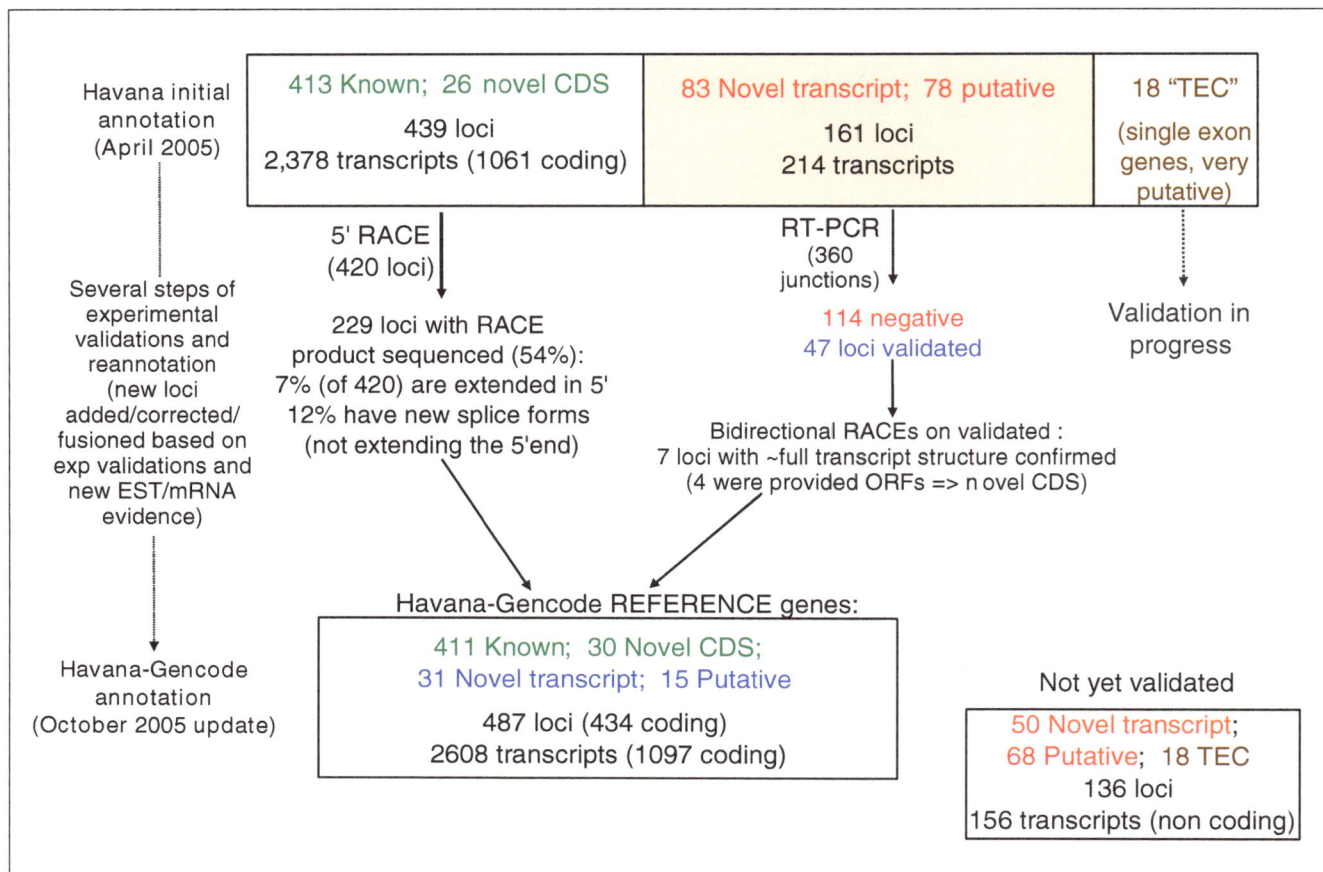

Figure 2
Experimental validation of HAVANA annotation. 'Known' and 'Novel_CDS' were submitted to 5' RACE, and 'Novel transcript' and 'Putative' loci were submitted to RT-PCR on all their exon junctions, followed by bi-directional RACE. Several steps of reannotation were performed during the process of experimental verification: the figure shows the update of the annotation between the first release in April 2005 and the release from October 2005.

expected result, given the fact that many alternative splice forms in GENCODE are likely to have a restricted expression pattern, and may not be represented in the 24 tissues tested.

Figure 2 summarizes the process of annotation, experimental validation and reannotation that has occurred since the original release of the GENCODE annotation in April and its current update in October 2005.

Assessing completeness of the GENCODE annotation
To examine whether the manual annotation had missed any coding loci, RT-PCR reactions in 24 tissues were also carried out for splice junctions from all those gene objects predicted by a panel of automated gene prediction algorithms before the E-GASP workshop (Geneid [12], Genescan [13], Twinscan [14], SGP [15], Fgenesh [16], Exonify [17], Acembly [18] Ecgene [19], Ensembl EST [20]) that lie outside a HAVANA annotated gene in 13 of the 44 ENCODE regions (corresponding to the training regions for which the annotations were released before the E-GASP predictions submission deadline). Of the 1,215 exon pairs tested, only 14 (1.2%)

produced a positive result, 9 of which perfectly predicted exon boundaries and 5 with displaced exon boundaries (8 other positive RT-PCRs were falling in 2 pseudogene loci). Among the 14 positive validated junctions, 8 were new splice forms internal to annotated loci, 4 were new splice forms extending annotated loci, and only 2 were completely intergenic to any annotation. These results suggest that the GENCODE gene set was relatively complete. It was then updated to include the new splice forms/loci suggested by these experiments.

To further assess the completeness of the GENCODE annotation, we have compared it with other publicly available and widely used human gene sets: RefSeq [21] and ENSEMBL [22]. These gene sets were downloaded from the UCSC genome browser in November 2005. Table 1 shows the overlap between these sets and GENCODE by at least one bp: 99% of RefSeq, and 94% of ENSEMBL exons overlap GENCODE exons. In contrast, only 80% and 84% of the GENCODE exons overlap RefSeq and ENSEMBL exons, respectively.

Table 1

Analysis of RefSeq and ENSEMBL ENCODE annotation compared with GENCODE

	RefSeq	ENSEMBL
No. (unique) exons	3,984	4,734
No. transcripts	577	738
No. exons overlapping GENCODE exons (%)	5,118 (98.6%)	4,469 (94.4%)
No. transcripts overlapping GENCODE (%)	567 (98.3%)	675 (91.5%)
No. GENCODE exons overlapped (total = 8,865) (%)	7,084 (80.0%)	7,450 (84.0%)
No. GENCODE transcripts overlapped (total = 2,608) (%)	2,327 (89.2%)	2,395 (91.8%)

Figure 3 illustrates the comparisons at exact exon/intron level. Although the exact agreement between GENCODE on the one hand, and RefSeq and ENSEMBL on the other, is lower than when considering one base overlap, the same trend is observed: 84% (3,361/3,984) of RefSeq and 76% (3,584/4,734) of ENSEMBL exons are included in the GENCODE set, but only about 40% of the GENCODE exons are included in RefSeq or ENSEMBL.

As illustrated by Figure 3, the exact agreement is larger for exons than for introns, which suggests that the disagreements are mostly found at the terminal exons, which is also reflected in the fact that the agreement is also larger for the subset of coding than for the set of all exons. In summary, the comparison shows that GENCODE contains most of the features from RefSeq and ENSEMBL but has more unique exons than the two sets, which is reflected by its high number of alternative splice forms.

Investigation of ENCODE regions that are problematic for automatic annotation

The gene prediction algorithms that performed most successfully in the E-GASP evaluation workshop when compared to the manual annotation were the ones that used alignments of expressed sequences to produce their gene predictions (see [7] in this issue). However, even the most successful methods of automated gene prediction achieved a maximum sensitivity of 70% at the gene level (where at least one coding transcript exon/intron structure was correctly predicted) and 45% at the transcript level (where all alternatively spliced variants were correctly predicted). There are several reasons for this. Some incidences of missed genes could be explained by the lack of high identity transcript evidence; for example, many of the olfactory receptor genes in ENm009 (Figure 4f) lack good transcript and protein support [23]. Another example is the ANKRD43 locus in ENr221, where partial coverage of the gene with human mRNA produces truncated automated predictions. However,

cross-species evidence supports an extended protein-coding gene (Figure 4c). In other cases, predictors fail to make a correct prediction even though a full length transcript with perfect sequence identity is present in the databases (for example, Pairagon at the TRIM22 locus in ENm009; Figure 4b). There are also examples where the predictions differ from the manual annotation gene structure, even though they use the same supporting evidence, because of problems with automated alignment (for example, Ensembl and Pairagon at the MAP3K1 locus in ENr221; Figure 4a). A problem that appears to be associated with tandem duplicated gene clusters is the linking together of adjacent loci. The predicted transcript uses consecutive exons from more than one locus, for example for a six exon gene taking exons 1 and 2 from locus A, 3, 4 and 5 from locus B and 6 from locus C. Because the equivalent exons of the different copies of the gene are very similar (often identical), the resulting predicted transcript is an elongated structure usually covering multiple loci (for example, AceView at the HBG1/HBG2 loci in ENm009).

Another observation is that there are predictions that have an identical intron/exon structure to the manual annotation but have a different CDS. In such cases, the CDS has either a 5' extension, that is, completely matches the GENCODE CDS but uses an upstream translation initiation codon (most often non-ATG; for example, AceView at the SEPT8 locus in ENr221 and approximately 41% of AceView have a non-ATG start), or has an entirely different CDS in a different frame. The latter often results in unusual structures, with multi-exon 5' and/or 3' UTRs that are at odds with rules governing re-initiation [24] and nonsense mediated decay (NMD) [25] (for example, Pairagon at the AC008937.5 locus in ENr221 and AceView at the IFNAR2 locus in ENm005; Figure 4e). Many of the predictors suffer from reduced specificity as a result of over-prediction of CDSs at loci where manual annotation does not identify any CDS that can be confidently assigned. These fall into two types; the first includes CDS predicted at pseudogene loci, often where the pseudogene suffers from small but significant disablements (for example, Ensembl at the AC08730.14 locus in ENm009; Figure 4d); and the second includes the 'rule-breaking' types of CDSs described above (AceView at the AC008937.2 in ENr221). Almost all the predictors (with AceView the notable exception) under-predict coding (and non-coding) splice variants, most predicting one transcript per gene.

GENCODE annotation uses only primary evidence; no predictions or RefSeq entries are used to support gene structures. This has the effect of reducing the risk of propagating any errors that may be present in the databases. The gene set annotated by GENCODE is supported using evidence from all available sources, human and non-human mRNAs, ESTs and proteins. The use of non-human evidence is supported by our analysis of four exons not present in our first pass annotation identified by the UNCOVER algorithm

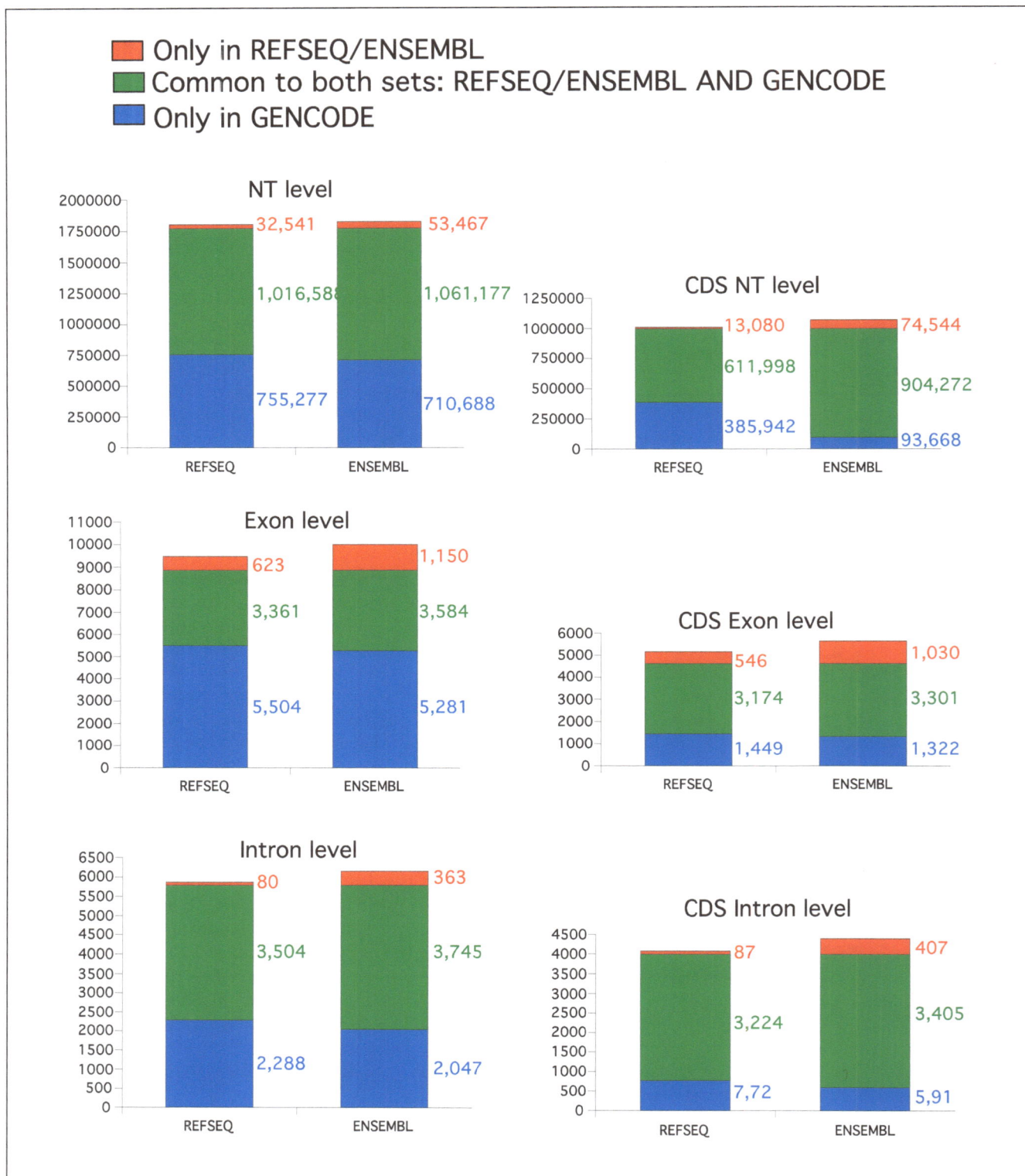

Figure 3
Comparison of GENCODE transcript annotation with RefSeq and ENSEMBL. The exact agreement between GENCODE and RefSeq and GENCODE and ENSEMBL exons, introns, and nucleotides (NT) for the full transcripts or only the coding parts of the transcripts (CDS) is represented: in blue is the fraction found only in GENCODE, in green the fraction common between GENCODE and the other set (RefSeq or ENSEMBL) and in red the fraction found only in the other set (RefSeq or ENSEMBL) but not in GENCODE. The RefSeq set only contained the curated transcripts tagged with the NM prefix.

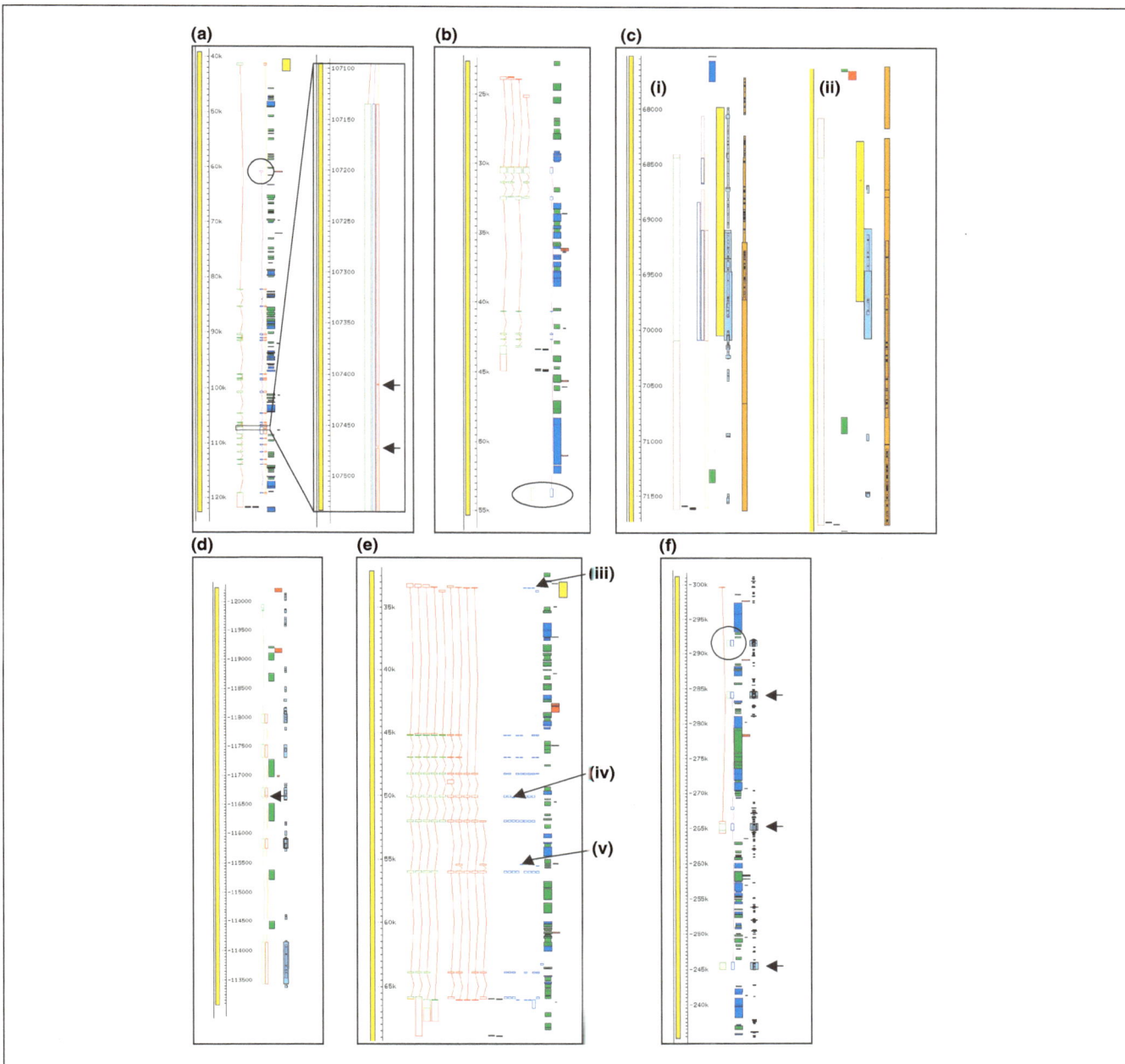

Figure 4

Comparison of GENCODE annotation with automated gene prediction methods. Viewed in Fmap of Acedb. Panel A shows the *MAPK1* gene in ENr221. The GENCODE annotated gene structure is represented in green and red, the circled region highlights the different first exon identified by Pairagon (dark pink/blue) and the expanded region shows tiny introns (indicated by arrows) predicted by Ensembl (orange/red). Panel B shows the TRIM22 locus in ENm009. The structure predicted by Pairagon differs from the GENCODE structure and incorporates an unprocessed pseudogene as the final exon (circled). Panel C shows the human ANKRD43 locus in ENr221 for which AceView (light pink/blue), Pairagon and Ensembl all predict a shorter CDS than GENCODE. C ii shows the mouse ANKRD43 locus in which the upstream ATG is conserved. Panel D shows the GENCODE unprocessed pseudogene locus AC087380.14 at which Ensembl predicts a coding gene. The arrow indicates a tiny intron introduced into the prediction to splice around an in-frame premature stop codon. Panel E shows the IFNAR2 locus in ENm005 with GENCODE coding (red/green) and non-coding (all red) variants and AceView predictions. The AceView CDSs differ from GENCODE in several respects; arrow 'a' indicates several transcripts that have their CDS extended to the start of the prediction upstream of the GENCODE CDS start; arrow 'b' indicates a CDS starting in exon 5 despite the presence of an upstream ATG, which would seem to preclude (re-)initiation from this site; and arrow 'c' indicates a predicted stop codon in the fourth from last exon, which would be likely to make this transcript a target from Nonsense-mediated decay (NMD). GENCODE annotation incorporates all these variants but keeps them as transcripts as CDSs cannot be assigned with certainty. Panel F shows part of the olfactory receptor (OR) cluster in ENm009. Here Pairagon predicts a coding gene at the pseudogene locus OR52Z1P and a multi-exon gene that links separate OR loci (pseudogene locus OR51A1P, coding loci OR52A1 and OR52A5), indicated by arrows.

[26], two of which are only supported by non-human EST evidence. The identification of a rare splice variant in the C16orf35 gene at the alpha globin locus is also facilitated using mouse EST evidence (J Hughes, personal communication). Importantly, manual annotation allows context to be taken into account when making a decision about difficult gene regions, which includes consulting literature and various web resources.

Conclusions

The E-GASP workshop as part of the ENCODE project has highlighted the need for a high quality reference gene set that can be used to improve and validate prediction algorithms, as well as a scaffold for further experimentation. RT-PCR and 5' RACE of predicted exons outside the GENCODE annotation has currently not revealed additional multi-exon protein-coding genes. However, the experimental validation continually adds evidence for more splice variants. In addition, other technologies such as mapping RNA to tiling arrays [27], cap analysis gene expression (CAGE) tags [28], and gene identification signature (GIS) ditags [29] indicate there is transcriptional activity outside the regions currently annotated by the GENCODE consortium. Therefore, the annotation will be continually evolving to represent the complete transcriptional landscape of the ENCODE regions.

Materials and methods
Annotation pipeline and software

Before the process of manual annotation begins, an automated analysis pipeline for similarity searches and *ab initio* predictions is run. The searches are run on a computer farm and stored in an Ensembl MySQL database using a modified Ensembl analysis pipeline system [30]. All searches and prediction algorithms, except CpG island prediction (see cpgreport in the EMBOSS application suite [31]) are run on repeat masked sequence. RepeatMasker [32] is used to mask interspersed repeats, followed by Tandem repeats finder [33] to mask tandem repeats. Nucleotide sequence databases are searched with wuBLASTN [34], and significant hits are re-aligned to the unmasked genomic sequence using est2genome [35]. The Uniprot protein database [36] is searched with wuBLASTX, and the accession numbers of significant hits are looked up in the Pfam database [37]. The hidden Markov models for Pfam protein domains are aligned against the genomic sequence using Genewise [38] to provide annotation of protein domains. We also run a number of *ab initio* prediction algorithms: Genescan [13] and Fgenesh [16] for genes, tRNAscan [39] to find tRNAgenes and Eponine TSS [40], to predict transcription start sites. Annotation assessed at the E-GASP workshop used data from searches of the 24th August 2004 of dbEST, vertebrate mRNA sequences from release 80 of the EMBL nucleotide database and protein sequences from version 2.4 of Swiss-Prot/TrEMBL.

Once the automated analysis is complete, the annotator uses a Perl/Tk based graphical interface, called 'otterlace', developed in-house to edit annotation data held in a separate MySQL database system [41]. The interface displays a rich, interactive graphical view of the genomic region, showing features like database matches, gene predictions, and transcripts created by the annotators. Gapped alignments of nucleotide and protein blast hits to the genomic sequence are viewed and explored using the 'Blixem' alignment viewer [42]. Additionally, the 'Dotter' dot plot tool [42] is used for showing the pair-wise alignments of unmasked sequence, thus revealing the location of exons that are occasionally missed by the automated blast searches because of their small size and/or match to repeat-masked sequence. The interface provides a number of tools that the annotator uses to build genes and edit annotations: adding transcripts, exon coordinates, translation regions, gene names and descriptions, remarks and polyadenlyation signals and sites.

Rapid amplification of cDNA ends

Both 5' and 3' RACE were performed on 12 human poly(A)⁺ RNAs (brain, heart, kidney, spleen, liver, colon, small intestine, muscle, lung, stomach, testis, placenta) using the BD SMART™ RACE cDNA amplification kit (BD BioScience-Clontech Catalogue No.634914, Mountain View, CA 95043, USA). Double-stranded cDNA synthesis and adaptor ligations to the synthesized cDNA were done according to the manufacturer's instructions. RACE fragments were separated on agarose gels and one or two strong single bands per gene purified and sequenced directly. Thus, successful RACE reactions appearing as a smear on the agarose gel would be discarded, therefore producing an approximate 54% success rate.

RT-PCR

Similar amounts of 24 human cDNAs (brain, heart, kidney, spleen, liver, colon, small intestine, muscle, lung, stomach, testis, placenta, skin, PBLs, bone marrow, fetal brain, fetal liver, fetal kidney, fetal heart, fetal lung, thymus, pancreas, mammary glands, prostate; final dilution 1,000×) were mixed with JumpStart REDTaq ReadyMix (Sigma, St Louis, MO, USA) and 4 ng/μl primers (Sigma-Genosys, St Louis, MO, USA)) with a BioMek 2000 robot (Beckman, Fullerton, CA, USA) as described and modified [43-45]. The 10 first cycles of PCR amplification were performed with a touchdown annealing temperature decreasing from 60°C to 50°C; the annealing temperature of the next 30 cycles was 50°C. Amplimers were separated on 'Ready to Run' precast gels (Pfizer, New York, NY, USA) and sequenced. This procedure was used to experimentally assay 1,215 exon-exon junctions of human genes predicted by five *ab initio* and four EST-based methods outside of HAVANA objects and 83 HAVANA novel and 78 putative transcripts (see Results and discussion for details).

Acknowledgments

This work was supported by grants from the Childcare and Désirée and Niels Yde Foundations, the European Union, the Swiss National Science Foundation and the NCCR Frontiers in Genetics, from the NHGRI ENCODE Project, and from the Spanish Ministry of Education and Science.

This article has been published as part of *Genome Biology* Volume 7, Supplement 1, 2006: EGASP '05. The full contents of the supplement are available online at http://genomebiology.com/supplements/7/S1.

References

1. International Human Genome Sequencing Consortium: **Finishing the euchromatic sequence of the human genome.** *Nature* 2004, **431:**931-945.
2. Mattick JS: **Non-coding RNAs: the architects of eukaryotic complexity.** *EMBO Rep* 2001, **2:**986-991.
3. Bartel DP: **MicroRNAs: genomics, biogenesis, mechanism, and function.** *Cell* 2004, **116:**281-297.
4. ENCODE project consortium: **The ENCODE (ENCyclopedia Of DNA Elements) Project.** *Science* 2004, **306:**636-640.
5. **GENCODE Consortium** [http://genome.imim.es/gencode]
6. **HAVANA Team** [http://www.sanger.ac.uk/HGP/havana/]
7. Guigo R, Flicek P, Abril J, Reymond A, Lagarde J, Denoeud F, Antonarakis S, Ashburner M, Bajic VB, Birney E, *et al*: **EGASP. The human ENCODE genome assessment project.** *Genome Biology* 2006, **7(Suppl 1):**S2
8. Deloukas P, Matthews LH, Ashurst J, Burton J, Gilbert JG, Jones M, Stavrides G, Almeida JP, Babbage AK, Bagguley CL, *et al*: **The DNA sequence and comparative analysis of human chromosome 20.** *Nature* 2001, **414:**865-871.
9. **Entrez Gene** [http://www.ncbi.nlm.nih.gov/entrez/query.fcgi?db=gene]
10. **HUGO Gene Nomenclature Committee** [http://www.gene.ucl.ac.uk/nomenclature/]
11. Will CL, Luhrmann R: **Splicing of a rare class of introns by the U12-dependent spliceosome.** *Biol Chem* 2005, **386:**713-724.
12. Parra G, Blanco E, Guigo R: **GeneID in Drosophila.** *Genome Res* 2000, **10:**511-515.
13. Burge C, Karlin S: **Prediction of complete gene structures in human genomic DNA.** *J Mol Biol* 1997, **268:**78-94.
14. Wang M, Buhler J, Brent MR: **The effects of evolutionary distance on TWINSCAN, an algorithm for pair-wise comparative gene prediction.** *Cold Spring Harb Symp Quant Biol* 2003, **68:**125-130.
15. Wiehe T, Gebauer-Jung S, Mitchell-Olds T, Guigo R: **SGP-1: prediction and validation of homologous genes based on sequence alignments.** *Genome Res* 2001, **11:**1574-1583.
16. Salamov AA, Solovyev VV: **Ab initio gene finding in Drosophila genomic DNA.** *Genome Res* 2000, **10:**516-522.
17. Siepel A, Haussler D: **Computational identification of evolutionarily conserved exons.** *Proc. 8th Annual Int'l Conf. on Research in Computational Biology.* pp 177-186, 2005 RECOMB'04, March 27-31, 2004, San Diego, California, USA.
18. **Acembly** [http://www.ncbi.nih.gov/IEB/Research/Acembly/index.html]
19. Kim P, Kim N, Lee Y, Kim B, Shin Y, Lee S: **ECgene: genome annotation for alternative splicing.** *Nucleic Acids Res* 2005, **33 (Database issue):**D75-79.
20. Eyras E, Caccamo M, Curwen V, Clamp M: **ESTGenes: alternative splicing from ESTs in Ensembl.** *Genome Res* 2004, **14:**976-987.
21. Pruitt KD, Tatusova T, Maglott DR: **NCBI Reference Sequence (RefSeq): a curated non-redundant sequence database of genomes, transcripts and proteins.** *Nucleic Acids Res* 2005, **33 (Database issue):**D501-504.
22. Birney E, Andrews TD, Bevan P, Caccamo M, Chen Y, Clarke L, Coates G, Cuff J, Curwen V, Cutts T, *et al.*: **An overview of Ensembl.** *Genome Res* 2004, **14:**925-928.
23. UCSC genome browser [http://genome.cse.ucsc.edu/ENCODE]
24. Kozak M: **Emerging links between initiation of translation and human diseases.** *Mamm Genome* 2002, **13:**401-410.
25. Lewis BP, Green RE, Brenner SE: **Evidence for the widespread coupling of alternative splicing and nonsense-mediated mRNA decay in humans.** *Proc Natl Acad Sci USA* 2003, **100**(1): 189-192.
26. Ohler U, Shomron N, Burge CB: **Recognition of unknown conserved alternatively spliced exons.** *PLoS Comput Biol* 2005, **1:** 113-122.
27. Kapranov P, Drenkow J, Cheng J, Long J, Helt G, Dike S, Gingeras TR: **Examples of the complex architecture of the human transcriptome revealed by RACE and high-density tiling arrays.** *Genome Res* 2005, **15:**987-997.
28. Shiraki T, Kondo S, Katayama S, Waki K, Kasukawa T, Kawaji H, Kodzius R, Watahiki A, Nakamura M, Arakawa T, *et al.*: **Cap analysis gene expression for high-throughput analysis of transcriptional starting point and identification of promoter usage.** *Proc Natl Acad Sci USA* 2003, **100:**15776-15781.
29. Ng P, Wei CL, Sung WK, Chiu KP, Lipovich L, Ang CC, Gupta S, Shahab A, Ridwan A, Wong CH, *et al.*: **Gene identification signature (GIS) analysis for transcriptome characterization and genome annotation.** *Nat Methods* 2005, **2:**105-111.
30. Potter SC, Clarke L, Curwen V, Keenan S, Mongin E, Searle SM, Stabenau A, Storey R, Clamp M: **The Ensembl analysis pipeline.** *Genome Res* 2004, **14:**934-941.
31. Rice P, Longden I, Bleasby A: **EMBOSS: the European Molecular Biology Open Software Suite.** *Trends Genet* 2000, **16:**276-277.
32. **RepeatMasker** [http://www.repeatmasker.org/]
33. Benson G: **Tandem repeats finder: a program to analyze DNA sequences.** *Nucleic Acids Res* 1999, **27:**573-580.
34. **wuBLASTN** [http://blast.wustl.edu]
35. Mott R: **EST_GENOME: a program to align spliced DNA sequences to unspliced genomic DNA.** *Comput Appl Biosci* 1997, **13:**477-478.
36. **Uniprot Protein Database** [http://www.uniprot.org]
37. Bateman A, Coin L, Durbin R, Finn RD, Hollich V, Griffiths-Jones S, Khanna A, Marshall M, Moxon S, Sonnhammer EL, *et al.*: **The Pfam protein families database.** *Nucleic Acids Res* 2004, **32(Database issue):**D138-141.
38. Birney E, Clamp M, Durbin R: **GeneWise and Genomewise.** *Genome Res* 2004, **14:**988-995.
39. Lowe TM, Eddy SR: **tRNAscan-SE: a program for improved detection of transfer RNA genes in genomic sequence.** *Nucleic Acids Res* 1997, **25:**955-964.
40. Down TA, Hubbard TJ: **Computational detection and location of transcription start sites in mammalian genomic DNA.** *Genome Res* 2002, **12:**458-461.
41. Searle SM, Gilbert J, Iyer V, Clamp M: **The otter annotation system.** *Genome Res* 2004, **14:**963-970.
42. Sonnhammer EL, Wootton JC: **Integrated graphical analysis of protein sequence features predicted from sequence composition.** *Proteins* 2001, **45:**262-273.
43. Reymond A, Friedli M, Henrichsen CN, Chapot F, Deutsch S, Ucla C, Rossier C, Lyle R, Guipponi M, Antonarakis SE: **From PREDs and open reading frames to cDNA isolation: Revisiting the human chromosome 21 transcription map.** *Genomics* 2001, **78:**46-54.
44. Reymond A, Camargo AA, Deutsch S, Stevenson BJ, Parmigiani RB, Ucla C, Bettoni F, Rossier C, Lyle R, Guipponi M, *et al.*: **Nineteen additional unpredicted transcripts from human chromosome 21.** *Genomics* 2002, **79:**824-832.
45. Guigo R, Dermitzakis ET, Agarwal P, Ponting CP, Parra G, Reymond A, Abril JF, Keibler E, Lyle R, Ucla C, *et al.*: **Comparison of mouse and human genomes followed by experimental verification yields an estimated 1,019 additional genes.** *Proc Natl Acad Sci USA* 2003, **100:**1140-1145.

Research

Open Access

Pairagon+N-SCAN_EST: a model-based gene annotation pipeline

Manimozhiyan Arumugam, Chaochun Wei, Randall H Brown and Michael R Brent

Address: Laboratory for Computational Genomics and Department of Computer Science, Washington University, One Brookings Drive, St. Louis, MO 63130, USA

Correspondence: Michael R Brent. E-mail: brent@cse.wustl.edu

Published: 7 August 2006

Genome Biology 2006, **7(Suppl 1)**:S5

The electronic version of this article is the complete one and can be found online at http://genomebiology.com/2006/7/S1/S5

Abstract

Background: This paper describes Pairagon+N-SCAN_EST, a gene annotation pipeline that uses only native alignments. For each expressed sequence it chooses the best genomic alignment. Systems like ENSEMBL and ExoGean rely on *trans* alignments, in which expressed sequences are aligned to the genomic loci of putative homologs. *Trans* alignments contain a high proportion of mismatches, gaps, and/or apparently unspliceable introns, compared to alignments of cDNA sequences to their native loci. The Pairagon+N-SCAN_EST pipeline's first stage is Pairagon, a cDNA-to-genome alignment program based on a PairHMM probability model. This model relies on prior knowledge, such as the fact that introns must begin with GT, GC, or AT and end with AG or AC. It produces very precise alignments of high quality cDNA sequences. In the genomic regions between Pairagon's cDNA alignments, the pipeline combines EST alignments with de novo gene prediction by using N-SCAN_EST. N-SCAN_EST is based on a generalized HMM probability model augmented with a phylogenetic conservation model and EST alignments. It can predict complete transcripts by extending or merging EST alignments, but it can also predict genes in regions without EST alignments. Because they are based on probability models, both Pairagon and N-SCAN_EST can be trained automatically for new genomes and data sets.

Results: On the ENCODE regions of the human genome, Pairagon+N-SCAN_EST was as accurate as any other system tested in the EGASP assessment, including ENSEMBL and ExoGean.

Conclusions: With sufficient mRNA/EST evidence, genome annotation without trans alignments can compete successfully with systems like ENSEMBL and ExoGean, which use trans alignments.

Background

There are three fundamental approaches to automated construction of exon-intron structure for protein-coding genes: **native alignment** – alignment of expressed sequences (including high quality cDNA sequences, expressed sequence tags (ESTs), and protein sequences) to the loci from which they were transcribed; **trans alignment** – non-native alignment of expressed sequences to loci that could potentially express similar sequences (can be within or between species); and **de novo** – prediction using the sequences of one or more genomes as the only inputs (no expressed sequences).

Native alignments of full insert, high quality cDNA sequences are the unquestioned gold standard in high-throughput annotation. However, even a concerted, high-budget effort to sequence cDNA libraries produces a full-open reading frame (ORF) sequence for only about 50% to 60% of loci in a mammalian genome [1]. Thus, trans alignments have played a key role in producing the most trusted genome predictions, including the ENSEMBL predictions (sometimes termed 'evidence based') that have been used in the first published analyses of many new genome sequences. Nonetheless, the evidence they provide for expression is circumstantial rather than direct – for example, the annotated genomic locus may represent a pseudogene derived from the true genomic source of the expressed sequence. Even when a trans alignment identifies a functional homologous gene locus, the alignments tend to be inaccurate in their details unless the expressed sequence is highly similar to the genomic sequence [2, 3].

De novo predictions have always been viewed with some suspicion. This suspicion derives in part from the tendency of gene predictors developed in the 1990s to predict far too many false positive genes and exons. It may also result, in part, from the fact that one cannot point to the evidence supporting *de novo* predictions – a large ensemble of individually weak statistical patterns – the way one can point to a single expressed sequence. Nonetheless, statistical evidence is biological evidence, with a track record extending back to Gregor Mendel.

If *de novo* prediction were indeed inaccurate, relying heavily on trans alignments would make sense when analyzing a genome for which few EST or cDNA sequences are available. However, the rapidly increasing accuracy of *de novo* prediction and the large number of very high quality cDNA sequences available for human suggest the possibility that high quality annotations might be produced without using trans alignments. A system that does not use trans alignments might be more accurate than one that does, since all alignments would have near 100% identity. Even if its accuracy were merely equal to that of a system using trans alignments, the evidence supporting each prediction might be considered more direct.

To build an annotation pipeline without trans alignment, we combined a number of tools that have been recently developed in our lab. These tools include Pairagon, a cDNA-to-genome aligner, N-SCAN_EST [4], a multi-genome gene predictor capable of taking guidance from EST alignments, and PPFINDER [5], a program for eliminating pseudogenes from sets of predicted protein-coding genes.

Pairagon uses a PairHMM to produce native cDNA alignments

To produce the best possible alignments of high quality cDNA sequences, we used Pairagon, a cDNA-to-genome aligner that is based on a pairHMM probability model [6]. A pairHMM is a hidden Markov model (HMM) whose states emit alignment columns. In our case, the columns contain

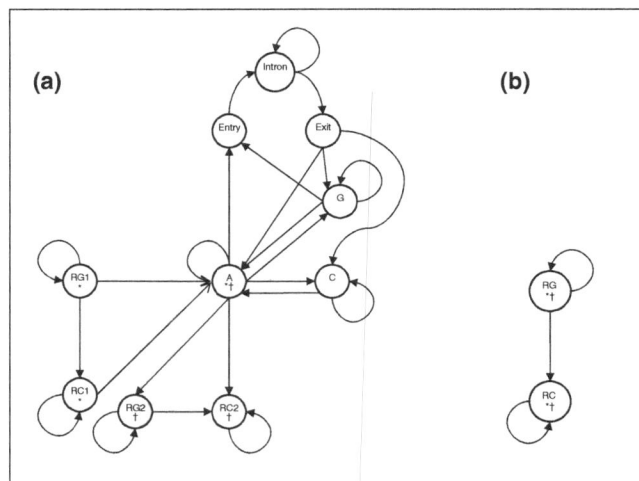

Figure 1
PairHMM state diagrams of Pairagon. **(a)** Alignment model and **(b)** Null model. RG1 and RG2 are unaligned genomic sequences in the 5' and 3' ends, respectively; RC1 and RC2 are unaligned cDNA sequences in the 5' and 3' ends, respectively; A, aligned; Entry corresponds to the first two bases of an intron; Exit corresponds to the last two bases of an intron; G, genomic insertion; C, cDNA insertion; RG and RC are random genomic and cDNA sequences, respectively. States that can start an alignment are marked with an asterisk and states that can end an alignment are marked with a dagger.

either a match between the two sequences, a mismatch, an insertion in the genome, a deletion in the genome, or an intron base in the genome (Figure 1). The particular pairHMM model we developed is 'strong', in the sense that it enforces prior biological and statistical knowledge rather than letting the data at hand dictate the alignment even when it is at odds with prior knowledge. In particular, our model only produces introns with plausible splice site sequences: GT-AG, GC-AG, AT-AC (AT-AG and other extremely rare U12 intron types [7] are not currently allowed). Furthermore, the probabilities of introns, matches, mismatches, genome insertions, and genome deletions are estimated from alignments of high quality cDNA sequences produced by BLAT [8] and the relative probabilities of the three intron types are derived from prior knowledge.

In order to make Pairagon run faster, we ran ungapped BLASTN as a preprocessing step and used the long alignments it produced to seed exon alignments (Figure 2, left side). For more details on Pairagon and its heuristics, see Materials and methods.

Our strategy was to use alignments of expressed sequences directly only when very high quality sequences were available. Thus, we applied Pairagon only to full ORF Mammalian Gene Collection (MGC) sequences [1, 9, 10] and human RefSeq mRNAs [11].

N-SCAN_EST threads complete gene structures through EST alignments

In the genomic regions between Pairagon's cDNA alignments, we combined EST alignments with *de novo* gene prediction by using N-SCAN_EST [4]. N-SCAN_EST is based on N-SCAN [12, 13], a multi-genome *de novo* gene predictor, which was the most accurate *de novo* predictor in the EGASP assessment [14] by every measure except nucleotide sensitivity. (*De novo* includes both the 'ab initio' and 'multi-genome' assessment categories.) N-SCAN_EST is a version of N-SCAN that takes guidance from EST alignments. Specifically, it takes as input a representation of EST alignments that we call ESTseq, by analogy to the 'conservation sequence' used in TWINSCAN (a three-character alphabet representing genome sequence conservation between two species) [15, 16]. N-SCAN_EST takes guidance from EST alignments, but it does not follow them blindly. Instead, it also considers the DNA sequence of the target genome and the evolutionary conservation information provided by alignments of the target genome with the genomes of other organisms. It predicts complete transcripts by extending or merging EST alignments or by building gene structures in which some exon regions are supported by EST evidence while others are not. We have shown elsewhere that this approach increases sensitivity and specificity not only for the genes that have EST support, but even for those that do not [4].

Pairagon+N-SCAN_EST annotates genomes without using *trans* alignment

To apply N-SCAN_EST, we downloaded human ESTs from dbEST and aligned them to the human genome using BLAT [8] (Figure 2, right side). We also downloaded alignments of the human, mouse, rat, and chicken genomes produced by MULTIZ [17] from the University of California Sant Cruz (UCSC) genome browser. These EST and genomic alignments were input to N-SCAN_EST. N-SCAN_EST was run on human genomic sequence that had been masked with PPFINDER, our processed pseudogene masker [5]. After the final round of N-SCAN_EST, all predicted transcripts that overlapped Pairagon alignments were removed. The remaining transcripts (one per locus) were combined with the Pairagon alignments to produce the final gene set.

In the remainder of this paper we present accuracy statistics for both the EGASP version of the pipeline and an updated version and analyze the relative contributions of Pairagon versus N-SCAN_EST. We then examine a series of examples where our pipeline gave a revealing result, whether correct or incorrect. Finally, we draw some lessons about how the pipeline could be improved in the future.

Results and discussion

RefSeq and MGC cDNA sequences mapped to the ENCODE regions were downloaded from the UCSC Genome Browser

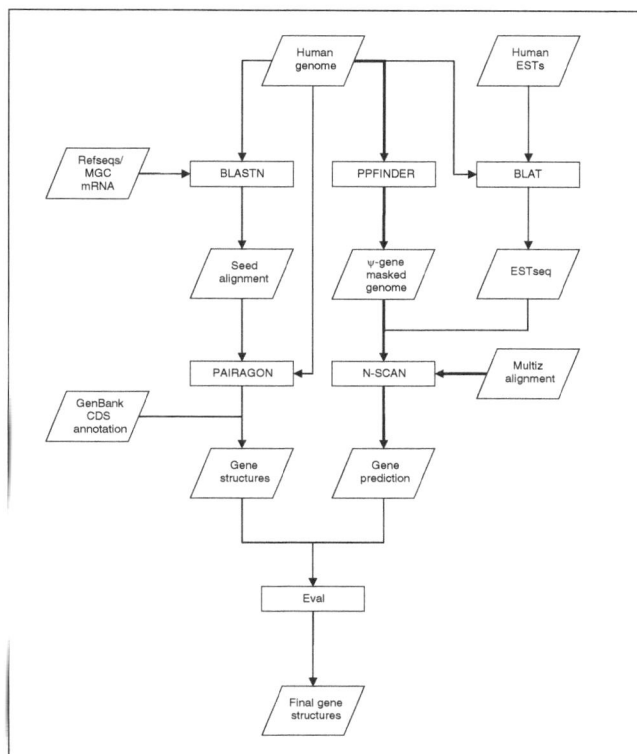

Figure 2
Block diagram of the Pairagon+N-SCAN_EST pipeline. The bold arrows mark the section of the flowchart corresponding to N-SCAN gene prediction.

and alignments were generated using the Stepping Stone implementation of Pairagon v0.5 as described in Materials and methods. GenBank's coding sequence (CDS) annotations of these cDNA sequences were used to produce 451 aligned transcripts annotated with GenBank ORFs (141 from MGC sequences and 310 from RefSeq sequences). Merging identical gene structures and removing inconsistent structures (for example, gap in the coding region leading to a frame shift in the genome) yielded 413 unique gene structures. N-SCAN_EST predictions were generated as described in Materials and methods. The 94 N-SCAN_EST predictions that did not overlap the 413 Pairagon gene structures were added to the gene set. We obtained seven gene structures by aligning sequences from our RT-PCR experiments. Two of these did not overlap the existing set and were included in our submission to the 'any evidence' category. We do not discuss this set in detail because it is almost identical to the submission to the 'mRNA/EST evidence' category. The accuracy statistics for this set can be found in the EGASP assessment report [14].

The official assessment of Pairagon+N-SCAN_EST shows high accuracy

Table 1 compares the coding region prediction accuracy measures of three submissions to the EGASP 'mRNA/EST

Table 1

Prediction accuracy measures of mRNA/EST evidence based gene prediction methods

Name	NSn	NSp	ESn	ESp	ExT	TSn	TSp	TrG	GSn	GSp
Pairagon + N-SCAN_EST	87.6	92.8	76.6	**89.0**	7.2	39.3	**60.6**	1.3	69.6	61.7
ENSEMBL	**90.2**	92.0	77.5	82.7	7.8	39.8	54.6	1.5	**71.6**	67.3
ExoGean	84.2	**94.3**	**79.3**	83.5	9.8	**42.5**	52.4	2.3	63.2	**80.8**

The highest value for each measure is in bold. The columns are Nucleotide sensitivity (NSn) and specificity (NSp), Exon sensitivity (ESn) and specificity (ESp), Exons per transcript (ExT), Transcript sensitivity (TSn) and specificity (TSp), Transcripts per gene (TrG), and Gene sensitivity (GSn) and specificity (GSp).

evidence' category at the gene, transcript, exon and nucleotide levels. Pairagon+N-SCAN_EST (Pairagon+N) is optimized for high accuracy in predicting exact exons and transcripts, so we will focus our analysis on those columns of Table 1. By both measures, ExoGean is the most sensitive of the three programs and Pairagon+N is the most specific; ENSEMBL is intermediate except in exact exon specificity, where it falls below the other two. None of the programs completely dominates any other, although one might argue that Pairagon+N has a slight edge, since the margin by which its specificity exceeds that of the second best program is substantially larger than the margin by which its sensitivity falls below the others. In absolute numbers, our pipeline identifies almost the same number of correct Gencode transcript structures as ENSEMBL (255 versus 258, respectively), and 21 fewer than ExoGean, but we have many fewer incorrect transcripts (149 versus 205 from ENSEMBL and 237 from ExoGean). Their gene accuracy measures are slightly better than ours because ENSEMBL and ExoGean predict more transcripts per gene locus on average. Predicting more transcripts at a locus increases the chance that at least one of them is correct, yielding a true positive by the gene measure, while no penalty in false positives is incurred for the additional incorrect transcripts. This is arguably a flaw in the gene level measure when applied to systems that can predict more than one transcript per locus.

Pairagon's cDNA alignments are highly accurate

The individual accuracies of Pairagon and N-SCAN_EST gene structures in the submission are given in Table 2. Pairagon's nucleotide and exon specificities are 98.8% and 96.1%, respectively. Pairagon is also very accurate in identifying splice sites – we estimated that 98.3% of the introns that Pairagon identified have supporting evidence in the Gencode reference genes. When there is high quality mRNA evidence, more than three-fourths of transcript structures predicted by Pairagon are correct.

Identifying the correct splice boundaries is the crucial step in cDNA-to-genome alignment, and here Pairagon proves to be extremely accurate. Out of the 1,834 introns Pairagon predicted (both within and outside coding regions), only 22

introns from 15 transcript structures were not supported by HAVANA annotation. Three of them (from a single transcript) matched the introns of a Gencode gene labeled 'putative' and eight of them were a result of using incorrect seed exons from BLASTN (discussed in detail below). The remaining 11 were from Refseq cDNAs that have no evidence in HAVANA annotation. Two of the eleven aligned to the reference genome with numerous mismatches.

There are 22 unique GC-AG introns in the protein coding part of the HAVANA annotation. Pairagon correctly identifies 12 of these. The remaining 10 are missed because they did not have supporting Refseq or MGC cDNA sequence. When other systems prefer a GT dinucleotide, especially if it occurs close to the actual GC donor site, Pairagon gets the GC splice boundaries correct. Figure 3 shows one such example where ENSEMBL, Augustus and ExonHunter choose an incorrect GT donor site that is four nucleotides downstream of the correct GC donor, which Pairagon chooses. There are two unique AT-AC splice sites in the annotation and Pairagon correctly identifies both of them. Among the methods that use mRNA/EST evidence, AceView identifies the two introns and ENSEMBL identifies one of them. There are also two AT-AG introns with one supporting Gencode annotation each, and only AceView predicts them. Pairagon's splice boundary model prevents it from identifying these introns.

In the Stepping Stone implementation of Pairagon, the accuracy of the final alignment depends on how well the seed exons are mapped in the genome (see Materials and methods and Figure 4 for details). Figure 5 shows an example where the first 112 bases (forming an exon) of the cDNA can be mapped to either of two tandem duplicates that are identical in those 112 bases. Because we chose to use BLASTN parameter topComboN=1, which does not return alignments of a query segment to more than one location in the genome, BLASTN aligned the exon arbitrarily to the locus farther from the rest of the alignment. As a result, Pairagon placed the exon in the same general region, while the annotation maps it to the nearer locus. One possible way to address this problem would be to follow Zhang and Gish [18], who report using topComboN=4 to generate multiple

Table 2

Individual prediction accuracies of Pairagon alignments and N-SCAN_EST predictions in the submission

Name	NSn	NSp	ESn	ESp	ExT	TSn	TSp	TrG	GSn	GSp
Pairagon	71.5	98.8	66.8	96.1	3.1	37.9	76.5	1.4	66.9	84.2
N-SCAN_EST	84.9	91.1	72.2	84.5	3.1	18.3	38.7	1.0	38.2	38.7
N-SCAN_EST*	16.0	73.0	9.8	59.1	4.1	1.4	8.5	1.0	2.7	8.5
N-SCAN_EST†	72.2	73.0	57.6	59.1	4.1	8.3	8.5	1.0	12.7	8.5

*N-SCAN_EST predictions not overlapping Pairagon alignments (sensitivities measured against all Gencode genes). †N-SCAN_EST predictions not overlapping Pairagon alignments (sensitivities measured against Gencode genes not overlapping Pairagon alignments). Columns are defined as in Table 1.

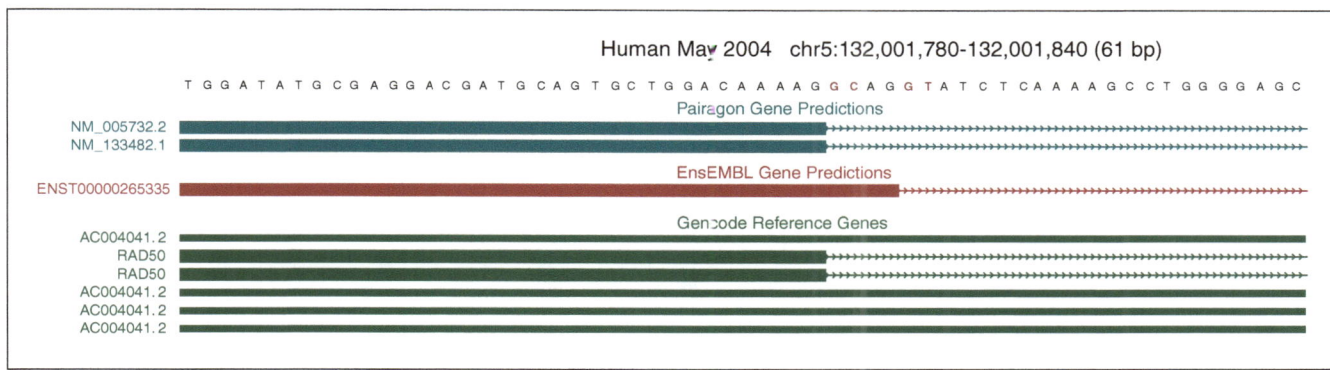

Figure 3
An annotated GC donor site that ENSEMBL misses. There is a GT dinucleotide four nucleotides downstream of the GC donor site (both dinucleotides are marked brown in the sequence). Pairagon identifies the correct donor site. (Screen shot obtained from UCSC Genome Browser web site [23].)

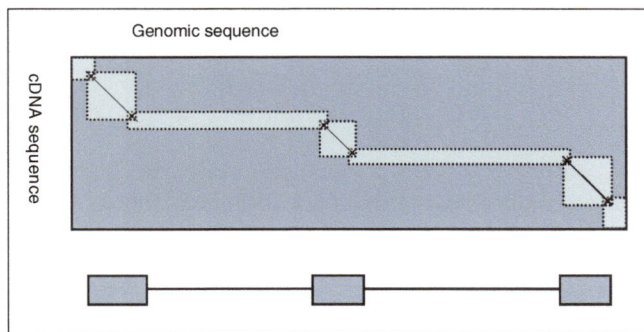

Figure 4
Generating the search subspace given three high-scoring segment pairs (HSPs) in the Stepping Stone algorithm. The three diagonal lines represent the three HSPs. The stars represent alignment pins. The lighter blue areas represent the search subspaces that are actually used in the heuristic method. The optimal algorithm uses the entire rectangle in blue. The block diagram shows the optimal spliced alignment where blue boxes represent an exon and the thin lines represent an intron.

combinations of high-scoring segment pairs (HSPs) as seed alignments for their cDNA-to-genome alignment program, EXALIN. We can then superimpose the search subspaces obtained from the possible HSP combinations. Using this approach, Pairagon would choose the correct alignment for the example in Figure 5 because, all other things being equal, it favors shorter introns over longer ones.

Pairagon's accuracy has improved since the official evaluation

Since the EGASP assessment, we have made several improvements to both Pairagon's probability model and its implementation. We have retrained Pairagon using its own alignments of 20,594 MGC cDNA sequences to 21,249 loci on the human genome. Several bug-fixes and optimizations have resulted in a faster and more robust program with lower memory requirements. Table 3 lists the accuracy measures of the current version of Pairagon (v0.95) when aligning the same cDNA sequences used for the assessment. Pairagon v0.95 shows improvement in all accuracy measures. It now identifies 22 more correct Gencode transcripts and 152 more correct exons with a small improvement in specificity as well. Thus, the accuracy of our pipeline using Pairagon v0.95 is substantially better than that of the version submitted for the assessment, which was already as good as, or slightly better than, that of the other entrants. Of course, other systems have likely improved as a result of this exercise, too.

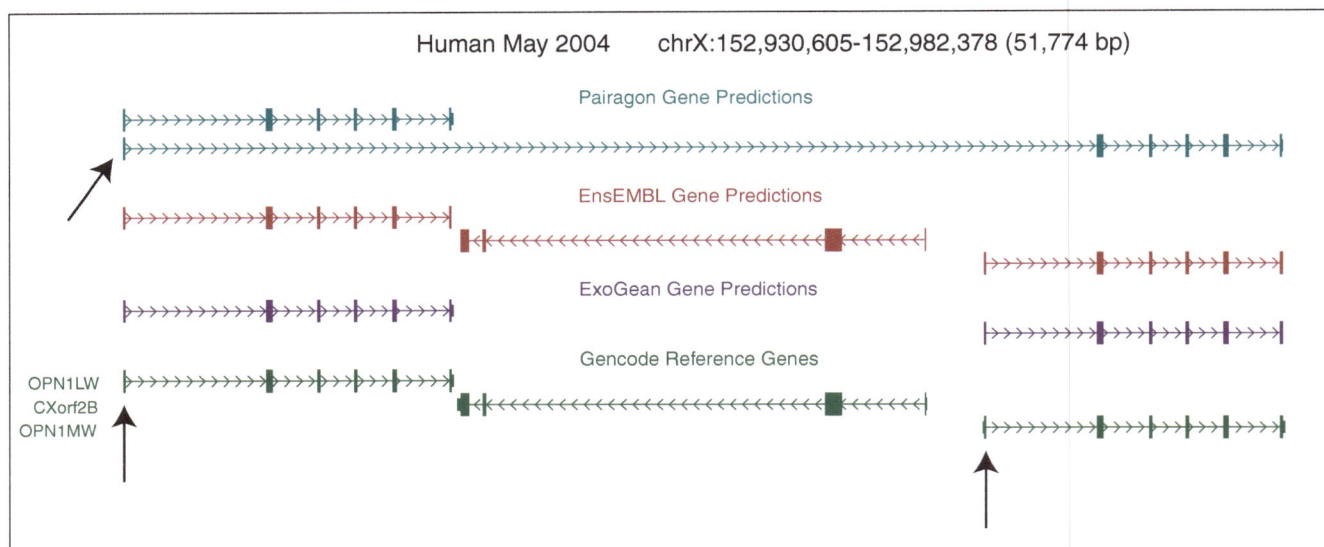

Figure 5

An incorrect alignment from Pairagon. The seed alignment from BLASTN aligned the 112-base exon at a location about 30 kb upstream (arrow in Pairagon gene prediction) instead of the annotated location (arrows in Gencode reference genes). Both alignments for that exon are 100% identical. (Screen shot obtained from UCSC Genome Browser web site [23].)

Table 3

Prediction accuracies of improved Pairagon alignments and Pairagon+N-SCAN_EST gene structures

Name	NSn	NSp	ESn	ESp	ExT	TSn	TSp	TrG	GSn	GSp
Pairagon v0.95	78.8	99.2	72.7	96.5	8.4	41.3	77.0	1.4	71.3	84.3
Pairagon v0.0.95 + N-SCAN_EST	89.9	92.5	79.0	88.9	7.6	42.4	63.3	1.3	73.3	64.5

Columns are defined as in Table 1.

A lack of biological evidence raises questions about ORF annotation

Identifying the coding region in (even) a full-length mRNA is an extremely difficult problem. NCBI and HAVANA do not always agree in their CDS annotations of mRNA sequences, even if they agree on the exon-intron structures. Because we relied on the CDS annotations from NCBI, a few of our gene predictions are incorrect according to HAVANA, although the underlying alignment is correct. For example, GenBank's annotated translation start sites for cDNA sequences BC001940 and NM_001004759.1 are 798 bases downstream and 81 bases upstream of HAVANA's annotated translation start sites in Gencode genes AC005538.1-001 and AC011711.3-001, respectively. A few more of our ORF predictions obtained from correct alignments are labeled incorrect because HAVANA has not made any CDS annotations on the exon-intron structures yet. For example, the exon-intron structure of our gene NM_181879.1 from aligning a reviewed RefSeq mRNA NM_181879.1 matches that of Gencode reference gene AC008984.1-003, which does not have a CDS annotation. Since the biological evidence supporting the GenBank ORF annotations, if any, is not available for evaluation, we might do better by using a modified version of N-SCAN to predict ORFs on aligned cDNA sequences.

N-SCAN_EST performs well on complete GENCODE test regions

After the release of the HAVANA annotations, we found that N-SCAN_EST predictions used to fill the gaps between Pairagon alignments had a very high proportion of incorrect genes – the gene/transcript specificity of the original N-SCAN_EST predictions was 8.5% in regions that did not overlap Pairagon alignments (gene and transcript specificity are the same for programs that predict only one transcript per locus). However, this is due largely to the fact that there are high quality cDNA sequences covering most of the real genes in the ENCODE regions. When these are not used and N-SCAN_EST's predictions on the complete GENCODE test regions are evaluated, their specificity is 38.7% (Table 2).

Table 4

Prediction accuracy measures of multiple-genome based gene prediction methods

Name	NSn	NSp	ESn	ESp	ExT	TSn	TSp	TrG	GSn	GSp
Augustus-dual	**88.9**	80.2	63.1	69.1	6.1	12.3	18.6	1.0	26.0	18.6
N-SCAN	85.4	**89.0**	**67.7**	**82.1**	8.0	**17.0**	**36.7**	1.0	**35.5**	**36.7**
Twinscan-MARS	84.3	74.1	65.6	61.7	8.6	15.9	15.1	1.7	33.5	24.9
Saga	52.5	81.4	38.8	50.7	5.6	2.2	3.4	1.0	4.4	3.4
Dogfish	64.8	88.2	53.1	77.3	8.7	5.1	14.6	1.0	10.8	14.6

The highest value for each measure is in bold. Columns are defined as in Table 1.

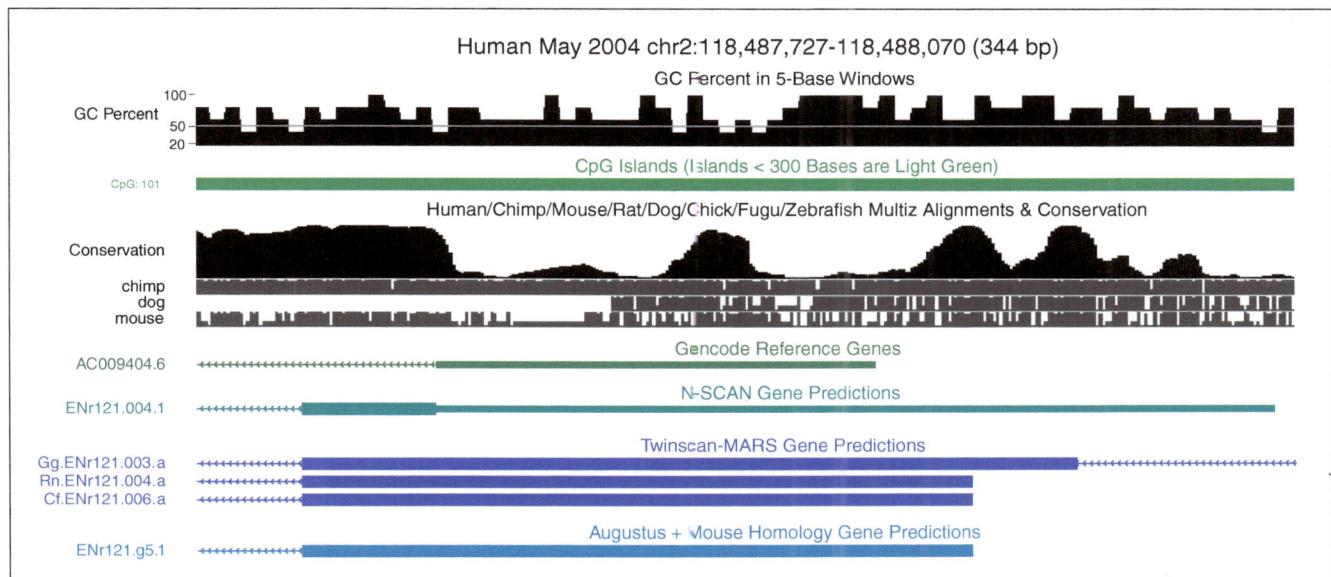

Figure 6
Initial exon of a gene where N-SCAN correctly discriminates coding region from the 5' UTR. Other gene prediction systems predict longer coding regions due to the high G+C content of the region. (Screen shot obtained from UCSC Genome Browser web site [23].)

In the ENCODE regions, the accuracy of N-SCAN_EST is due in large part to the accuracy of N-SCAN itself (this may not hold in less gene-dense regions). Table 4 compares the five submissions to the Dual or Multiple Genome category of EGASP that score the highest on exons, transcripts, and genes. N-SCAN scores the highest in all categories except for nucleotide sensitivity. In terms of exon specificity, N-SCAN is 4.8% better than the next best system (Dogfish) and in transcript specificity 18% better than the next best system (Augustus-dual). For transcript and exon sensitivity, N-SCAN is 4.7% and 4.6% better, respectively, than any other system except TWINSCAN-MARS. N-SCAN outperforms TWINSCAN-MARS by about 1% transcript sensitivity and 2% exon sensitivity. TWINSCAN-MARS has relatively high sensitivity in part because it predicts several transcripts per gene, for which it pays a price in specificity. Even with the hit it takes in specificity, TWINSCAN-MARS is among the top

three performers, especially at the transcript level. This may be explained, in part, by the fact that N-SCAN and TWINSCAN-MARS share nearly identical models for DNA sequence [16], although their conservation models are quite different.

N-SCAN's ability to explicitly model untranslated regions (UTRs) [12, 13, 19] facilitates the distinction between coding and non-coding exons. Figure 6 illustrates this advantage of N-SCAN when compared to other dual- or multiple-genome gene predictors on Gencode reference gene AC009404.6. Only the N-SCAN prediction agrees with the Gencode reference gene; N-SCAN's ability to model 5' UTR content is the key. The 168 base-pair (bp) region upstream of the annotated start codon lies within a 1,012 bp CpG island (annotated on the UCSC Genome Browser CpG-island track). The 67% G+C content of this 168 bp region is very

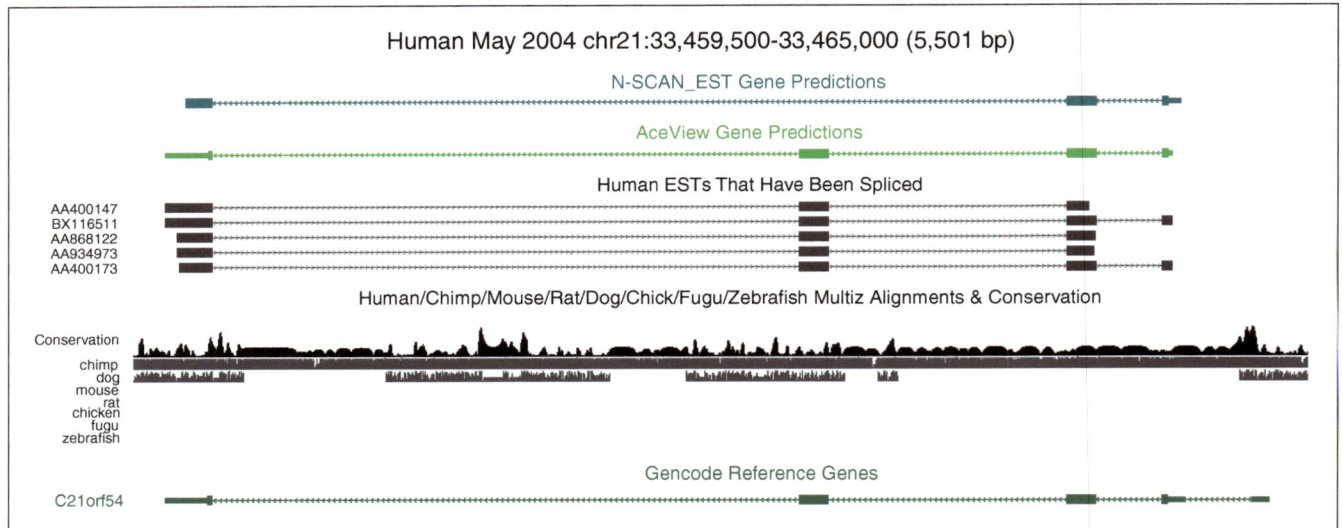

Figure 7
A gene where N-SCAN_EST predicts three out of the four exons right. All other programs except AceView do not predict anything in that locus. N-SCAN_EST missed an exon even though there is EST evidence for it. We believe that lack of conservation overwhelmed the EST evidence for that exon. (Screen shot obtained from UCSC Genome Browser web site [23].)

high compared to typical intronic and intergenic regions and even high compared to most exonic regions. However, this is not unusual for a region of this size within a CpG island. Without explicit 5' UTR-content modeling, however, it is more likely to be predicted as a coding region rather than as a 5' UTR, intronic, or intergenic region. For example, Augustus + Mouse Homology and TWINSCAN-MARS annotate this region as coding. N-SCAN's modeling of DNA content and conserved sequence for 5' UTR regions facilitates the correct categorization of this region.

When the genome sequence and conservation do not provide sufficient information about the coding potential of a gene locus, EST evidence can be very useful in gene prediction. Figure 7 shows a gene where N-SCAN_EST predicts three out of four exons correctly while both ENSEMBL and N-SCAN do not predict any gene in the region. In fact, N-SCAN_EST is one of only two gene predictors that predict any gene in this locus. There are high quality EST alignments supporting this gene, such as BX116511 with a 100% identical alignment of 583 bases, which aid N-SCAN_EST in predicting this gene even though the conservation rate of the coding regions is low. This low conservation may explain why N-SCAN failed to predict a gene; likewise, the extremely low genomic conservation in Exon 3 may explain why even N-SCAN_EST missed this exon.

Conclusions

The results of this exercise have demonstrated two things. First, this careful community assessment has been very valuable, particularly for the way in which it uncovered

weaknesses in, and inspired improvements to, Pairagon and other systems. Second, genome annotation without trans alignments can compete successfully with systems like ENSEMBL and ExoGean, which use *trans* alignments, under certain circumstances. However, annotation accuracy in the EGASP assessment is determined largely by the accuracy with which high quality native cDNA sequences can be aligned, and secondarily by the accuracy with which HAVANA's ORF calls on those cDNA sequences, or lack thereof, can be anticipated. We cannot extrapolate the results of this exercise to situations in which fewer full length cDNAs and/or fewer ESTs are available. In such situations, the accuracy of our pipeline would depend more on N-SCAN and N-SCAN_EST, while the accuracy of ENSEMBL would depend more on *trans* alignments. In future assessments, it would be worthwhile to assess prediction pipelines under a range of scenarios between the two evaluated this time –freedom to use all available native cDNA and prohibition against using any. In particular, the selective elimination of cDNA and EST sequences from the available pool would shed light on the tradeoffs among different approaches under a range of situations of practical significance (see [4] for such a study on Pairagon+N-SCAN_EST).

Materials and methods
Pairagon gene predictions
The state diagram of Pairagon's pairHMM model for cDNA-to-genome alignment is given in Figure 1. The different states model different alignment columns as follows: matches and mismatches are modeled by state A; intron is modeled by a

loop consisting of Entry, Intron and Exit; insertion and gap in genome are modeled by states G and C, respectively. Four additional states – RG1, RC1, RG2 and RC2 modeling unaligned genomic and cDNA sequences – were added to facilitate local alignment. Although, for simplicity, Figure 1 shows only one loop modeling introns, our model contains two such loops. One of them requires GT or GC at the splice donor site and AG at the splice acceptor site. The other requires AT and AC at those sites, respectively. Each state can emit the different columns of a cDNA-to-genome alignment with certain probabilities (emission probabilities). For each state there is also a probability of staying in that state or transitioning to different states (transition probabilities). These probabilities can be estimated using maximum likelihood from example alignments.

We implemented the Viterbi algorithm, an optimal dynamic programming algorithm for finding the most probable alignment between two sequences, in C. Although it produced accurate alignments, the time and space complexity for optimally aligning two sequences increases in proportion to the product of the sizes of the input sequences, imposing limitations on the size of the input sequences. Therefore, we adapted the Stepping Stone algorithm [20], a heuristic modification to the optimal algorithm. Stepping Stone relies on faster seeded alignment programs like BLASTN to identify regions of high identity between the cDNA and the genomic sequence (diagonal lines in Figure 4). It restricts the optimal dynamic programming algorithm to regions close to the approximate exons that the seed alignments correspond to (light blue region in Figure 4).

Pairagon v0.5 was trained using 15,766 BLAT alignments of 15,297 MGC [1, 9, 10] cDNA sequences to the human genome build NCBI35 (May 2005). Transition probabilities between the states were estimated from the alignments using maximum likelihood. Because this was a bootstrap procedure, and BLAT does not pay careful attention to splice sites, we assigned reasonable estimates for probabilities of GT-AG, GC-AG and AT-AC splice site combinations (98.9%, 1.0% and 0.1%, respectively). All bases were equally probable in states RG1, RC1, RG2, RC2, G, C and Intron. The probability of a match in the aligned state was estimated using maximum likelihood and was evenly distributed among the four possible combinations. Similarly, the probability of a mismatch in the aligned state was distributed among the 12 possible combinations.

Ungapped local alignments between the cDNA sequences and the unmasked ENCODE regions were generated using BLASTN [21] with parameters M=1 N=-3. These approximate seed exons were then used by the Stepping Stone implementation of Pairagon v0.5 to generate an alignment. GenBank CDS annotations of the cDNA sequences were used to convert these alignments into gene structures.

N-SCAN gene predictions

The genome sequence was masked for putative processed pseudogenes using PPFINDER [5]. N-SCAN gene predictions were then obtained as explained in [12,13].

N-SCAN_EST gene predictions

Human ESTs, downloaded from dbEST on 20 January 2005, were aligned to whole human genome (build NCBI35) by BLAT [8]. For each EST sequence, the alignment with the greatest number of bases matching the genome was selected. Alignments with at least 98% of the bases in the entire EST matching the genome were chosen to generate an ESTseq for each chromosome. ESTseq parameters were estimated from regions corresponding to a set of cleaned Refseq annotations containing 17,798 transcripts. An additional 1,000 bases on either side of the genes were used to train intergenic regions. The genome sequence was masked for putative processed pseudogenes using PPFINDER [5]. ESTseqs corresponding to the ENCODE regions were obtained by cutting the relevant sections out of the chromosomal ESTseq, and N-SCAN_EST was then used to predict genes.

Pairagon+N-SCAN_EST pipeline

A block diagram showing the steps involved in generating Pairagon gene structures and N-SCAN_EST gene predictions, and combining them is given in Figure 2. Because multiple mRNA sequences are available for some genes, identical Pairagon gene structures are merged into one gene. N-SCAN_EST predictions are added to the final set if they do not overlap the merged Pairagon gene structures. We used the Eval software package [22] for finding these overlapping genes.

Acknowledgments

We are grateful to Jeltje van Baren for help with her PPFINDER software for detection of processed pseudogenes in gene annotation sets. Thanks also to the organizers of the GENCODE evaluation, including especially Roderic Guigó and Paul Flicek. This work was supported in part by grants U01 HG003150 (ENCODE) and R01 HG02278 from the National Human Genome Research Institute and by Contract N01-CO-12400 from the National Cancer Institute (Mammalian Gene Collection).

This article has been published as part of Genome Biology Volume 7, Supplement 1, 2006: EGASP '05. The full contents of the supplement are available online at http://genomebiology.com/supplements/7/S1.

References

1. The MGC Project Team: **The status, quality, and expansion of the NIH full-length cDNA project: The Mammalian Gene Collection (MGC).** Genome Res 2004, 14:2121-2127.
2. Brent MR: **Genome annotation past, present and future: How to define an ORF at each locus.** Genome Res 2005, 15:1777-1786.
3. Birney E, Clamp M, Durbin R: **GeneWise and Genomewise.** Genome Res 2004, 14:988-995.
4. Wei C, Brent MR: **Integrating EST alignments and de novo gene prediction using TWINSCAN.** BMC Bioinformatics 2006, In Press.
5. van Baren MJ, Brent MR: **Iterative gene prediction and pseudogene removal improves genome annotation.** Genome Res 2006, 16:678-685.

6.   Durbin R, Eddy SR, Krogh A, Mitchison G: *Biological Sequence Analysis: Probabilistic Models of Proteins and Nucleic Acids.* Cambridge, UK: Cambridge University Press; 1998.
7.   Levine A, Durbin R: **A computational scan for U12-dependent introns in the human genome sequence.** *Nucleic Acids Res* 2001, **29:**4006-4013.
8.   Kent WJ: **BLAT - the BLAST-like alignment tool.** *Genome Res* 2002, **12:**656-664.
9.   Strausberg RL, Feingold EA, Grouse LH, Derge JG, Klausner RD, Collins FS, Wagner L, Shenmen CM, Schuler GD, Altschul SF *et al.*: **Generation and initial analysis of more than 15,000 full-length human and mouse cDNA sequences.** *Proc Natl Acad Sci USA* 2002, **99:**16899-16903.
10. Strausberg RL, Feingold EA, Klausner RD, Collins FS: **The mammalian gene collection.** *Science* 1999, **286:**455-457.
11. Pruitt KD, Tatusova T, Maglott DR. **NCBI Reference Sequence (RefSeq): a curated non-redundant sequence database of genomes, transcripts and proteins.** *Nucleic Acids Res* 2005, **33(Database issue):**D501-D504.
12. Gross SS, Brent MR: **Using multiple alignments to improve gene prediction.** In *Research in Computational Molecular Biology, 9th Annual International Conference, RECOMB 2005, Cambridge, MA, USA, May 14-18, 2005, Proceedings.* Edited by Miyano S, Mesirov JP, Kasif S, Istrail S, Pevzner PA, Waterman MS. Cambridge: Springer; 2005:374-388.
13. Gross SS, Brent MR: **Using multiple alignments to improve gene prediction.** *J Comput Biol* 2006, **13:**379-393.
14. Guigo R, Flicek P, Abril JF, Reymond A, Lagarde J, Denoeud F, Antonarkis S, Ashburner M, Bajic VB, Birney E, *et al.*: **EGASP: The ENCODE Genome Annotation Assessment Project.** *Genome Biology* 2006, **7** (Suppl 1) :S2.
15. Flicek P, Keibler E, Hu P, Korf I, Brent MR: **Leveraging the mouse genome for gene prediction in human: from whole-genome shotgun reads to a global synteny map.** *Genome Res* 2003, **13:** 46-54.
16. Korf I, Flicek P, Duan D, Brent MR: **Integrating genomic homology into gene structure prediction.** *Bioinformatics* 2001, **17 (Suppl 1):**S140-S148.
17. Blanchette M, Kent WJ, Riemer C, Elnitski L, Smit AF, Roskin KM, Baertsch R, Rosenbloom K, Clawson H, Green ED, *et al.*: **Aligning multiple genomic sequences with the threaded blockset aligner.** *Genome Res* 2004, **14:**708-715.
18. Zhang M, Gish W: **Improved spliced alignment from an information theoretic approach.** *Bioinformatics* 2006, **22(1):**13-20.
19. Brown RH, Gross SS, Brent MR: **Begin at the beginning: predicting genes with 5' UTRs.** *Genome Res* 2005, **15:**742-747.
20. Meyer IM, Durbin R: **Comparative ab initio prediction of gene structures using pair HMMs.** *Bioinformatics* 2002, **18:**1309-1318.
21. Altschul SF, Gish W, Miller W, Myers EW, Lipman DJ: **Basic local alignment search tool.** *J Mol Biol* 1990, **215:**403-410.
22. Keibler E, Brent MR: **Eval: a software package for analysis of genome annotations.** *BMC Bioinformatics* 2003, **4:**50.
23. **UCSC Genome Browser** [http://genome.ucsc.edu]

Research

Vertebrate gene finding from multiple-species alignments using a two-level strategy

David Carter and Richard Durbin

Address: Wellcome Trust Sanger Institute, Wellcome Trust Genome Campus, Hinxton, Cambridge CB10 1SA, UK.

Correspondence: David Carter. Email: dmc@sanger.ac.uk

Published: 7 August 2006

Genome *Biology* 2006, **7(Suppl 1)**:S6

The electronic version of this article is the complete one and can be found online at http://genomebiology.com/2006/7/S1/S6

Abstract

Background: One way in which the accuracy of gene structure prediction in vertebrate DNA sequences can be improved is by analyzing alignments with multiple related species, since functional regions of genes tend to be more conserved.

Results: We describe DOGFISH, a vertebrate gene finder consisting of a cleanly separated site classifier and structure predictor. The classifier scores potential splice sites and other features, using sequence alignments between multiple vertebrate species, while the structure predictor hypothesizes coding transcripts by combining these scores using a simple model of gene structure. This also identifies and assigns confidence scores to possible additional exons. Performance is assessed on the ENCODE regions. We predict transcripts and exons across the whole human genome, and identify over 10,000 high confidence new coding exons not in the Ensembl gene set.

Conclusions: We present a practical multiple species gene prediction method. Accuracy improves as additional species, up to at least eight, are introduced. The novel predictions of the whole-genome scan should support efficient experimental verification.

Background

Gene finding can usefully be viewed as a two-level task. At the lower or local level there is a classification task: one of assigning probability estimates to potential features such as splice sites and coding start and stop sites on the basis of sequence information associated with each potential feature. At the higher or global level, on the other hand, we have a structure-building task: finding the most probable way(s) to combine potential features into exons, transcripts and genes. Classification and structure building are very different tasks, and although a gene finder can be based on a single formalism, such as hidden Markov models (HMMs) [1,2], there is no reason to assume that the same technique will be optimal for both tasks. Although HMMs seem to offer a good basis for structure building, they impose independence assumptions that are not particularly well suited to feature classification; formalisms such as neural networks [3,4], maximum entropy modeling [5], Bayesian networks [6-8], support vector machines [9-11] and relevance vector machines (RVMs) [12-14] provide alternative approaches with potential benefits.

Gene finders have conventionally analyzed a single sequence [2,15-17] or, more recently, alignments between sequences for two species [18-25]. In the past year or two, gene finders processing alignments of more than two species have begun to appear [26-31]. In principle at least, the additional information provided by extra species should lead to improved predictions, but it is far from trivial to extend existing formalisms to make the best use of it.

In parallel with systems processing only genomic data, gene finders have been developed to use expressed sequence tag (EST), cDNA and protein sequences [32-36]; these can achieve better overall accuracy than systems using multiple-species alignments, but they are effective only where the sequences in question have been detected.

The gene finder described in this paper, DOGFISH (for 'detection of genomic features in sequence homologies'), is based on the above observations. It predicts gene structures in the sequence for a target species based on alignments with one or more informant species. At the global, structure-building level it employs a fairly conventional HMM. Its two main novelties lie at the local, classification level. At this level, it analyses multiple-species alignments (of eight species in the work reported here), passing the results up to the HMM for structure building. In this way, it avoids having to deal with the complexities of multiple-species alignments and the HMM formalism in the same tightly coupled framework. To do the classification, it uses a cascade of relevance vector machines to derive a single probability estimate from many thousands of individual scores based on particular aspects of the aligned sequences around a feature of interest. The HMM sees only the predictions of the classifier, not the genomic sequences or alignments, resulting in some useful simplifications.

Results and discussion
In this section, we present results first for classification of individual splice sites and start and stop codons, and then for HMM-based gene finding on the ENCODE test regions using the outputs of the classifier.

Classifier results
As explained in more detail in the Materials and methods section, DOGFISH's classifier consists of two main components, which adopt respectively a 'vertical' and a 'horizontal' view of alignments of multiple species around each feature of interest (see Figure 1 for an example alignment). The vertical component applies a separate evolutionary model to each column in an alignment, explicitly modeling mutations but taking only very limited account of the context in which the column occurs. The horizontal model is complementary: it uses Markov models and nucleotide tuple frequencies to assess the aligned sequence for each species as a possible instance of the feature under consideration without reference to the other species, and then combines the results to produce a single estimate. Thus, in contrast to the vertical model, it analyses context as thoroughly as possible but ignores mutations. Since both kinds of information are

important, one might expect each component to perform well on its own, and a combination of the two to do better still.

We trained DOGFISH to detect genes in the human genomic sequence on the basis of the University of California, Santa Cruz (UCSC) MultiZ alignments [37] with seven other species. We used the multi-way alignments with mouse, rat, dog, chicken, zebrafish and fugu, discarding chimp from the original set because it did not improve results, and adding in the separately available human-frog pairwise alignments. All sequences were soft repeat masked using RepeatMasker [38]. The classifier was trained and evaluated using all the Vega annotations for human (nine chromosomes, downloaded August 2005), excluding those for all 44 ENCODE regions and for positions 100M to 110M of chromosome 9, a region with typical gene density that we used for various tuning purposes. We did not use the 18-species ENCODE comparative sequences [39], which were only available for the ENCODE regions, covering 1% of the human genome, for two reasons. Firstly, this quantity of sequence would not be enough to train fully the thousands of parameters in the classifier. Secondly, we wanted to run the system on the whole human genome, for which the UCSC alignments were the most comprehensive available.

At the local level, DOGFISH assigns a probability estimate to every potential splice site, start codon and stop codon in a genomic region to be analyzed and, for splice sites only, a probability distribution over the possible coding phases. A potential splice site is defined here as any AG or GT dinucleotide; GC splice donors and U12 splice sites are too rare to be accurately detected. In what follows, by a 'true' acceptor site we mean any AG splice site, while a 'decoy' is an AG that is not a splice site. True and decoy donor splice sites and start and stop codons are defined similarly.

We evaluated a number of variants of the classifier on a specially constructed 'challenging' set of candidate sites. The probability of including a site in this set, irrespective of whether true or decoy, was a strongly increasing function of the score assigned to it by a first version of the classifier that was itself trained on randomly selected sites. Such a challenging set is necessary to achieve clearly distinct performance figures; if sites are randomly selected from the genome, the classification task is too easy, at least for splice sites, and many versions of the classifier score close to 100%.

Comparing classifier components
We evaluated performance using the horizontal component alone, the vertical component alone, and both together. As well as the scores derived from each of the horizontal and/or vertical components, we used one further value in all the experiments. This was derived from a simple 'presence' component that just returns a score depending on the set of species aligned to a site, irrespective of the content of the

```
hs  TGGGTGGGCACGTGTGACGCTGGTCCCCTCTCCTTTGTAGCTGTGGTGCACCTGCCATAAGAAGTCCTTGGTGGAAACAG
mm  TAGAC-------TGTGATCTCAATCAATTTTCCTCTGCAGCTGTGGTGTACCTGCCACAAGAAGTCATTGGTGAAAACAG
rn  TAGAC-----TGTGTGGCCTGGATCACTTCTCTTCTGCAGCTGTGGTGCACGTACCACAAGCAGTCTTTGGTGAAAACGG
cf  GGGCG-----------------TCCTTCCTCCCCGGCAGCTCTGGTGCACCAGCCACAAGAAGTCGCTGGTGAAGGCGG
gg  ------------GTGAGGCT--CGGCTTTTTATCTCCAGCTATGGTCCACGTTTCATGAAAGACATCTGGTGAAGGAAG
dr  ----------------------------TTTCTTTCACAGCTCTGGTGCACCTTCCATGAAAAGGCCTTGGTGAAAGGAG
fr  TGAAT-----CATGAGACGTTGACGTCTTCTTTTTTGTAGCTGTGGTCCACTTTCCACAAAAAGTCCATGGTCAAGGAGA
                                                    ·  ·  ·  ·  ·  ·  ·  ·  ·  ·  ·  ·  ·  ·  ·
```

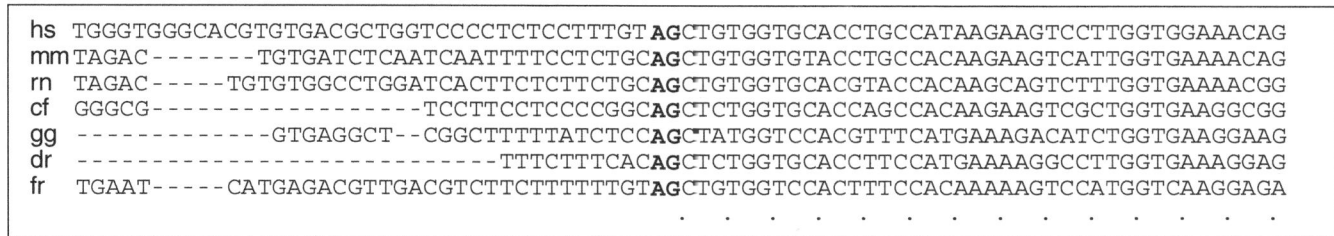

Figure 1
Alignment for a coding splice acceptor site. The figure shows the central part of a typical alignment window used by the classifier component of DOGFISH. Codon boundaries on the exon side of the splice site are indicated with dots. This site has an alignment with all species except frog: hs; *Homo sapiens*: mm; *Mus musculus*: rn; *Rattus norvegicus*: cf; *Canis familiaris*: gg; *Gallus gallus*: dr; *Danio rerio*: fr; *Fugu rubripes*. The AG dinucleotide for the acceptor site itself is shown in bold.

Table 1

Prediction accuracies for vertical and horizontal components

	Acceptors	Donors	Starts	Stops
Train set size	204,021	221,421	7,571	25,071
Eval set size	52,605	57,179	1,805	6,162
%True sites	14.05	13.01	16.68	8.08
F scores (%)				
Presence	52.72	48.77	39.70	34.64
Vertical	82.01	81.00	55.70	49.25
Horizontal	84.36	84.43	57.01	48.22
Both	84.86	84.60	58.22	49.60
ENCODE CI	63.18	65.86	27.44	14.67
ENCODE GF	80.23	81.38	42.47	50.49
100-ROC (%)				
Presence	12.41	12.66	20.62	23.98
Vertical	2.46	2.52	14.49	12.76
Horizontal	1.81	1.58	12.48	11.77
Both	1.74	1.54	10.41	10.90
ENCODE CI	0.99	0.61	9.14	10.49

The table shows the F score (geometric mean of sensitivity and specificity, which are close to each other) for various classifier components. The test set for the presence, vertical, horizontal and 'both' conditions is 'challenging' data; we show results for a mixture of the classifiers trained on challenging and randomly selected data. The 'ENCODE CI' and 'ENCODE GF' lines are for the 31 ENCODE test regions, using classifier scores and gene-finder scores, respectively. The table also shows the 100%-ROC (receiver operating characteristic) error value for each condition. This error value is the probability that if a true instance and a decoy are selected at random, the classifier will give the decoy a higher score than the true instance.

alignment; thus, a site that aligns with many informants is likely to score higher. This favors true sites because the true splice sites in our challenging set align with locations in an average of 5.6 out of our seven informant genomes, compared to 2.0 for decoys, while for start and stop sites the corresponding figures are 3.8 for true sites and 1.6 for decoys.

We also evaluated the full classifier against all potential sites in the roughly 21.5 Mb of the 31 ENCODE test regions. In genomic regions, decoy sites were thousands of times more numerous than true ones, rather than just a few times as in our main evaluation set. This serves as a 'reality check' that our main set, despite its challenging nature, is not artificially easy. For comparison, we also evaluated the site estimates output by the full gene finder; these values are based partly on the classifier estimates but also on the availability of nearby sites to make up legal gene structures.

F-score and receiver operating characteristic (ROC) error values are shown in Table 1 for each condition. These results can be summarized as follows. Firstly, start and stop codons are much harder to detect than splice sites. Secondly, for splice sites, presence scores alone are much better than random: the F scores in the 'Presence' line of the table are well over the small percentage of true sites in the evaluation set, which would be the F scores expected from a random-choice strategy. Thirdly, adding either the vertical or the horizontal component improves performance markedly over using the presence component alone. Fourthly, for splice sites, the horizontal component alone is better than the vertical component alone. Fifthly, using both the horizontal and the vertical component is consistently, but only slightly, better than using the horizontal alone. Sixthly, classifier results on the ENCODE regions confirm that performance is good on whole genomic regions, where decoys outnumber true sites by thousands to one. (The simultaneous decrease in both F score and ROC error rate is a consequence of these regions having far more, but on average easier, decoys than the main test set; see Materials and methods.) Finally, not surprisingly, the full gene finder is much more accurate than the classifier alone on the ENCODE regions.

Analysis of classification errors
The errors in classification on the challenging test set are broken down by site type in Table 2. For this table, we set

Table 2

Error rates broken down by site type

	Acceptors	Donors	Starts	Stops
True sites, false reject percentages				
Overall	4.45	4.08	17.61	15.86
Coding	4.34	3.88	17.61	15.86
Non-coding	7.48	10.00	NA	NA
Decoy sites, false accept percentages				
Overall	4.45	4.08	17.35	15.84
Coding	2.34	1.08	23.19	2.92
Non-coding intra	2.54	2.21	12.17	16.58
Non-coding inter	8.00	8.31	18.05	16.07

The table shows the proportion (in the challenging test set) of various site types that received an incorrect classification. The classification threshold is adjusted to achieve roughly equal proportions of false positives and false negatives. NA: not applicable.

Table 3

Phase prediction error rates on coding splice sites

	Acceptors	Donors
Vertical	3.84	3.02
Horizontal	5.17	4.79
Both	1.99	1.60

The table shows the percentage (in the challenging test set) of coding splice sites for which the coding phase that was assigned the highest probability was not the annotated phase.

acceptance thresholds so that false positives balance false negatives. (We treat a decoy as 'coding' not only if it falls within a coding region of the genome but also if it is within 50 bases of a coding region. In the latter case, it will generally have 50 or more coding positions within the 200-nucleotide region described in Materials and methods, making it in that regard more similar to a true coding site, which usually has 100, than to a true non-coding site, which usually has none.)

Not surprisingly, the figures indicate that non-coding splice sites are harder to detect (have a higher error rate) than coding ones. However, we were initially surprised that intergenic splice site decoys (which are by definition non-coding) should have a much higher error rate than intragenic non-coding or even coding ones. This could be due either to suppression of non-functional splice sites inside transcripts or to non-annotated exons outside annotated transcripts. We found no evidence of suppression (decoys inside and outside transcripts were similarly distributed) but we did find evidence for unannotated exons.

If substantial numbers of exons are present in a region, one would expect high-scoring candidate acceptor (A) sites to alternate with high-scoring donors (D) more often than chance would predict. Therefore, we looked at the highest-scoring N acceptor candidates and the highest-scoring N donors, for various values of N. If no exons are present, we would expect neighboring AD and DA pairs on the same strand to occur no more often than AAs or DDs. However, if there are exons, then as N rises, we expect Δ, the excess of ADs and DAs over AAs and DDs, to rise as genuine splice site pairs enter the set, then to fall again as the pattern is destroyed by lower-scoring, mostly decoy sites.

We looked at how Δ varied with N on human chromosome 13. This chromosome was selected because, in proportion to its length, it had contributed the smallest number of sites to the top-scoring 2% of intergenic decoys to the test set and, therefore, seemed likely to contain the fewest unannotated exons. Even on this chromosome, we found Δ rising to a highly significant level and then falling again, as predicted. At maximum, we found a total of 2,062 AD and DA pairs in the chromosome 13 intergenic regions, compared to 1,712 AA and DD pairs, giving Δ = 350. The corresponding Δ value for intragenic regions was 4,773 - 1,658 = 3,115. The Vega annotation of chromosome 13 contains about 3,000 internal (bounded by an acceptor and a donor) exons, which would suggest there are around 3,000 × 350/3,115 = 337 exons still be to be annotated. We return to the implications of this later.

Splice site phase determination

We have seen that using the vertical component in addition to the horizontal one does not improve splice site detection by more than a small amount. However, this is not the case for the task of determining splice site phases. For coding splice sites, the error rate (percent incorrect) for the various combinations is given in Table 3; we take a prediction as correct if the true phase is the one assigned the highest probability. These results show that for phase determination, the vertical component is superior to the horizontal. This would appear to be because the vertical component explicitly looks for patterns of amino acid conservation, which are a more powerful indicator of phase than the per-species nucleotide preferences detected by the horizontal component. However, using both vertical and horizontal is much better than using vertical alone, suggesting that the horizontal component, with its wider view of context, is picking up phase-indicating contextual effects wider than individual codons, even though it does not compare sequences so is blind to patterns of mutation.

The effect of additional species

Finally, we tested one of the assumptions behind this work, that the more informant species are used, the better the classifier works. We evaluated the configuration of the system containing the species-presence and horizontal components of the classifier trained on challenging data; this is almost as

Table 4

Prediction accuracies for different numbers of species

	Acceptors	Donors	Starts	Stops
Train set size	204,021	221,421	7,571	25,071
Eval set size	52,605	57,179	1,805	6,162
F scores (%)				
Human only	66.78	67.25	35.34	22.20
Human+mouse	80.67	82.74	43.38	30.57
All 4 mammals	82.53	83.99	44.02	31.88
All 8 species	84.31	84.82	51.45	34.93
100-ROC (%)				
Human only	5.22	4.31	18.30	20.03
Human+mouse	2.45	1.93	13.18	15.54
All 4 mammals	2.21	1.81	11.77	14.75
All 8 species	1.76	1.54	10.53	11.68

The table shows the F score (geometric mean of sensitivity and specificity) and ROC error rate (area not under the ROC curve) for the horizontal component of Classifier Two trained on different numbers of informant species and running on the challenging evaluation (Eval) set. All scores are percentages.

accurate as the full system in classifying splice sites. We made available one species (human), two (human and mouse), four (the mammals) and all eight; Table 4 shows the results. As expected, the greatest gain comes from the first additional species, mouse. However, more gains are apparent as further species are added, with non-mammal species apparently just as useful overall as additional mammals.

Gene finder results

We combined the classification results into gene structures using an HMM as described in Materials and methods. In the evaluation here, we focus on exon performance as the primary indicator.

Table 5 gives sensitivity and specificity results at the nucleotide, exon and transcript level on the 31 ENCODE testing regions, for DOGFISH-1, the version available at the time of the ENCODE competitive evaluation in May 2005, and for DOGFISH-2, the current version. Although the latter version was developed after the detailed annotations of the testing regions were released, no nucleotide sequences, alignments or annotations for any of these regions were used in any way in developing any version of DOGFISH.

DOGFISH-2, the current version of the system, is described throughout this paper. The most important differences between DOGFISH-1 and DOGFISH-2 are as follows. Firstly, although DOGFISH-1 constructed coding-phase-specific models within the horizontal and vertical components, the RVM cascade did not maintain separate per-phase hypothe-

Table 5

Exon and transcript accuracies

	DOGFISH-1	DOGFISH-2
Exon sensitivity	53.11	63.68
Exon specificity	77.34	84.90
Transcript sensitivity	5.08	8.94
Transcript specificity	14.61	33.12

The table shows percentage sensitivity and specificity at the exon and transcript levels for the workshop version, DOGFISH-1, and the current version, DOGFISH-2.

ses during its later data reduction. This both decreased the accuracy of its estimates and meant it was unable to pass phase information on to the HMM. Secondly, DOGFISH-1's HMM component was less sophisticated than that of DOGFISH-2, and in particular did not use N-best lists [2] (see Materials and methods) to mitigate the negative effects of using exon and intron length penalties. Thirdly, the training set used for DOGFISH-1's classifier was not constructed systematically to include difficult decoys and, therefore, the classifier was less well-matched to the needs of the gene finder.

Error analysis

The gene-finding results for both DOGFISH-1 and DOGFISH-2 are derived from the single best-scoring HMM path; thus only one transcript per gene is predicted, a bias that is reflected in sensitivity scores being rather lower than specificity. In fact, the excess of false-negative over false-positive exon detection errors made by DOGFISH-2 on the ENCODE test set is almost exactly equal to the number of alternative exons in the reference annotation; these account for half of all exon errors.

The next most important source of errors is the classifier's poorer performance on start and stop codons than on splice sites. The overall exon sensitivity of 63.68% in fact breaks down to around 73% for internal exons and only 37% for external (initial and terminal) ones, while specificity (84.90% overall) is 87% for internal exons and 73% for external. This difference directly accounts for about a quarter of all the exon errors, and has an additional knock-on effect in the form of increased numbers of errors in internal exons adjacent to external ones, accounting for a further 20% of the errors. Most of the final 5% of errors can be traced to imperfect classifier estimates on splice sites.

This analysis suggests a number of ways in which DOGFISH could be improved. Firstly, by explictly modeling splicing signals not currently handled, such as enhancers and repressors; this could be done by applying independently derived information to train weight matrices for such signals, which can be longer than the six-nucleotide patterns

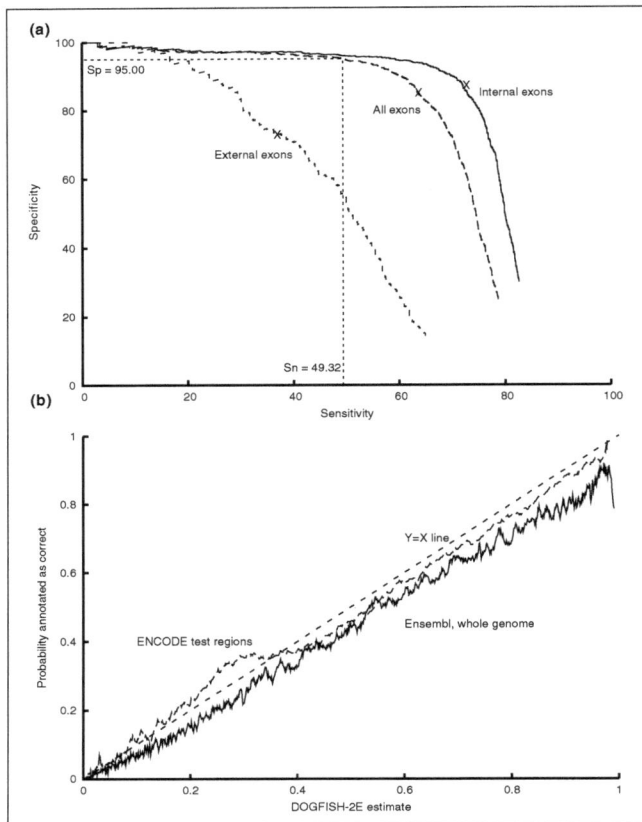

Figure 2
DOGFISH-2E results. **(a)** Sensitivity and specificity for DOGFISH-2E output. The figure shows plots for specificity against specificity on the ENCODE test regions as the acceptance probability threshold is varied for internal exons, external (initial and terminal) exons, and all exons together. 'X' is used to mark the DOGFISH-2 sensitivity and specificity values, and the specificity value of 95% for almost 50% sensitivity is highlighted. **(b)** Probability of annotation as a function of DOGFISH-2E estimate. The figure shows DOGFISH-2E probability estimates on the *x* axis and, on the *y* axis, the probability that a site a DOGFISH-2E estimate of the given magnitude is annotated in ENCODE and Ensembl, respectively. The Y=X line is shown for comparison.

processed by the current method. Secondly, by an explicit treatment of the specific characteristics of alternative exons [40]; including high-scoring exons not on the HMM's best path as suggested in [41] did not work well. Thirdly, by better modeling of untranslated regions [7,28]. Fourthly, by using alignments with more informant species, both closely related and more distant. Fifthly, by improving accuracy on start and stop codons.

Of these, there is reason to hope for good progress from applying variants of DOGFISH's existing machinery to the first four problems; but we have already devoted substantial effort to the last issue, start and stop codons, and it is not clear to us how much better accuracy could be obtained for these features. The difficulty seems to be that despite the known consensi around these sites, interspecies conserva-

tion is not as strong as for splice sites and so a multiple-alignment based method cannot predict them as accurately.

Exon probability estimates
The gene finder HMM assigns a score to every candidate site and exon. Using these scores, we trained separate relevance vector machines (RVMs) for initial, internal and terminal exons to estimate the probability of correctness of each candidate coding exon. By setting the threshold for acceptance, we were able to trade off sensitivity against specificity. We call this version of the system DOGFISH-2E, since it predicts individual exons with no requirement that they make up correct transcripts; this could indicate additional exons incompatible with the most likely gene structure, and also allows low-scoring exons (even when on the best path) to be discarded. Figure 2a shows the behavior on the ENCODE test regions for internal exons, external exons (initial and terminal individually show similar behavior) and all exons together. The points corresponding to DOGFISH-2 are shown there as crosses; note also that close to 50% of all exons are predicted with specificity 95% or better.

Whole-genome scan
We ran DOGFISH-2E over the whole human genome (excluding chromosome Y because of its overlap with X), estimating probabilities for over 1.3 million candidate exons, and looked at how these estimates correlated with whether each exon was among the 181,475 coding exons in the Ensembl database (downloaded 9th November 2005). We found that the probability of an exon being present in Ensembl was very well modeled by its DOGFISH-2E estimate multiplied by 0.889 (compare Ensembl's 0.775 sensitivity against the ENCODE annotations; see companion paper in this supplement). For DOGFISH-2E on the ENCODE test data, the corresponding factor was 1.001, though the relationship was less linear (Figure 2b). It seems likely from the difference between the factors that substantial numbers of exons are missing from Ensembl.

DOGFISH-2E assigns an estimate of 0.95 or greater to 99,369 exons over the whole genome. On the ENCODE test data, 95.9% of exons scoring over this threshold are annotated as correct; thus, it seems reasonable to assume that $0.959 \times 99,369 = 95,295$ of the whole-genome predictions are correct. Of the 99,369, only 88,385 are annotated in Ensembl as coding exons, with 10,984 either absent altogether or, in a minority (15%) of cases, annotated as non-coding. Even if we assume that all of the 88,385 are correct, we are left with an expected 95,295 - 88,385 = 6,910 correct predictions among the 10,984 additional ones, giving a specificity of 62.9%. Adding 6,910 new coding exons to Ensembl's existing total of 181,475 would increase the number by 3.8%.

These results, together with the pattern of alternation of high-scoring 'decoy' acceptor and donor splice sites in regions annotated as intergenic in Vega, lead us to conclude it would be fruitful to use high-scoring DOGFISH-2E predictions to guide experiments searching for new coding exons. It would also be interesting to investigate how far these 'missing' exons overlap with existing EST data and with so-called transfrags [42].

Conclusions

Distinguishing two levels of the task of gene finding allows separate strategies to be applied at each level, allowing us to make good use of the information present in multiple alignments without the system becoming unmanageably complex. The current accuracy of DOGFISH is comparable to that of the best published gene finders that use multiple-species alignments (see other papers in this supplement), confirming that a two-level approach can yield good results.

Perhaps surprisingly, vertical (evolutionary) models do not appear to offer much advantage over combining the results of horizontal ones when it is a matter of distinguishing true sites from decoys; however, they are useful for determining phase, a task that is important for guiding the gene finder, since a phase mismatch can help rule out an otherwise promising exon.

The strategy of using multiple species pays off: we have demonstrated that the more species are used, the more accurately splice sites can be detected. It remains to be verified whether this effect will continue to apply if more than eight species, or different species, are used, but Table 4 does not suggest that saturation has been reached. Furthermore, adding more closely related informants as their genomes become available should also improve performance, since 3.4% of confirmed coding splice sites in our data set have no alignments at all, and a further 3.5% only align to one other species.

Three useful resources arise from this work. The first is the challenging data set used to train Classifier Two, which we offer for use for training and testing both single- and multiple-species feature classifiers. The second is the single-species subpart of the horizontal component, which is a strong single-sequence classifier in itself. The third is a set of predictions of splice sites, exons and genes obtained by running DOGFISH over the whole human genome, which will enable experimental effort to be concentrated on predictions that are not part of known genes; we estimate that if the highest-scoring 50% of these extra predictions are selected, over 60% of them will be correct.

Materials and methods

In this section, we devote most attention to DOGFISH's classifier, which contains most of the novel aspects of the system. We finish with a description of the structure-building HMM, focusing on the way it uses classifier outputs and the respects in which it differs from conventional HMM technology.

Classification methods

The main mechanism that DOGFISH uses in its classifier is the Biojava [43] implementation of the RVM [12,14], a robust and accurate new classification technology that dispenses with many of the independence assumptions inherent in HMMs. An RVM is a trainable device for mapping any number of input scores (which may or may not themselves represent probabilities) to a single output probability. In contrast to most other classification methods, when the mapping is trained, a few inputs typically receive high weights (are viewed as 'relevant'), a few more get low ones, and many are assigned a weight of zero, on the basis that they do not offer any further useful information once the other inputs, with which they may be correlated, have been taken into account. The tendency of RVMs presented with many inputs to select only a few of them as relevant leads to good robustness, greater transparency than some alternative techniques, and some efficiency gains because the values of zero-weighted inputs do not need to be calculated.

DOGFISH applies a cascade of RVMs to carry out a stage by stage reduction of many thousands of scores, each derived from one small facet of an alignment around a site of interest, to a single estimate of the probability that the site is a true instance of a particular feature such as an acceptor splice site.

DOGFISH classifies a feature by looking at a 200-nucleotide window centered on it. Each column of the window contains a target-species nucleotide and, for each informant species, either a gap character or a nucleotide from that species. The window is much wider than the known consensus of a dozen or so base-pairs around splice sites; however, this choice makes it possible to detect not only these consensi but also coding phases and transitions between introns and exons and between non-coding and coding regions, both of which are marked by distinctive patterns of conservation and divergence in the alignments. Doing this removes most of the need for an explicit model of coding sequence in the HMM, which is able as a consequence to avoid looking at nucleotides altogether and work simply on the classifier output scores.

The inner 78 positions of a classifier window, for a typical phase-zero acceptor site, are shown in Figure 1. Sequences from seven species are aligned here, with species identifiers shown to the left; the top one is the human sequence, and the frog sequence is missing. The AG dinucleotide at the site itself is shown in bold, and codon boundaries are indicated by dots under the alignment. Characteristically for this type of site, we see much better alignment on the exon (downstream) side than the intron side; a polypyrimidine tract just upstream of the site, clearly present in all species but with poor inter-species alignment at the nucleotide level; and, on the exon side, at least close to the splice site, more mutations in codon-final positions. The classifier uses all this information not only to distinguish true sites from decoys, but also, for the case of splice sites, to determine coding phase.

'Vertical' and 'horizontal' perspectives

There are many ways in which a classifier could be trained on such a data structure, but two are clearly worth pursuing. As discussed briefly above, we call them vertical and horizontal approaches according to which dimension of the window they treat as primary.

In the vertical approach, we look primarily at the columns of the window, each of which contains the target-species nucleotide at a particular offset from the (candidate) site in question and its alignment, if any, with each informant species. We apply offset-dependent evolutionary models to derive a score for each column having arisen at that offset from a feature of the type under consideration (for example, 17 bases upstream of a phase-zero splice donor). We then, secondarily, look at the horizontal dimension, combining the per-offset scores resulting from the primary step into a single estimate.

By contrast, in the horizontal approach, we first treat the sequence for each species as a potential instance of the feature in question and derive an estimate of the probability that it is indeed one. We then, secondarily, combine these species-specific estimates together (making suitable allowance for one or more species being absent altogether) into a single estimate.

Each approach has its strengths and weaknesses. The vertical approach involves an explicit treatment of mutation at a given position but, because of the complexities of evolutionary models, it can take only limited account of contextual influences between neighboring positions [29,44, 45]. In contrast, the strength of the horizontal approach is a thorough treatment of just these influences, at the price of ignoring the relationships between aligned nucleotides. The complementary nature of these two approaches means there is reason to hope that a combined approach will do better than either one on its own.

We accordingly combine the two components on an equal basis, in the following sense. For each window to be evaluated, the horizontal component makes eight predictions (one for each available species) that are then combined into a single one. We therefore implemented the vertical component also to make eight predictions by dividing the 200-nucleotide window into 8 subwindows of 25 bases, and combining each set of 25 column-specific scores to produce a single value. We then combine the 16

resulting values (one horizontal for each species, and one vertical for each 25 base-pair subwindow) into a single estimate.

For our vertical component, we use the PAL phylogenetic analysis package [46], selecting the generalized time-reversible model of mutation [47]. We train separate sets of models on sets of true and decoy candidate sites and on sites of different coding phases. We also distinguish intragenic from intergenic decoy sites, giving us nine 'site types' for acceptors (phases zero, one and two true sites, non-coding true, phases zero, one and two decoys, and two types of non-coding decoy), nine for donors, and six each for starts and stops (since true instances can only be phase zero). Within each 25 base-pair subwindow, we divide the training data differently depending on whether that subwindow represents a coding or non-coding region in the target species. For a 25 base-pair non-coding region, we train each offset with a separate model, yielding 25 models. For a coding region, we train separately for each codon position of each amino acid or stop codon, yielding $3 \times (20 + 1) = 63$ models. Thus, in total, over all subwindows and site types, we trained over 2,700 evolutionary models for each kind of splice site and over 1,800 for both coding starts and stops. This was possible because of the tens of thousands of training examples available to us, each containing information at every offset.

Because PAL models only mutations and not gaps, we included in the vertical model a simple gap model that applied an RVM to the counts of gaps, and ungapped runs of nucleotides, of particular lengths in particular parts of the window. For example, one such feature would be number of gaps of length 4 to 15 starting (in any species) at an offset between 0 and 25 to the right of the center of the window. In subsequent processing, the estimate derived from the gap model was treated just like each of the eight estimates for 25-base subwindows.

For the horizontal component, we again train separate sets of models for each site type. We analyze each sequence in two ways. Firstly, we estimate the likelihood of each nucleotide using position-specific weight matrices [1], using a context length of up to six nucleotides; smoothing is achieved by only using a longer context when the distribution of its predicted target nucleotide is significantly different on the training data from that given by a shorter context. Secondly, we look for the words of length six or less whose frequency of occurrence over given parts of the window varied most between training sets. For example, the triplet TCT is much more common in the 20 bases upstream of true acceptor sites than of decoy AGs because of the presence of the polypyrimidine tract in true acceptors. To detect coding biases, we counted both overall occurrences of this type and occurrences starting at offsets differing by a multiple of three.

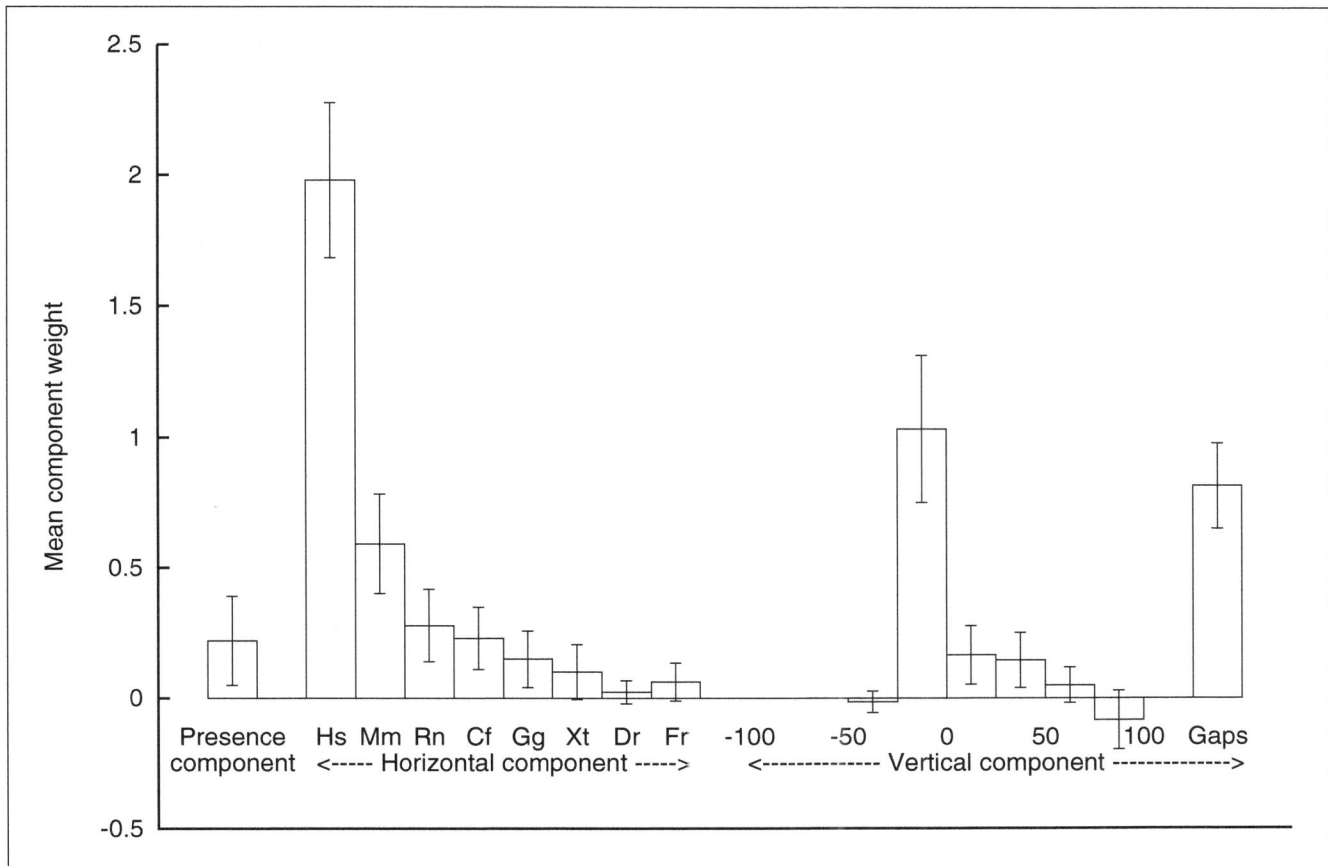

Figure 3
Mean RVM weights for horizontal and vertical component inputs. The figure shows the means, with $p = 0.05$ two-tail error bars, for weights assigned to inputs by acceptor site-type-pair RVMs in Classifier Two, averaging over all 20 pairings of decoy with true site types. The presence component has a single score. Two-letter abbreviations are used for the species-specific scores output by the horizontal component, while the vertical-component quantities are for eight 25 base-pair subregions (only six of which ever get non-zero scores) with one gap score. Species abbreviations are as in Figure 1.

Combining estimates using relevance vector machines

Each component thus yields several hundred different scores on each candidate site for each hypothesized site type. We reduce these to the final true-site and phase probability estimates for a site as follows.

First, we considered each pair of possible types for the site, for example, phase-zero true with phase-two decoy, taking logs of ratios of corresponding estimates in both the horizontal and vertical components. For each pair, we train a RVM on the scores from the horizontal component (using target-species sequences), and one RVM for each of the subwindows in the vertical component. Each of these RVMs selects anything from a handful of its inputs to nearly all of them as 'relevant', and maps from those scores to a single output. Site types are considered in pairs rather than all together because an efficient approximation for the optimization process involved in training the RVM is only known for the case of two classes, not multiple ones.

By this stage, for each pair, we have eight RVM output scores from the horizontal component (one for each species present in the alignment, with suitable trained defaults used where species were absent), and nine from the vertical component (one for each subwindow and one for the gap model). Next, we train another RVM to combine these scores (plus that of the 'species-presence' component) into a single estimate for the probability that the given instance represents one of our current pair of site types rather than the other.

Each kind of splice site, as we have seen, has nine types, yielding $9 \times 8/2 = 36$ different pairs, and coding starts and stops have six, yielding 15 pairs. Our next step is thus to train a further RVM to make the true versus decoy distinction on the basis of all decoy-and-true site type pairs. For splice sites, we also train one to predict the probability of each phase among true sites on the basis of all true-true pairs. The outputs of the phase RVMs are then normalized so that they sum to one in the probability domain.

Finally, for reasons explained below, we run two separate instances of the classifier trained on two different data sets, and average their results together; we could have trained RVMs to do this step, too, but we found that performance was quite insensitive to the weights used.

Figure 3 illustrates one stage of the data-reduction process, showing how one presence, eight horizontal and nine vertical scores are weighted. The values given are means over all 20 (5 decoy types by 4 true) acceptor site RVMs for Classifier Two, with $p = 0.05$ error bars on the means. Each RVM input is separately prenormalized to have a standard deviation of one, so that the weights are directly comparable. It can be seen that the weights given to horizontal-component scores decrease with evolutionary distance from human. The vertical component gives a lot of weight to its gap model and to the subregion from 0 to 25 bases upstream of the AG, and some weight to those 0 to 25 and 25 to 50 downstream, but effectively none to any others, since their weights are either always zero or are on average indistinguishable from zero. The presence component makes a positive but small contribution.

Rational choice of training data

Choosing appropriate training data for the local level of DOGFISH is an important and non-trivial issue, because there are thousands of times more decoy sites in a genome (in the sense of specific di- and trinucleotides) than true ones. Training a classifier with many parameters usually gives best results with many thousands of true sites, which implies using a significant portion of the genome, containing many millions of decoys. Processing all those decoys in training may not be practically feasible; and even if it is, doing so may well, as pointed out in [48], result in a classifier that rejects every item.

The imbalance can be reduced or even eliminated by procedures such as random sampling of decoys [6] or only considering regions known to be relatively rich in true instances, such as the coding extents of genes ([49], resulting in a decoy-to-true ratio of around 100). However, random sampling is likely to leave the classifier somewhat undertrained on the more difficult decoys, only a few of which will be selected for training; and annotation-based region selection will systematically exclude whole classes of decoys, many of which may be difficult ones (compare the large proportion of intergenic false positives in Table 2). Both procedures represent a partial mismatch with the requirements of the gene finder, which has to process whole genomic regions and is especially likely to be misled by poor classifier estimates on the more challenging decoys. Therefore, although we do need the classifier to reliably recognize the easier decoys that form the vast majority of the sites it will encounter, we also need it to be well-trained on challenging ones.

We therefore train and run two versions of the classifier, and give the gene finder the average of their estimates. Classifier

One is trained using a large set of true sites and randomly sampled decoys. The training set for Classifier Two is constructed by running the classifier one and the gene finder over the whole Vega portion of the genome. To do this in reasonable time, we run Classifier One in a 'lite' mode in which the horizontal component only examines the target (human) sequence, and the vertical component is replaced by a much simpler one based on counting occurrences of codons and amino acids in different site types.

We then create a training set for the Classifier Two by a highly non-uniform random selection process, favoring high-scoring sites from the output of the first-pass HMM, irrespective of whether they are true and decoy, but without excluding low-scoring ones altogether. Crucially, this selection process does not rely on any form of annotation. The result is a set consisting of nearly all the true sites that have a reasonable chance of being detected by DOGFISH, and several times as many decoys, most of which are challenging ones. Around 20% of true splice sites and 65% of true start and stop codons are omitted, along with the vast majority of decoys, because they score low as a result of aligning with few species and/or not reflecting the consensus sequence well.

Training Classifier Two on this set has the effect of tuning it to the hardest kinds of decisions that the second-pass HMM will ask it to make. Furthermore, we believe that this training set is of interest in its own right as a challenging testbed for genomic feature classification, since it is enriched for difficult (that is, realistic, from a gene-finding perspective) decoy cases rather than being made artificially easy by being enriched for true sites on the basis of existing annotations.

To train each classifier instance, we first divided the data into 10 roughly equal-sized portions, P1 to P10. P1 and P2 were used to train the underlying horizontal and vertical models (Markov, word-based and evolutionary); P3 to P6 to train the intra-component RVMs; P7 and P8 to train the site-type-pair RVMs; P9 to train the RVMs to produce the final estimates; and P10 (taken from challenging, second-pass data set for both classifier instances, not just for Classifier Two) for evaluation. The classifier results given in this paper are for two evaluation runs, in one of which P9 and P10 were exchanged. The gene-finding results instead used both P9 and P10 together to train the final RVMs; there was no need to hold either of them out, as the entire data set under discussion here is disjoint from all the ENCODE regions.

To avoid the training and evaluation sets being too similar to each other and thereby artificially boosting the accuracy scores, we allocated sites to portions not at random but so as to ensure that as far as possible, paralogs were allocated to the same portion. First, all sites (true and decoy) from within the same gene were put in the same portion. Second, genes were clustered so that as far as possible, two genes that both (partially) aligned to the same piece of informant sequence were also put in the same portion.

The global level: structure building using HMMs

Most of the complexity of DOGFISH is, as we have seen, located in its local-level classifier, allowing the global-level HMM component to be relatively simple. The system works as follows. Every potential splice site and start and stop codon on both strands in the target sequence is handed to the classifier, which, as we have seen, returns an estimate of the probability that the site is a true instance of the feature in question, accompanied, for splice sites only, by a probability distribution among the four possible coding phases (zero, one, two and non-coding) conditioned on the site being a true one. The HMM sees only these estimates, not the DNA sequences themselves, and searches for the best-scoring combinations of sites that are consistent with (its model of) the structure of protein-coding genes. Before this search is carried out, the site scores undergo linear transformations, with different parameters for splice sites and for start/stop sites; parameters for these transformations were optimized on the 13 ENCODE training regions for evaluation on the 31 testing regions.

The HMM's topology imposes several simplifications on biological reality. Firstly, no attempt is made to model transcription start sites and polyadenylation sites. Instead, a gene starts either with a start codon (for the case where coding starts in the first exon) or with a non-coding splice donor (the end of the first exon where coding starts in some later exon). Similarly, it ends with either a stop codon or a non-coding splice acceptor. Secondly, non-coding transcripts are excluded for the same reason. Thirdly, genes with a single coding exon are handled, but are not treated specially despite evidence [50] that they should be: such genes often arise from reverse transcription of mature mRNAs, so that their single exon tends to be as long as several exons in the more common kinds of genes. As a result, few are predicted. Fourthly, no provision is made for overlapping or embedded genes, on either the same or opposite strands, although alternative paths through the lattice can be pulled out once the HMM has run. Fifthly, no provision is made for start and stop codons interrupted by introns, largely because of the difficulties of training the classifier on sufficient numbers of these relatively rare cases. Sixthly, as stated earlier, only AG acceptor sites and GT donors are considered, for similar reasons.

The first of these simplifications is applied because transcription start sites and polyadenylation sites are notoriously hard to model accurately and in most cases are not even known precisely. Each of the other simplifications makes the overall model simpler, excluding various rare and, therefore, hard-to-train cases; we believe that these decisions make an overall positive contribution to accuracy by ruling out many

false positives, at the admitted cost of also excluding a relatively small number of correct structures.

The only respect in which DOGFISH's HMM departs from the basic technology is that it explicitly models the observed distributions of exon and intron lengths, penalizing very short introns and exons. These penalties can be applied only to complete hypothesized exons and introns, not to partial ones, with the consequence that the algorithm is no longer quite sound: the overall least-cost path is no longer guaranteed to be found. To mitigate this effect, we maintain at each position a N-best list [2] of the best few path continuations in each direction, rather than just the best one. We have found $N = 5$ maintains reasonable efficiency while excluding few if any correct and (theoretically) highest-probability paths. Accuracy is much improved overall by modeling lengths; for example, if they are not modeled, many more very short exons and introns (lengths less than 20 and 50 nucleotides, respectively) are accepted than really occur.

The DOGFISH-2E exon probability estimates were derived by training three separate RVMs, for initial, internal and terminal exon candidates, respectively. The inputs to each RVM were the scores assigned by the gene finder to the sites and each end of the exon; the log of the length of the exon; and the 'competition score', the difference between the HMM score for the exon itself and that of the best-scoring overlapping exon. The competition score is positive for exons on the best-scoring path and negative for all others; the RVMs for internal and terminal exons used it almost to the exclusion of all the other inputs, while the initial-exon RVM mainly favored the minimum of the two end-site scores. For evaluating DOGFISH-2E on ENCODE test data (Figure 2a), we trained only on the ENCODE training regions, while for the whole-genome scan we used RVMs trained on all the ENCODE data; the resulting differences appeared to be minimal.

Software

The DOGFISH comparative gene finder software and its predictions on the human genome are available under the GNU public license at [51].

Author contributions

D.C. and R.D. designed the experiments and completed this paper; D.C. wrote the software, carried out the experiments and wrote the first draft of the paper.

Acknowledgements

We are grateful to the UCSC Genome Bioinformatics Group, who created the multiz8way annotation [52], and to the providers of the original sequence data listed at [53]; to the teams involved in the VEGA project [54]; to Alexei Drummond and Korbinian Strimmer for making the PAL code available; to Thomas Down for providing RVM training code [55] and accompanying insightful discussions; to Paul Flicek and Evan Keibler for their gene-prediction evaluation code; to Avril Coghlan and Jean-Karim Hériché for detailed comments on a draft of this paper; to two anonymous referees for their comments; to the organizers of the

ENCODE gene prediction workshop; to the Sanger Institute's computer systems group; and to the Wellcome Trust for funding.

This article has been published as part of *Genome Biology* Volume 7, Supplement 1, 2006: EGASP '05. The full contents of the supplement are available online at http://genomebiology.com/supplements/7/S1.

References

1. Durbin R, Eddy S, Krogh A, Mitchison G: *Biological Sequence Analysis.* Cambridge: Cambridge University Press; 1998.
2. Krogh A: **Two methods for improving performance of an HMM and their application for gene finding.** *Proc Int Conf Intell Syst Mol Biol* 1997, **5:**179-186.
3. Hawkins J, Bodén M: **The applicability of recurrent neural networks for biological sequence analysis.** *IEEE ACM Trans Comput Biol Bioinformatics* 2005, **2:**243-253.
4. Hebsgaard S, Korning P, Tolstrup N, Engelbrecht J, Rouze P, Brunak S: **Splice site prediction in** *Arabidopsis thaliana* **DNA by combining local and global sequence information.** *Nucl Acids Res* 1996, **24:**3439-3452.
5. Yeo G, Burge C: **Maximum entropy modeling of short sequence motifs with applications to RNA splicing signals.** *J Comput Biol* 2004, **11:**377-394.
6. Castelo R, Guigó R: **Splice site identification by idlBNs.** *Bioinformatics* 2004, **20(Suppl 1):**i69-i76.
7. Eden E, Brunak S: **Analysis and recognition of 5' UTR intron splice sites in human pre-mRNA.** *Nucl Acids Res* 2004, **32:**1131-1142.
8. Ho L, Rajapakse J: **Splice site detection with a higher-order Markov model implemented on a neural network.** *Genome Informatics* 2003, **14:**64-72.
9. Burges C: **A tutorial on support vector machines for pattern recognition.** *Data Mining Knowledge Discovery* 1998, **2:**121-167.
10. Zhang XH, Heller K, Hefter I, Leslie C, Chasin L: **Sequence information for the splicing of human pre-mRNA identified by support vector machine classification.** *Genome Res* 2003, **13:**2637-2650.
11. Sonnenburg S, Rätsch G, Jagota A, Müller KR: **New methods for splice site recognition.** In *Proceedings of the International Conference on Artificial Neural Networks: August 2002; Madrid.* Edited by Dorronsoro J. Springer, Berlin. 2002:329-336
12. Tipping ME: **Sparse Bayesian learning and the relevance vector machine.** *J Machine Learning Res* 2001, **1:**211-244.
13. Down T, Hubbard T: **Computational detection and location of transcription start sites in mammalian genomic DNA.** *Genome Res* 2002, **12:**458-461.
14. Down T, Hubbard T: **Relevance vector machines for classifying points and regions in biological sequences.** *Quantitative Biology Archive* 2003, [http://arxiv.org/abs/q-bio.GN/0312006].
15. Burge C, Karlin S: **Prediction of complete gene structures in human genomic DNA.** *J Mol Biol* 1997, **268:**78-94.
16. Stanke M, Waack S: **Gene prediction with a hidden Markov model and a new intron submodel.** *Bioinformatics* 2003, **19(Suppl 2):**ii215-ii225.
17. Majoros W, Pertea M, Antonescu C, Salzberg S: **GlimmerM, Exonomy and Unveil: three** *ab initio* **eukaryotic genefinders.** *Nucl Acids Res* 2003, **31:**3601-3604.
18. Korf I, Flicek P, Duan D, Brent M: **Integrating genomic homology into gene structure prediction.** *Bioinformatics* 2001, **17(Suppl 1):**S140-S148.
19. Meyer I, Durbin R: **Comparative** *ab initio* **prediction of gene structures using pair HMMs.** *Bioinformatics* 2002, **18:**1309-1318.
20. Batzoglou S, Pachter L, Mesirov J, Berger B, Lander E: **Human and mouse gene structure: comparative analysis and application to exon prediction.** *Genome Res* 2000, **10:**950-958.
21. Wiehe T, Gebauer-Jung S, Mitchell-Olds T, Guigó R: **SGP-1: prediction and validation of homologous genes based on sequence alignment.** *Genome Res* 2001, **11:**1574-1583.
22. Parra G, Agarwal P, Abril J, Wiehe T, Fickett J, Guigó R: **Comparative gene prediction in human and mouse.** *Genome Res* 2003, **13:**108-117.
23. Alexandersson M, Cawley S, Pachter L: **SLAM: cross-species gene finding and alignment with a generalized pair hidden Markov model.** *Genome Res* 2003, **13:**496-502.
24. Taher L, Rinner O, Garg S, Sczyrba A, Brudno M, Batzoglou S, Morgenstern B: **AGenDA: homology-based gene prediction.** *Bioinformatics* 2003, **19:**1575-1577.
25. Ovcharenko I, Boffelli D, Loots G: **eShadow: a tool for comparing closely related sequences.** *Genome Res* 2004, **14:**1191-1198.
26. Kellis M, Patterson N, Birren B, Berger B, Lander E: **Methods in comparative genomics: genome correspondence, gene identification and regulatory motif discovery.** *J Comput Biol* 2004, **11:**319-355.
27. Chatterji S, Pachter L: **Multiple organism gene finding by collapsed Gibbs sampling.** In *Proceedings of the eighth annual international conference on research in computational molecular biology: 27-31 March 2004; San Diego.* ACM Press, New York; 2004:187-193.
28. Brown R, Gross S, Brent M: **Begin at the beginning: predicting genes with 5' UTRs.** *Genome Res* 2005, **15:**742-747.
29. McAuliffe J, Pachter L, Jordan M: **Multiple-sequence functional annotation and the generalized hidden Markov phylogeny.** *Bioinformatics* 2004, **20:**1850-1860.
30. Hobolth A, Jensen J: **Applications of hidden Markov models for comparative gene structure prediction.** *J Comput Biology* 2005, **12:**186-203.
31. Siepel A, Haussler D: **Computational identification of evolutionarily conserved exons.** In *Proceedings of the eighth annual international conference on research in computational molecular biology: 27-31 March 2004; San Diego,* ACM Press, New York; 2004: 177-186
32. Meyer I, Durbin R: **Gene structure conservation aids similarity based gene prediction.** *Nucl Acids Res* 2004, **32:**776-783.
33. Allen J, Pertea M, Salzberg S: **Computational gene prediction using multiple sources of evidence.** *Genome Res* 2004, **14:**142-148.
34. Issac B, Raghava G: **EGPred: prediction of eukaryotic genes using ab initio methods after combining with sequence similarity approaches.** *Genome Res* 2004, **14:**1756-1766.
35. Birney E, Clamp M, Durbin R: **GeneWise and GenomeWise.** *Genome Res* 2004, **14:**988-995.
36. Brejová B, Brown E, Li M, Vinar T: **ExonHunter: a comprehensive approach to gene finding.** *Bioinformatics* 2005, **21(Suppl 1):**i57-i65.
37. Blanchette M, Kent W, Riemer C, Elnitski L, Smit A, Roskin K, Baertsch R, Rosenbloom K, Clawson H, Green E, Haussler D, Miller W: **Aligning multiple genomic sequences with the threaded blockset aligner.** *Genome Res* 2004, **14:**708-715.
38. **RepeatMasker Open-3.0** [http://www.repeatmasker.org]
39. Blakesley R, Hansen N, Mullikin J, Thomas P, McDowell J, Maskeri B, Young A, Benjamin B, Brooks S, Coleman B, *et al.*: **An intermediate grade of finished genomic sequence suitable for comparative analyses.** *Genome Res* 2004, **14:**2235-2244.
40. Sorek R, Ast G: **Intronic sequences flanking alternatively spliced exons are conserved between human and mouse.** *Genome Res* 2003, **13:**1631-1637.
41. Brent M, Guigó R: **Recent advances in gene structure prediction.** *Curr Opin Struct Biol* 2004, **14:**264-272.
42. Kapranov P, Drenkow J, Cheng J, Long J, Helt G, Dike S, Gingeras T: **Examples of the complex architecture of the human transcriptome revealed by RACE and high-density tiling arrays.** *Genome Res* 2005, **15:**987-997.
43. Mangalam H: **The Bio* toolkits - a brief overview.** *Briefings Bioinformatics* 2002, **3:**296-302.
44. Siepel A, Haussler D: **Phylogenetic estimation of context-dependent substitution rates by maximum likelihood.** *Mol Biol Evol* 2004, **21:**468-488.
45. Lunter G, Hein J: **A nucleotide substitution model with nearest-neighbour interactions.** *Bioinformatics* 2004, **20(Suppl 1):**i216-i223.
46. Drummond A, Strimmer K: **PAL: an object-oriented programming library for molecular evolution and phylogenetics.** *Bioinformatics* 2001, **17:**662-663.
47. Liò P, Goldman N: **Models of molecular evolution and phylogeny.** *Genome Res* 1998, **8:**1233-1244.
48. Yoon K, Kwek S: **A filtering approach to splice site predictions in human genes.** *Adv Bioinformatics Applications* 2005, **8:** 447-458
49. Pertea M, Lin X, Salzberg S: **GeneSplicer: a new computational method for splice site prediction.** *Nucl Acids Res* 2001, **29:**1185-1190.
50. Chen C, Gentles A, Jurka J, Karlin S: **Genes, pseudogenes and Alu sequence organization across human chromosomes 21 and 22.** *Proc Natl Acad Sci USA* 2002, **99:**2930-2935.
51. **DOGFISH** [http://www.sanger.ac.uk/Software/analysis/dogfish]
52. **UCSC Genome Browser** [http://genome.ucsc.edu]
53. **UCSC Genome Browser** [http://genome.ucsc.edu/goldenPath/credits.html]
54. **Vega Genome Browser** [http://vega.sanger.ac.uk]
55. **BioJava** [http://www.biojava.org]

Research

Exogean: a framework for annotating protein-coding genes in eukaryotic genomic DNA

Sarah Djebali*†, Franck Delaplace† and Hugues Roest Crollius*

Addresses: *Dyogen Lab, CNRS UMR8541, Ecole Normale Supérieure, 46 rue d'Ulm, 75005 Paris, France.
†IBISC Lab, CNRS FRE2873, Université d'Evry Val d'Essonne, Genopole, 523 place des terrasses de l'Agora, 91000 Evry, France.

Correspondence: Hugues Roest Crollius. Email: hrc@ens.fr

Published: 7 August 2006

Genome *Biology* 2006, **7(Suppl 1)**:S7

The electronic version of this article is the complete one and can be found online at http://genomebiology.com/2006/7/S1/S7

Abstract

Background: Accurate and automatic gene identification in eukaryotic genomic DNA is more than ever of crucial importance to efficiently exploit the large volume of assembled genome sequences available to the community. Automatic methods have always been considered less reliable than human expertise. This is illustrated in the EGASP project, where reference annotations against which all automatic methods are measured are generated by human annotators and experimentally verified. We hypothesized that replicating the accuracy of human annotators in an automatic method could be achieved by formalizing the rules and decisions that they use, in a mathematical formalism.

Results: We have developed Exogean, a flexible framework based on directed acyclic colored multigraphs (DACMs) that can represent biological objects (for example, mRNA, ESTs, protein alignments, exons) and relationships between them. Graphs are analyzed to process the information according to rules that replicate those used by human annotators. Simple individual starting objects given as input to Exogean are thus combined and synthesized into complex objects such as protein coding transcripts.

Conclusions: We show here, in the context of the EGASP project, that Exogean is currently the method that best reproduces protein coding gene annotations from human experts, in terms of identifying at least one exact coding sequence per gene. We discuss current limitations of the method and several avenues for improvement.

Background

Ideally, the process of annotating protein coding genes (hereby referred to as 'genes') in a region of genomic DNA involves locating the exact external and internal boundaries of all the genes it includes and, for each, finding all the possible transcript variants. In practice, achieving this is very difficult in eukaryotic genomes for many reasons. First, eukaryotic genes are generally composed of a succession of exons and introns, which makes their structure complex and highly variable. Second, genes cover only a small fraction of eukaryotic genomes (30% in mammals) and exons cover an even lower fraction (1% to 2% in mammals). Third, some eukaryotic genomes contain many pseudogenes, which are non-functional copies of genes sometimes nested within genes and with similar compositions. Finally, each gene may give rise to many different transcripts, often with minor variations, a mechanism that modulates the function or the spatial or temporal availability of the corresponding protein. Despite these difficulties, precise gene annotation is crucial for biomedical research: it is a basic requirement to link genotype and phenotypes in human and model species and generally to focus the work of biologists and bio-informaticians on an essential functional part of the genome. Forty eukaryotic genome sequences have now been completed and each is commonly tens of millions or even billions of nucleotides long: annotating genes in this massive amount of data undoubtedly requires mathematical models.

Mathematical models have been proposed to automatically locate genes in genomic DNA, either by similarity to expressed or evolutionary conserved sequences, or by capturing our current biological understanding of genes in statistical algorithms, or a combination of these methods. Despite tremendous advances, automatic gene annotations are still considered predictions that require validation by human experts, particularly when expensive and time consuming experimental work will be based upon them. This paradigm is exemplified in the ENCODE Genome Annotation Assessment Project (EGASP) competition [1], where the reference against which all the automatic methods are measured is a set of annotations experimentally verified and manually curated by human experts. Such high quality reference gene annotations (also including those collated in the Vega repository [2]) are generated by humans based on a number of resources: cross-species sequence alignments, mRNA sequences, *ab initio* predictions, and so on. Generally, with the aid of sophisticated annotation and curation software tools, these resources are reviewed on a gene-by-gene basis using strict rules rooted in a deep knowledge of both the data at hand and the biology associated with gene expression (transcription, splicing, translation, and so on).

Based on these observations, we were interested in designing an automatic annotation method that explicitly establishes the same relationships between biological objects, and applies the same rules, as human experts. In computer science, some such rules can be assimilated to heuristics of the type '*if (X) then (Y) else (Z)*'. The automatic annotation of protein coding genes may, therefore, appear deceivingly simple and be reduced to coding the rules extracted from biological expertise into a set of heuristics and to applying them to the experimental evidence. In the field of gene annotation, however, encoding human expertise is rather a problem of untangling the body of evidence that experts build to elect a sequence to the status of protein coding gene. This body of evidence can be viewed as a complex network of relationships between DNA, mRNA and protein sequences. These relationships are difficult to formalize because biological knowledge cannot (yet) be assimilated to a structured list of observations based on a controlled vocabulary. It is instead a rich and heterogeneous set of often unconnected observations. It is also in constant evolution and may, therefore, vary from one set of experts to another. It does not always follow strict logical rules and instead may rely on arbitrary variables. In the context of

gene annotation, the latter is often a consequence of a lack of knowledge on specific aspects of gene structure and biology. For instance, a transcript with an annotated coding region of less than 100 amino acids is often considered too short and not classified as coding for a protein.

Our first objective was, therefore, to design a formal framework within which rules and biological objects may be represented and manipulated using computers to produce gene annotations. We then identified a number of rules and biological objects used by human annotators and integrated them into the framework. The resulting software tool is called Exogean for EXpert On GEne Annotation. We believe that this strategy is currently the only possibility to approach the level of completeness and accuracy reached by human experts. Here we show, in the context of the EGASP competition of the ENCODE project, that Exogean already performs better than any other automatic method in identifying at least one exact coding sequence per gene.

Results
The Exogean method
Human annotators manipulate and integrate information stemming from multiple heterogeneous sources (for example, *ab initio* predictions, mRNA and expressed sequence tag (EST) alignments, protein alignments). Each source has specific properties and is thus treated with specific rules. For instance, mRNA sequences from the same species that is being annotated should align to the genome with high similarity (98% to 100%) while protein sequences from a different species typically align less perfectly owing to base substitutions, insertions or deletions during evolution. Hence, mRNA alignments are dealt with using more stringent rules and are given more importance because they can be aligned perfectly, while protein alignments are typically treated with more caution. The different sources of data will, therefore, be represented differently, and will be processed by different heuristics.

To develop a flexible yet formal framework, we decomposed human expertise into heuristics of two independent types: the establishment of relationships between objects on the one hand; and the action of connecting and merging the objects based on these relationships on the other hand.

This independence between relationships and actions provides the flexibility required to solve a number of difficulties: heuristics change over time - the system must be able to easily adapt to these changes by modifying, adding or deleting heuristics; heuristics are applied to different sources of information (for example, different types of sequence alignment) - the system must be able to handle heterogeneous sources; heuristics are themselves of different types, whether they deal for instance with structural concepts (properties of aligned sequences) or on prioritizing

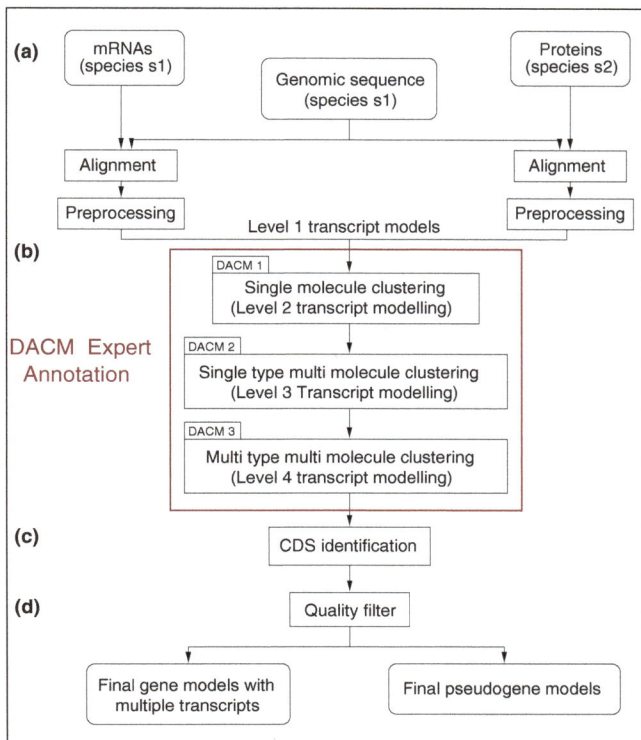

Figure 1

Overview of the Exogean methodology. Exogean annotations are based on protein and/or mRNA alignments. Protein sequences come from a different species than that of the DNA to be annotated (for example, mouse protein to annotate human genes), while the mRNAs originate from tissues of the species being annotated. **(a)** These alignments are first processed to remove artifacts using heuristics. **(b)** The core of the method is the directed acyclic multigraph (DACM) component, which builds transcripts of increasing complexity in three iterative steps using a second set of heuristics. **(c)** The transcripts are then examined to identify a complete or a partial coding sequence (CDS), and **(d)** a last filter applies a last set of rules to remove pseudogenes.

sources - the system must be able to handle heterogeneous heuristics.

In Exogean (Figures 1 and 2), transcripts are built from relatively simple objects (sequence alignments) into more complex structures. Throughout the manuscript, therefore, the term transcript model will designate transcripts at any level of complexity, from basic sequence alignments to the final structure that represents the predicted functional mRNA (Figure 2). Following this we define that heuristics of the type 'relationship' will always be established between two transcript models of the same level of complexity. The fact that relationships may be directional (can be represented as an arrow between two objects) greatly simplifies the actions (see below) and is based on the directionality of the DNA molecule and, hence, of transcription itself. Relationships are thus directed by default, unless otherwise specified.

To represent transcript models and relationships, and to apply actions on the former using the latter, we use directed acyclic coloured multigraphs (DACMs; Figures 1 and 2). In such graphs, nodes are transcript models and multiple edges between nodes are the relationships. In its current version, Exogean uses three DACMs, each with increasingly complex and accurate transcript models (nodes) and different relationships (edges). While DACM1 is built from the original sources of information given to Exogean (mRNA and/or protein alignments), its output will be the basis of DACM2, and DACM3 will be built from the output of DACM2. To proceed from one DACM to the next, Exogean performs a graph reduction. The first step in reducing a graph involves the definition of a set of relationships (edges) that will represent a certain path. Then Exogean finds all the paths of maximal length in the DACM, which results in combining the different nodes located on each of these paths. The nodes collected along a maximal path together form a more complex object (transcript model) that will be a new node ready for processing in the next DACM. Edges are then built between these nodes, and this constitutes a new DACM that, in turn, can be reduced. In summary, Exogean automates the annotation protocol followed by human experts by iteratively building edges (making relationships) and subsequently reducing DACMs (taking actions based on these relationships). Before and after the three core DACMs, Exogean also applies heuristics to respectively prepare the data for transcript modeling, and to identify coding sequence (CDS) within transcripts (Figure 1).

The EGASP assessment

Exogean is one of 20 automatic methods that were compared in the EGASP project [1] (see Materials and methods). In brief, each method predicted protein coding transcripts in 31 regions of the human genome totaling about 21 Mb. Independently, a group of experts (the Havana group at the Sanger Institute [3]) annotated the same regions using manual curation and experimental validation and identified 296 genes that were considered as reference against which all the automatic methods were compared. We refer to this set of genes as the GENCODE annotations [4]. The comparison between GENCODE annotations and Exogean predictions is summarized in Table 1. Except for the DNA sequence itself, the only source of information used by Exogean to predict transcripts were human mRNA and mouse protein sequence alignments (see Materials and methods). Exogean predictions are evaluated both if they overlap and if they exactly match a GENCODE annotation (see Materials and methods).

Overlapping predictions

Two standard measures to evaluate the accuracy of predictions against a reference are sensitivity (percentage of annotations identified) and specificity (percentage of predictions that identify an annotation). In the overlap evaluation, Exogean consistently detects GENCODE coding

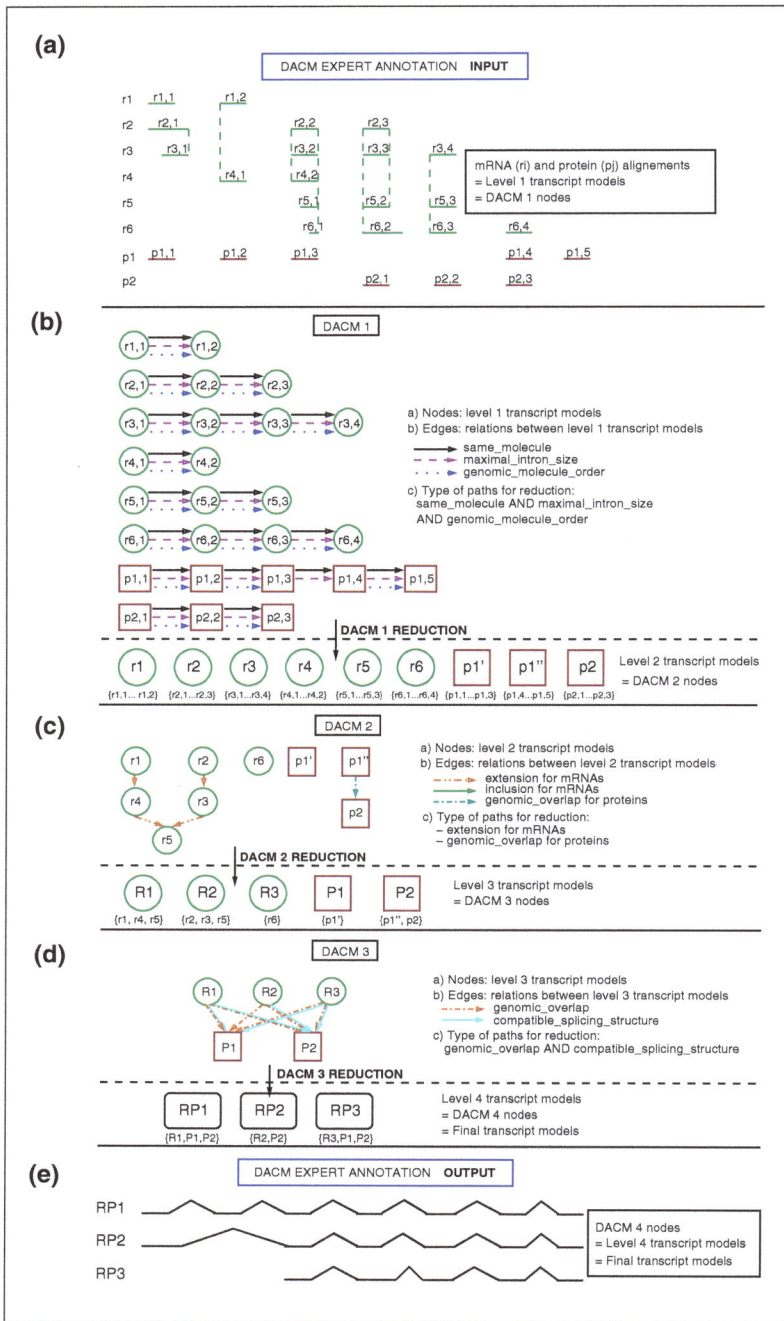

Figure 2

Example of transcript modeling from a set of protein and mRNA alignments using DACMs. **(a)** The DACM input are mRNA (r1...r6) and protein (p1, p2) sequences that have been aligned to a genomic sequence S. The individual local alignments are each a level 1 transcript model (L1TMs) and constitute the nodes of a graph DACM1. **(b)** This graph has three possible directed edges: same_molecule, maximal_intron_size, and genomic_molecule_order. Each corresponds to a different relationship that connects two nodes if they respectively: are alignments produced by the same mRNA or same protein; are separated by a distance smaller than a user defined threshold (for example, 75 kilobases); and are collinear on the molecule of origin (mRNA or protein) and the genomic DNA. There are nine maximal paths along the three combined edges, which reduce DACM1 into the nine nodes (r1 to r6 and p1',p1", p2) of a graph DACM2, each representing a level 2 transcript model (L2TM). Note that the reduction of DACM1 splits nodes p1,1 to p1,5 into two DACM2 nodes (p1' and p1") because of the absence of a genomic_molecule_order edge between p1,3 and p1,4. **(c)** DACM2 has three possible edges, inclusion, extension (for mRNAs) and genomic_overlap (for proteins), which respectively connect two nodes if: they overlap and their overlapping introns are identical; they overlap and their overlapping introns are identical but the second node also extends the first in 3'; and the span of the two nodes have overlapping genomic coordinates. The reduction follows either the 'extension' rule for mRNAs edges or the genomic_overlap protein edge and produces here the five nodes of graph DACM3 (mRNA nodes R1 to R3 and protein nodes P1 and P2), which represent level 3 transcript models (L3TMs). **(d)** DACM3 has two possible edges, genomic_overlap and compatible_splicing_structure, which connect (combines) protein and mRNA transcript models if they respectively have overlapping genomic coordinates and if the protein transcript model does not have any exons in introns of the mRNA transcript model. To reduce the graph, Exogean first identifies the path that contains both edges and from these, the reduction consists in grouping all nodes that are connected to the same RNA node. This generates the three nodes of a graph DACM4 (RP1 to RP3), which represent level 4 transcript models (L4TMs). These L4TMs are the final transcript models generated by the DACM expert annotation. **(e)** Graphical representation of the DACM expert annotation output: the final transcript models RP1 to RP3 are represented on the genomic sequence S. No information has been lost during the three graph reductions. Note that transcript models produced by the DACM component of Exogean are not yet final, and will be further examined and potentially extended when looking for splicing and start/stop signals.

nucleotides, exons, transcripts and CDS with more than 80% sensitivity and 94% specificity. Of particular interest is the identification of transcripts and genes, where Exogean predicts less than 3% false positives (15 transcripts out of 513 predicted) corresponding to 8 genes that do not overlap a Havana gene. We investigated in more details the reasons why Exogean predicted these sequences as genes, since some may potentially represent new CDS. Two predictions correspond to retro-transposable elements (one L1 and one LTR) that are both supported by at least one mRNA aligned at

these positions. One prediction corresponds to the H19 maternally imprinted non-coding RNA on chromosome 11, where Exogean nevertheless predicts a 356 amino acid protein sequence across its 5 exons spanning more than 50% of the length of the RNA. Another prediction is a GENCODE putative gene directly downstream of H19 supported by a single human placental mRNA that, upon manual inspection, displays no pseudogene characteristics but shows no similarity to any known protein. GENCODE 'putative' genes are not considered *bona fide* coding transcripts in this

Table 1

Exogean performances in identifying GENCODE coding sequences

Measure	Sensitivity (%)	Specificity (%)	Exogean correct	Exogean wrong	GENCODE missed	GENCODE total
Overlap						
Nucleotides	84.18	94.33	371,369	22,301	69,791	441,160
Exons	90.12	94.94	2,495	133	273	2764
Transcripts	89.37	97.08	498	15	151	649
Genes	82.09	96.35	211	8	53	296
Exact						
Nucleotides	84.18	94.33	371,369	22,301	69,791	441,160
Exons	79.34	83.45	2,193	435	571	2764
Transcripts	42.53	52.44	267	237	373	649
Genes	63.18	80.82	187	32	109	296

Table 2

Diagnostic for GENCODE genes missed by Exogean

Class of diagnostic	Number of cases	Description of the cause
1	33	Proteins matching multiple neighboring genes in clusters
2	8	GENCODE CDS size below 300 nucleotides
3	3	GENCODE CDS interrupted by the boundaries of the Encode region
4	2	Insufficient evidence to predict the gene
5	3	Protein evidence eliminated by Exogean
6	1	mRNA evidence eliminated by Exogean
7	2	Intron donor/acceptor sites not accepted by Exogean
8	1	mRNA evidence was misaligned with genomic DNA

evaluation, and it is thus considered an incorrect Exogean prediction. Finally, four predictions correspond to clear cases of pseudogenes that Exogean did not identify as such. In conclusion, across the 31 Encode regions tested, Exogean predicts only six true false positive genes that do not overlap GENCODE annotations, and two neighboring genes that are expressed in the form of RNAs but are either putative CDS or known non-coding functional RNA.

Conversely, Exogean misses 53 GENCODE protein-coding genes out of 296 (18%). Examination of each case revealed eight possible causes, listed in Table 2. The most prominent reason for which Exogean fails to predict a gene overlapping a GENCODE annotation is when such genes are organized in clusters (33 genes missed). In these situations, homologous mouse proteins invariably produce alignments to most genes in the cluster because they all share a high sequence similarity. If one such mouse protein bridges two or more

neighboring genes by producing alignments that are contiguous both in the protein sequence and in the genomic DNA, this contiguity will not be eliminated by Exogean provided it continues to comply with the other rules. Consequently, Exogean defined transcripts spanning the entire cluster and the CDS found in each prediction only covers one or perhaps two GENCODE annotations, resulting in most genes in the cluster being missed. Most cases (26 out of 33) concern the Encode region ENm009, which contains an olfactory receptor gene cluster. Clearly, the rules currently implemented in Exogean for exploiting protein alignment need revisiting to address such cases, which theoretically should not pose a major problem and thus provide a rich avenue for improvement. The other causes for false negative predictions each concern fewer cases (between 1 and 8). For instance, Havana annotated 42 transcripts in 29 genes with a CDS smaller than 300 nucleotides. Exogean currently does not predict CDS that would produce a protein with less than

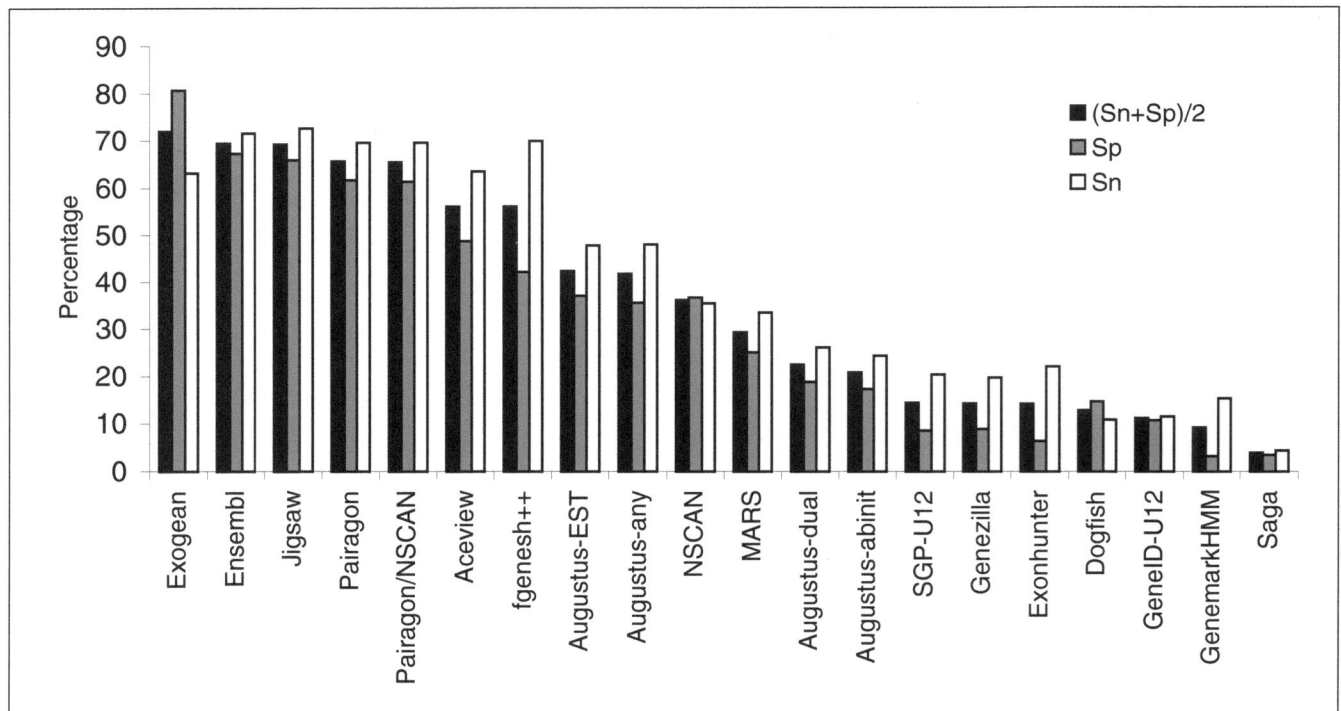

Figure 3
Exogean outperforms all other automatic methods confronted during EGASP. Histogram of the performances in terms of specificity (Sp, grey), sensitivity (Sn, white), and average between specificity and sensitivity ((Sn+Sn)/2, black) for all methods designed to predict protein-coding sequences in the EGASP competition. The values were provided by the EGASP organizers, and are based on predictions matching at least one coding sequence (CDS) exactly for each Havana annotated gene (known or novel). The methods are ranked left to right from best to worse average between Sn and Sp.

100 amino acids, which prevented predictions overlapping eight of these GENCODE genes. Three GENCODE genes were not predicted because they are interrupted by the limits of the Encode region. Since we filtered out as potentially unreliable any evidence that was truncated by the boundaries of Encode regions, Exogean was unable to predict transcripts in these genes. The remaining eight GENCODE genes that were not predicted by Exogean are due to rules implemented in the program that are slightly too stringent, resulting either in the elimination of some evidence or in the inability to identify rare forms of intron donor/acceptor sites. These rules can thus probably be refined further.

Exactly matching predictions
The difficulty in automatically annotating protein-coding genes in eukaryotic DNA lies not so much in identifying predictions that at least partially overlap the coding sequence of each real gene but rather in identifying the precise positions of the coding sequence of every transcript, that is, the start codon, all the internal exon boundaries if they exist, and the stop codon. In designing Exogean, we have focused on maintaining a high specificity in order to obtain a strong and reliable baseline annotation, with as few compromises as possible on sensitivity. This is reflected in Exogean's specificity in exact gene CDS predictions

(Table 1), which is higher than any other method by a large margin: no other method shows more than 70% specificity while Exogean shows more than 80% specificity (Figure 3). Does this come at the cost of a low sensitivity? A group of seven methods including Exogean show a distinctly better sensitivity (between 63% and 73%) than all the others (below 50%). Exogean's sensitivity (63%) is not, therefore, notably affected by the quest for a high specificity. In fact, four GENCODE genes are uniquely identified by Exogean (supported by both mouse proteins and human mRNAs) and by no other method from the same category. Altogether, based on the average between specificity and sensitivity (a standard measure to compare different methods [5,6]) for exact gene CDS predictions, Exogean comes in first position when ranking all the methods that participated in the EGASP competition (Figure 3).

Like Havana, Exogean is able to predict several alternative transcripts per gene: Exogean and Havana identify on average 2.34 and 2.19 isoforms per gene, respectively. Interestingly, while its sensitivity for detecting transcripts is the second highest across all methods, Exogean does not predict exactly matching transcripts with the same high specificity as for genes (Table 1). To explain this apparent contradiction, we were interested to see if a specific category

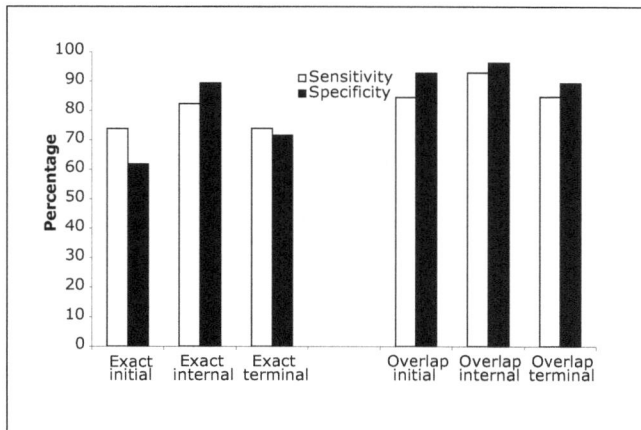

Figure 4
GENCODE and Exogean agree more often on exact boundaries of internal rather than external exons. Histogram of Exogean sensitivity and specificity in identifying the exact boundaries (left) and overlapping boundaries (right) of Havana initial, internal and terminal exons. While Exogean more or less predicts overlapping exons with similar specificities for initial, internal and terminal exons, this is not the case for exactly matching exons: Exogean internal exons are much more specific (89%) than external ones (62% and 72%).

of GENCODE transcripts is better predicted than others. Indeed, out of 267 exact transcripts predicted by Exogean, a remarkable 266 correspond to GENCODE transcripts that are complete, that is, where a start and a stop codon have been found. The single remaining exact Exogean transcript matches one of the 194 incomplete GENCODE transcripts. This result has two important consequences: the first is the conclusion that Exogean reproduces GENCODE annotations much better when the latter are complete transcripts. The second is the suggestion that a fraction of the complete Exogean transcripts overlapping incomplete GENCODE transcripts, and thus not showing exact matches, might be correct. Indeed, there are 23 GENCODE genes that only include incomplete transcripts, and complete Exogean transcripts overlap 11 of these 23 genes. It is thus not excluded that Exogean is able to completely and correctly annotate genes that were partially annotated by Havana. If this was the case, one would expect discrepancies between Exogean and GENCODE transcripts to occur more often at the end of transcripts, where arbitrary end points are more frequent. Figure 4 shows indeed that initial and terminal exons are less well identified exactly by Exogean than internal ones, although Exogean does overlap these external exons with the same sensitivity as internal ones.

An important factor that likely explains why Exogean transcripts are exact in complete GENCODE transcripts but not in partial ones is that Exogean only uses human mRNA and mouse protein alignments, while Havana also includes human ESTs among other additional sources. When a GENCODE annotation is only supported by ESTs, then

Exogean will often predict a different transcript or no transcript at all, depending on the conservation of the corresponding protein in mouse. This affects mainly incomplete GENCODE transcripts because ESTs typically cover only parts of complete transcripts of a given gene. Conversely, if mRNA evidence exists for a gene, then Exogean and Havana will both use it and are thus more likely to predict the same corresponding transcript, which is more likely to be complete because mRNAs tend to cover the entire length of transcripts. We are currently formulating heuristics that will also allow Exogean to take ESTs into account.

Finally, Exogean predictions show the highest average number of exons per transcript (9.8) compared to Havana (8.28) and all other methods (below 8.6). One factor contributing to this high figure is that Exogean predicts fewer transcripts with few exons than Havana (Figure 5). In contrast, Havana and Exogean predict a remarkably similar number of transcripts with many exons (more than 9 exons) and this is accompanied by a higher sensitivity in correct predictions (Figure 6) for these particular transcripts. Here also, the different sources of evidence used by the two strategies probably explain these observations: transcripts with many exons are more likely to be predicted based on mRNA alignments, while shorter transcripts probably reflect more EST-based alignments, simply because ESTs are generally shorter than mRNAs.

An improved version of Exogean (post-EGASP)
Since the EGASP experiment, we have addressed many of the limitations described above in a new version of Exogean. Major areas of improvement have focused on refining the rules to untangle protein alignments in clusters of paralogous genes, and in the definition of the CDS of transcripts. These rules have a direct positive impact on sensitivity, with very little consequence on specificity; sensitivity in exact GENCODE CDS identification increases to 72.64% and specificity remains essentially stable at 79.30%. The average between these two measures is 75.97%, which demonstrates a substantial improvement over the version used in EGASP (72.00%). Using this version, we tested the influence of the nature of the information provided to Exogean. Indeed, human mRNAs generally provide a high specificity and precise exon boundaries but only cover a subset of genes, while each mouse protein tends to identify a broad spectrum of genes in human (paralogs), albeit with fuzzy boundaries. The complementarity of these two sources of information is confirmed when each is used individually and then in combination. Human mRNAs alone provide very specific predictions for a substantial fraction of GENCODE CDSs (sensitivity and specificity for exact CDS prediction are 64.86% and 82.11%, respectively) and mouse proteins provide little sensitivity and little specificity on exact CDS predictions (17.23% and 50.00%, respectively). But combining both sources yields the performances described

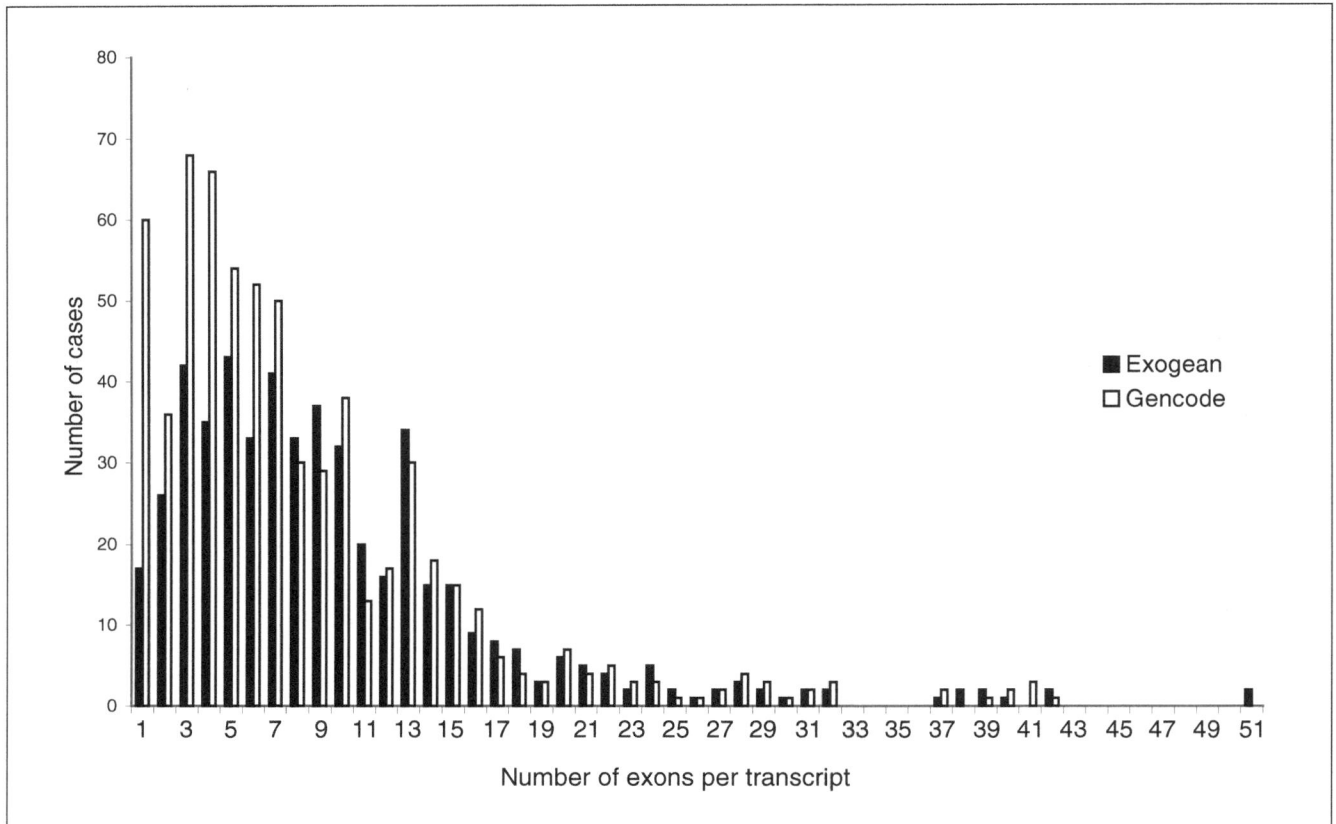

Figure 5
Histogram of the number of Exogean and Havana predicted transcripts as a function of their respective number of exons. For transcripts with large numbers of exons (more than 9), Exogean and Havana predict remarkably similar numbers of transcripts.

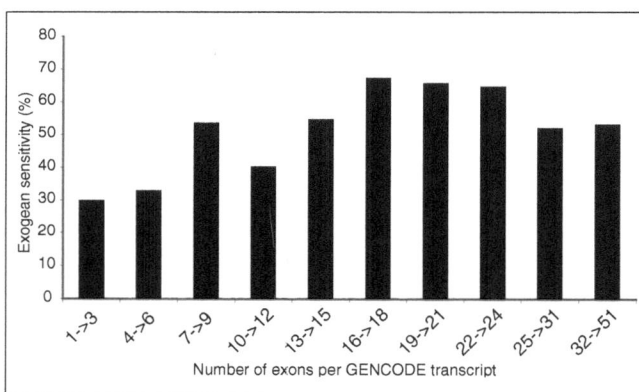

Figure 6
Histogram of Exogean's sensitivity in predicting exact GENCODE transcripts as a function of the number of exons in the latter. Clearly, GENCODE transcripts with larger numbers of exons are better identified than smaller transcripts.

above for the new version of Exogean, with a sharp increase in sensitivity and a minor decrease in specificity compared to the use of human mRNAs only. Of note, the sensitivity of

Exogean predictions increases by 8% when mouse proteins are added to human mRNAs. The reason stems from a rule that forbids the use of single exon transcript models based solely on human mRNAs, whereas many such genes are recovered by mouse proteins.

Discussion

Conclusions from previous studies aiming at comparing automatic annotation methods in eukaryotic genomes have often been limited by the availability of a large and reliable reference dataset. In this respect, the EGASP assessment project has been a unique opportunity to rigorously measure how well current strategies replicate meticulous and detailed protein coding gene annotations on a large and varied set of genomic regions in a blind test [1] (see Materials and methods). A commonly accepted standard for annotating genes is to consider that at least one coding transcript must be entirely and exactly identified [7]. Using this measure, Exogean is the method that currently best replicates reference annotations out of 20 methods tested in EGASP. In particular, Exogean is the most specific by a large measure (12% more than the next best), which reflects our initial objective when designing the method. In addition, out of all methods with good performances, Exogean is also the most consistent across the 31 ENCODE regions [1]. This suggests that Exogean would be the most likely method to reproduce its performances on a different set of human genomic regions. Because 8 out of 10 genes annotated by Exogean are correct, one possible use of Exogean is to accelerate annotations by human experts, especially since the methodology behind Exogean intuitively follows the same logic. To assist in this task, Exogean generates (in addition to the positions of transcripts and their sequences) information on each predicted gene and transcript that summarizes their structure, the evidence used, the problems and conflicts encountered and the solutions applied. Human experts may continue from there and use additional rules, resources and experiments to correct or confirm the automatic predictions.

While Exogean is specific, its sensitivity could be improved in several ways. First, the annotations that we produced for the EGASP assessment relied solely on two sources of alignments: human mRNA sequences and proteins predicted in the mouse genome. It is thus not unexpected that other participating methods that rely on a wider range of resources (human ESTs, mRNAs from other species, conserved genomic DNA) identify more genes, and we are currently designing rules to integrate some of these resources as well. We also show that current rules designed to exploit mouse protein alignments fail when human genes are in clusters, such as the olfactory receptor gene cluster. The EGASP experience was extremely useful in helping to uncover such limitations, many of which have been addressed in a new version of Exogean.

Automatic annotation methods have traditionally used statistical models to capture properties of genes and annotate them in genomic DNA, either alone [8,9] or in combination with evidence from other sources [10-13]. Exogean departs from these approaches in that it only relies on rules extracted from human expertise, and as such does not need to train on a set of known genes to 'learn' their statistical properties. Directed acyclic graphs (DAGs), the component used in the Exogean strategy to store and manipulate the information, have already been used in the context of gene annotation, albeit differently. The program AIR [14] uses DAGs where exons are nodes and edges are introns. ESTGenes [15] uses similar DAGs to Exogean but with a unique edge between nodes, whereas multiple edges are used in Exogean's multigraphs. One advantage of using DAGs as in ESTGenes or Exogean is the strict independence that can be maintained between the data and the heuristics applied to the data.

Translating human expertise for gene annotation into a computational framework could be generalized if an appropriate language was developed. The formalization that

is at the core of Exogean, namely the DACM algorithm, can be seen as a natural starting point for developing such a language. If successful, this approach could lead to a more general and expressive method to integrate any rule that biologists use to synthesize information about biological objects in order to create more complex objects. Such approaches could potentially be of great use in the future.

Materials and methods
The Exogean software
Exogean is written in Ocaml [16]. Precompiled executables are available for several platforms [17]. Exogean is currently able to annotate eukaryote protein coding genes based on alignments with proteins from a different species and/or mRNAs from the same species. Input files with the alignments must be provided in one of several possible formats (psl, gff or exf, the latter being a simple format developed for Exogean). For the EGASP assessment, Exogean used proteins from the mouse International Protein Index database (March 2005 version containing 42,799 protein sequences) and human mRNA from EMBL (March 2005 release, containing 213,695 mRNA sequences) aligned by BLAT [18]. If protein alignments are used, a fasta formatted file with the protein sequences must also be provided. Finally, a configuration file is required where a large number of parameters pertaining to the rules used by Exogean are specified. Given that the alignments are computed, Exogean is fast since the entire human genome is annotated in approximately 100 minutes on a single 3 GHz processor with 1 Gb memory. A formal description of the Exogean method will be described elsewhere.

The EGASP assessment
To place in their context the results described here, we briefly summarize the conditions of EGASP [1]. EGASP was organized by the GENCODE [19] group of the ENCODE project [20] and the aims were twofold: to evaluate how well automatic methods are able to reproduce manual and experimental gene annotation of the human genome; and to assess how complete our current knowledge is of the gene content of the human genome. A set of 31 regions of the human genome from the ENCODE project totaling 21.5 Mb of DNA were used by 14 groups to predict protein coding genes. After submitting the predictions, a workshop [21] was organized to confront the prediction and the annotations from the Havana group [3] at the Sanger Institute. The Havana group annotates genes by combining information from a variety of sources using human expertise and experimental validations (designated here as the GENCODE annotations [4]). All the figures for sensitivity and specificity used for Table 1 and Figure 3 are those provided by the EGASP organizers based on these comparisons. When necessary (Table 1, Figures 4 and 6, and text) additional results for Exogean were computed using the Eval software [22], kindly provided by the organizers. In the overlap mode,

any GENCODE annotation with boundaries that overlap an
Exogean prediction is counted as true positive.

Acknowledgments

We wish to thank Paul Flicek for kindly providing an updated version of
the Eval software, Matthieu Muffato for assistance with engineering
aspects and Fiona Francis for critical reading of the manuscript.

This article has been published as part of *Genome Biology* Volume 7,
Supplement 1, 2006: EGASP '05. The full contents of the supplement are
available online at http://genomebiology.com/supplements/7/S1.

References

1. Guigo R, Flicek P, Abril JF, Raymond A, Lagarde J, Denoeud F,
Antonarkis S, Ashburner M, Bajic VB, Birney E, *et al.*: **EGASP: The
ENCODE Genome Annotation Assessment Project.** *Genome
Biology* 2006, **7(Suppl 1):**S2
2. Ashurst JL, Chen CK, Gilbert JG, Jekosch K, Keenan S, Meidl P,
Searle SM, Stalker J, Storey R, Trevanion S, *et al:* **The Vertebrate
Genome Annotation (Vega) database.** *Nucleic Acids Res* 2005,
33(Database Issue):D459-465.
3. **The HAVANA Project** [http://www.sanger.ac.uk/HGP/havana/]
4. Harrow J, Denoeud F, Frankish A, Reymond A, Chen CK, Chrast J,
Lagarde J, Gilbert JGR, Storey R, Swarbreck D, *et al.*: **GENCODE:
The HAVANA-based reference gene annotation of
ENCODE.** *Genome Biology* 2006, **7(Suppl 1):**S4
5. Burset M, Guigo R: **Evaluation of gene structure prediction
programs.** *Genomics* 1996, **34:**353-367.
6. Rogic S, Mackworth AK, Ouellette FB: **Evaluation of gene-finding
programs on mammalian sequences.** *Genome Res* 2001, **11:**
817-832.
7. Brent MR: **Genome annotation past, present, and future:
How to define an ORF at each locus.** *Genome Res* 2005, **15:**
1777-1786.
8. Burge C, Karlin S: **Prediction of complete gene structures in
human genomic DNA.** *J Mol Biol* 1997, **268:**78-94.
9. Parra G, Blanco E, Guigo R: **GeneID in Drosophila.** *Genome Res*
2000, **10:**511-515.
10. Korf I, Flicek P, Duan D, Brent MR: **Integrating genomic homology into gene structure prediction.** *Bioinformatics* 2001, **17
(Suppl 1):**S140-148.
11. Meyer IM, Durbin R: **Comparative ab initio prediction of gene
structures using pair HMMs.** *Bioinformatics* 2002, **18:**1309-1318.
12. Parra G, Agarwal P, Abril JF, Wiehe T, Fickett JW, Guigo R: **Comparative gene prediction in human and mouse.** *Genome Res*
2003, **13:**108-117.
13. Yeh RF, Lim LP, Burge CB: **Computational inference of homologous gene structures in the human genome.** *Genome Res*
2001, **11:**803-816.
14. Florea L, Di Francesco V, Miller J, Turner R, Yao A, Harris M,
Walenz B, Mobarry C, Merkulov GV, Charlab R, *et al.*: **Gene and
alternative splicing annotation with AIR.** *Genome Res* 2005, **15:**
54-66.
15. Eyras E, Caccamo M, Curwen V, Clamp M: **ESTGenes: alternative
splicing from ESTs in Ensembl.** *Genome Res* 2004, **14:**976-987.
16. **OCAML Development Site** [http://caml.inria.fr/]
17. **Exogean Development Site** [http://www.biologie.ens.fr/dyogen/
exogean]
18. Kent WJ: **BLAT - the BLAST-like alignment tool.** *Genome Res*
2002, **12:**656-664.
19. **The GENCODE Project** [http://genome.imim.es/gencode/]
20. **The ENCODE Project** [http://genome.gov/10005107]
21. **The EGASP Project** [http://genome.imim.es/gencode/workshop/
meeting.html]
22. Keibler E, Brent MR: **Eval: a software package for analysis of
genome annotations.** *BMC Bioinformatics* 2003, **4:**50.

Research

Using several pair-wise informant sequences for *de novo* prediction of alternatively spliced transcripts

Paul Flicek* and Michael R Brent[†]

Addresses: *European Bioinformatics Institute, Wellcome Trust Genome Campus, Hinxton, Cambridge CB10 1SD, UK.
[†]Laboratory for Computational Genomics, Washington University, Saint Louis, MO 63130, USA.

Correspondence: Paul Flicek. Email: flicek@ebi.ac.uk

Published: 7 August 2006

Genome Biology 2006, **7(Suppl 1)**:S8

The electronic version of this article is the complete one and can be found online at http://genomebiology.com/2006/7/S1/S8

Abstract

Background: As part of the ENCODE Genome Annotation Assessment Project (EGASP), we developed the MARS extension to the Twinscan algorithm. MARS is designed to find human alternatively spliced transcripts that are conserved in only one or a limited number of extant species. MARS is able to use an arbitrary number of informant sequences and predicts a number of alternative transcripts at each gene locus.

Results: MARS uses the mouse, rat, dog, opossum, chicken, and frog genome sequences as pair-wise informant sources for Twinscan and combines the resulting transcript predictions into genes based on coding (CDS) region overlap. Based on the EGASP assessment, MARS is one of the more accurate dual-genome prediction programs. Compared to the GENCODE annotation, we find that predictive sensitivity increases, while specificity decreases, as more informant species are used. MARS correctly predicts alternatively spliced transcripts for 11 of the 236 multi-exon GENCODE genes that are alternatively spliced in the coding region of their transcripts. For these genes a total of 24 correct transcripts are predicted.

Conclusions: The MARS algorithm is able to predict alternatively spliced transcripts without the use of expressed sequence information, although the number of loci in which multiple predicted transcripts match multiple alternatively spliced transcripts in the GENCODE annotation is relatively small.

Background

Accurate prediction of protein-coding genes in mammals remains a challenging and active area of research [1]. In the past decade the most important advance in *de novo* gene prediction came with the initial availability of extensive human and mouse genomic sequences. Several gene prediction algorithms were introduced at that time that improved gene prediction by using the specific patterns of evolutionary conservation that are indicative of protein coding genes [2-4].

Dual-genome gene finding algorithms

All of the dual-genome (category 4) gene finders participating in EGASP rely on alignments to one or more informant genome sequences. For predicting human genes, dual-genome gene prediction algorithms most often use the mouse genome sequence as a source of evolutionary conservation information. This was originally a consequence of the early availability, with respect to other mammals, of the mouse genome sequence [5-8]. However, as additional genomes were sequenced, it became apparent that the evolutionarily divergence between human and mouse is near the point of optimal value for dual-genome gene prediction [9-11].

Twinscan is one of the most accurate *de novo* dual-genome gene prediction algorithms. It has proven effective for genome annotation in nematodes [12], plants [13], fungi [14], and mammals [6,15]. Recently, the gene-prediction program N-SCAN was introduced as a way to incorporate whole-genome multiple alignments into gene prediction [11]. Twinscan is a special case of the more general N-SCAN algorithm.

Both Twinscan and N-SCAN have focused on the prediction of the single mostly likely transcript in a given gene locus, although alternative splicing is now known to occur in a large majority of mammalian genes. In fact, Kan *et al.* [16] reported that nearly all genes with high expressed sequence tag (EST) coverage showed evidence of multiple splice forms. Even the well characterized human alpha globin cluster was recently shown to contain previously unknown, small, alternatively spliced exons [17]. Moreover, rare alternatively spliced transcripts can have important consequences in health and disease [18].

In an attempt both to address the problem of *de novo* prediction of alternatively spliced genes and to improve multi-genome *de novo* gene prediction, we developed the MARS ('Multiple Informants: Alternative Splices') extension to the Twinscan algorithm.

Almost all current methods for automatically annotating alternatively spliced transcripts rely on a rich EST database [19-21]. One of the few exceptions to an EST-based technique used a pair-hidden Markov model (pair-HMM) to successfully identify alternatively spliced exons conserved in human and mouse [22]. These conserved alternative splicing events are thought to be relatively rare [23]. MARS seeks to leverage the apparently more common situation that for some human genes only one splice variant appears to be conserved in another species [24,25]. One recently described example is the *Tfam* gene, which encodes a mitochondrial transcription factor and has a conserved alternative isoform in primates and rat, but not in mouse [26].

Description of the MARS algorithm
The MARS algorithm consists of two major steps. In the first step, transcript predictions are created from a number of different evolutionarily related informant sequences using Twinscan. For EGASP, MARS used the publicly available assemblies of the mouse (UCSC id mm5), rat (rn3), dog (canFam1), chicken (galGal1), frog (xenTro1), and opossum (monDom1) genomes as informant sources for Twinscan. These six informant sources make up the informant set. In the second step of the algorithm, the predicted transcripts based on each of the informant sources in the informant set

are collected into multi-transcript genes using coding (CDS) region overlap. We refer to gene predictions created this way as MARS genes. MARS genes may be created from any informant subset that contains two or more informant sources.

The predictions described in this paper are based on a version of the MARS algorithm that has been updated compared to the version of the algorithm used to create the predictions submitted to the EGASP workshop. The current predictions use each member of the informant set as a pairwise informant sequence for Twinscan, which is run once for each of the sequences in the informant set to generate transcript predictions based on each specific informant sequence (for example, a total of six times for the informant set described above). This set of transcript predictions is collected into MARS genes.

For the predictions submitted to the EGASP workshop and used in the official evaluation [1], the first step transcript predictions were based on probabilistic combinations of the mouse conservation model with the conservation model from each of the other informant sequences [27]. Briefly, this strategy defines a weighted average of the mouse conservation model with the conservation model of another informant source within the Twinscan probability model to produce the single best transcript predictions based on both informant sources simultaneously. We refer to this procedure as the 'full weight' method, and it is described in detail elsewhere [27]. Thus, the EGASP submissions were created from a set of transcripts based on running Twinscan five times with uniformly weighted averages of the probability models for mouse-rat, mouse-dog, mouse-chicken, mouse-frog, and mouse-opossum. This set of transcripts was collected into MARS genes as described above.

MARS currently predicts only the coding (CDS) regions of genes, thus all references to exons and transcripts are to coding exons and coding transcripts only.

Results
The results for the updated MARS algorithm differ from those reported in the EGASP summary because of the updates to the MARS algorithm that are described above. Compared to the submitted predictions, those produced from the updated MARS algorithm are more sensitive compared to the GENCODE annotation, but less specific at both the transcript and exon levels. A summary of the accuracy of the EGASP submission version of the MARS algorithm and the updated version described in this paper is given in Table 1. The updated predictions also include approximately twice as many coding transcripts per gene as the predictions submitted to EGASP. Because we made very limited use of the 13 EGASP training regions, we have chosen to present results here based on all 44 regions. These

Table 1

Submitted versus updated prediction characteristics

	ESn	ESp	TSn	TSp	GSn	GSp
Predictions submitted to EGASP	69.3%	65.8%	18.2%	17.8%	38.0%	28.3%
Updated MARS algorithm	74.4%	45.1%	19.4%	10.4%	40.6%	33.0%

A comparison of the predictive accuracy for the MARS genes submitted to the EGASP workshop and those produced by the updated MARS algorithm. The columns are sensitivity and specificity at the coding exon (ESn/ESp), coding transcript (TSn/TSp), and gene level (GSn/GSp).

Table 2

Pair-wise prediction characteristics

	Mouse	Rat	Dog	Chicken	Frog	Opossum
Predicted transcripts	486	476	530	431	422	467
Exons per transcript	7.55	7.62	6.82	11.02	11.28	8.54

The total number of predicted transcripts in the 44 ENCODE regions and the number of coding exons per transcript for each of the six informant sources in the MARS informant set.

Table 3

Aligned fraction of the ENCODE regions

	Mouse	Rat	Dog	Chicken	Frog	Opossum
Whole regions	15.2%	14.7%	28.8%	2.8%	2.0%	5.6%
Coding sequence	87.5%	85.0%	87.4%	53.0%	50.1%	76.1%

A comparison of the total fraction of bases aligned in the 44 ENCODE regions and the fraction of bases aligned in the coding portion of the GENCODE annotation for each of the informant sources in the MARS informant set. See Materials and methods for the alignment protocol.

considerations result in a slight difference between the EGASP evaluation of the submitted results and those displayed in Table 1, but do not materially change the results or the interpretation of them.

Transcript predictions from individual informant sources

For the set of all 44 ENCODE regions, each individual informant results in a similar number of predicted transcripts (Table 2). Informant sources at greater evolutionary distances tend to result in fewer, longer transcripts than informants within the mammalian lineage. However, the summary information from the ENCODE regions presented in Table 2 only hints at the diversity of predicted transcripts from the various informant sources. For example, mouse and rat shared a common ancestor approximately 25 million years ago and align similar fractions of the human genome using our alignment procedure (see Materials and methods and Table 3), but using these two rodent genome sequences as informant sources leads to a significantly different set of transcripts. In fact, the total number of predicted transcripts made using the mouse genome as the informant sequence is similar to the total number of predicted transcripts using the

rat genome as the informant sequence (486 and 476, respectively), but less than 50% (213) of these transcripts are predicted to have identical intron-exon structure. Similar results are seen on the human genome as a whole (data not shown).

The four mammalian informant sequences lead to more accurate predictions than either the frog or the chicken informant. Predictions based on the opossum informant sequence are slightly more accurate than those based on either mouse or rat (Figure 1). Compared to the rodent informant sequences the dog sequence aligns significantly more of the ENCODE regions, without additional alignment in the coding sequence. Conversely, the opossum aligns approximately one-third the total number of bases as the rodent sequences, while retaining alignment in 76% of the coding regions (Table 3).

Informative value of the pair-wise alignments

The alignment characteristics for each of the six informant sequences shown in Table 3 are primarily responsible for the characteristics of the pair-wise prediction sets shown in Table 2. To asses how the alignments affect the various

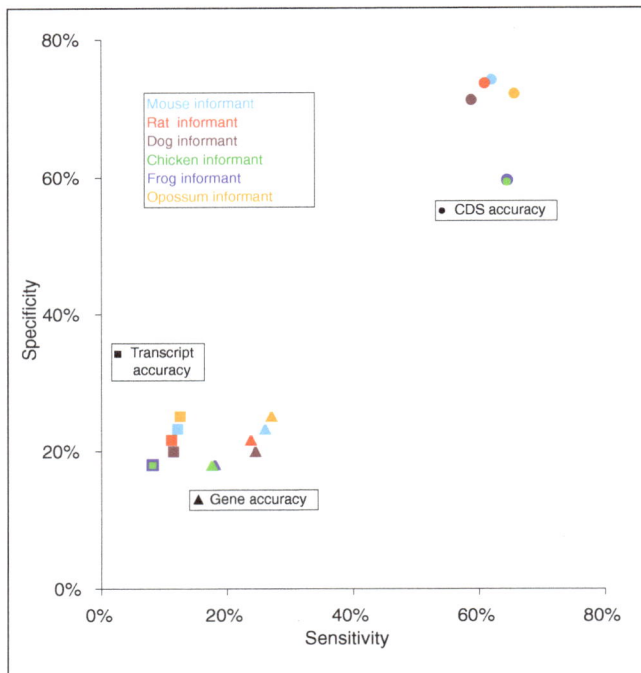

Figure 1
Pair-wise predictive accuracy for each of the six sequences in the informant set. The sensitivity and specificity, as compared to the GENCODE annotations, of Twinscan predictions based on the mouse (blue), rat (red), dog (brown), chicken (green), frog (purple), and opossum (orange) informant sequences. Gene level accuracy (triangles), transcript level accuracy (squares), and coding exon level accuracy (circles) are presented.

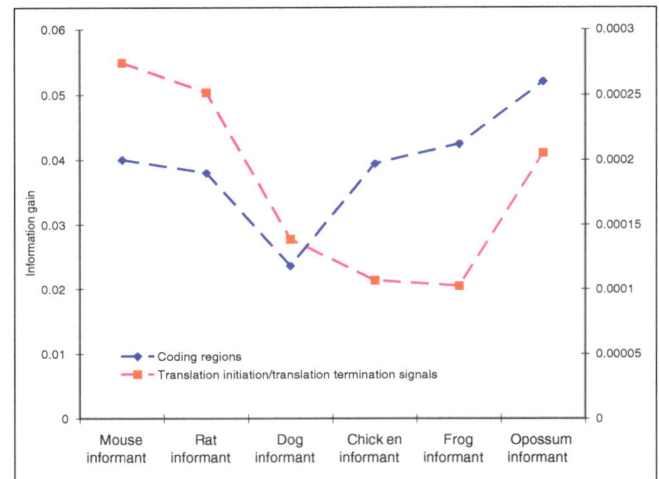

Figure 2
The information gain for the informant alignments with respect to the training set annotations for the six informant sequences. The information gain in the coding portion of the model is displayed in blue with the scale on the left side of the graph. The information gain for the translation initiation and termination signals is displayed in red with the scale on the right-hand side of the graph.

components of the Twinscan conservation model, we calculated the information gain of the alignments with respect to the training sequence annotations (see Figure 2 and Materials and methods). The difference in the number of exons per transcript is partially the effect of the amount of information available to the coding portion of the model and the translation initiation and termination signals (that is, the transcript ends). In cases such as mouse, rat, and opossum, where the information gain of the alignments with respect to the annotations is relatively high in both the coding regions and the transcript ends, the number of exons per transcript most closely resembles the annotation. When the information gain for the coding region portion of the conservation model is relatively high and the information gain for the transcript ends is relatively low, longer genes are predicted because the relative information gain of correct gene boundaries is low with respect to incorrect gene boundaries, thus the model is less inclined to end a transcript. In other words, for the case of the frog and the chicken informant sequence, it is more probable, under the model, for a gene to contain additional internal exons rather than boundary exons, which also contain the translation initiation or translation termination signals. This effect also leads to a greater number of exon predictions for the more distantly

related informant species. For the case of the dog informant, in which the information gain in both the coding regions and the transcript ends of the model is relatively low, genes are predicted with fewer exons than the annotation. The number of exons per transcript from the dog informant-based predictions is more similar to *ab initio* transcript predictions that do not use evolutionary conservation, such as those reported in group 2 of the EGASP experiment [1].

MARS genes predicted from informant sets

As MARS genes are created from an increasing number of informant sources, we see an increase in predictive sensitivity as the transcripts based on each additional informant sequence are added to the genes. At the same time, the gene specificity improves as addition of longer transcripts from non-mammalian informant sources leads to longer genes (Figure 3).

The predictive sensitivity of both the coding exons and complete coding transcripts also increases as the predictions based on each additional informant sequence are clustered together, but the specificity falls as the number of apparent false positive transcripts and CDS exons increases. The difference in the performance trend at the gene level and the transcript level is based on the definition of gene level accuracy, which rewards predicting at least one transcript correctly with no penalty for additional, incorrectly predicted transcripts.

Both the number of coding exons and the number of transcripts in each MARS gene increase with the size of the

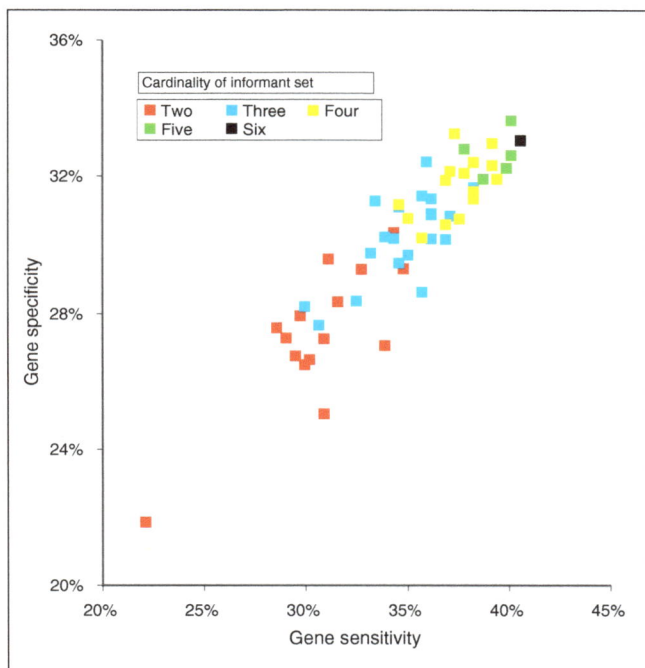

Figure 3
Gene accuracy versus informant subset size. The effect of informant subset size on gene level sensitivity and specificity compared to the GENCODE annotations.

informant set (Table 4). This increase corresponds partly to the usage of distantly related informant species in the informant set. For example, MARS genes predicted by the three species informant subset that includes mouse, rat, and dog average 8.07 exons per gene, while MARS genes from the frog, chicken, and opossum informant subset average 11.64 exons per gene.

We separately evaluated the subset of transcripts that are predicted based on at least two informant sequences. These transcripts are significantly more specific at all levels of the evaluation (Table 5), although the predictions are less sensitive (as expected). The set of transcripts common to only mammalian informant sources is more specific than the set common to all informant sources and only slightly less sensitive (Table 5).

Prediction of alternatively spliced transcripts
Using the informant set consisting of all six informant sources, MARS correctly predicts alternatively spliced transcripts for 11 of the 236 multi-exon GENCODE genes that are alternatively spliced in the coding region of their transcripts. For these 11 genes, a total of 24 (out of 59) correct transcripts are predicted (we observed that just 2 of these 11 genes accounted for 25 of the 59 coding transcripts: RP1-309K30.2 on ENr333 and RP4-696P19.3 on ENr334). Moreover, when compared to a set of 134 cassette (that is, skipped) coding exons from the GENCODE annotation, MARS predicted 85 of these exons correctly in at least one transcript, including 19 that are correctly predicted as cassette exons. MARS predicts a total of 247 cassette exons.

Table 4

Effect of increasing informant subset size

	Pair-wise prediction characteristics					
Informant set size	Two	Three	Four	Five	Six	Annotation
Average transcripts per gene	1.76	2.44	3.05	3.62	4.15	2.25
Average exons per transcript	9.31	9.67	9.94	10.16	10.35	8.64

The number of coding transcripts per gene and coding exons per transcripts increases with the cardinality of the informant set. The gene level accuracy also increases with informant set size (see Figure 3).

Table 5

Transcripts common to several informant sources

	Prediction accuracy for transcripts common to several informants					
	ESn	ESp	TSn	TSp	GSn	GSp
All transcripts	74.4%	45.1%	19.4%	10.4%	40.6%	33.0%
Common transcripts	43.0%	71.6%	15.1%	29.6%	32.0%	35.0%
Mammalian transcripts	40.0%	77.1%	14.8%	33.7%	31.3%	37.7%

A comparison of the predictive accuracy for all MARS genes, with those having at least two transcripts predicted with identical structure from more than one informant source across the entire informant set, and with those having two transcripts with identical structures from at least two mammalian informant sources. Columns are defined as in Table 1.

When all six informant sources are used simultaneously, the predictive sensitivity is at its highest. MARS predicts about twice as many unique coding transcripts (1,873) as exist in the reference GENCODE annotation (975).

Experimental verification

An important part of the EGASP experiment is the attempt to experimentally validate a subset of the computational predictions outside of the reference GENCODE annotation. As part of EGASP, Guigo *et al.* [1] selected a total of 47 exon pairs predicted by MARS for experimental confirmation by RT-PCR. Of these, 7 (15%) were found to be expressed in at least one tissue. Interestingly, although a number of the other EGASP gene-prediction methods also predicted as many as 4 of these exon pairs, these 7 were the only ones that could be confirmed in the EGASP experiment.

Discussion

One of the goals of the ENCODE pilot project was to develop new high-throughput methods to identify the functional elements in the human genome [28,29]. To address the continued need for *de novo* gene discovery, we have introduced the MARS method for prediction of alternatively spliced transcripts without the use of any expressed sequence information. MARS genes are built by combining the predicted transcripts from a number of informant species and are significantly more likely to contain correctly predicted transcripts than any individual informant. MARS performed effectively when compared to other dual-genome *de novo* gene prediction systems in EGASP [1] and is unique among the EGASP methods in its ability to predict alternatively spliced transcripts using only patterns found in pair-wise alignments between a target sequence and a set of informant sequences.

We have updated the MARS algorithm between the EGASP workshop and submission of this paper. The updated version of MARS correctly predicts multiple alternatively spliced transcripts at one additional locus compared to the submitted version. Additionally, the updated algorithm is more sensitive for all measures, although this increased sensitivity comes at a cost of a significant reduction in specificity at the transcript and exon levels compared to the version submitted to the EGASP workshop. Regardless, we feel the update is justified on theoretical grounds because the original submission gives too much consideration to the mouse informant to the detriment of other informant sources. A second source of error comes from the addition of transcripts from the two non-mammalian informant sources, which appears to have enriched the prediction set for false positive transcripts.

A number of gene-finding algorithms create consensus genes by combining sets of gene predictions and other information [13,30]. One example in the EGASP experiment is JIGSAW

[31], a program that uses 'any information' (EGASP category 1) to create gene structures. Much of the information used by JIGSAW is based on expressed sequences and is, therefore, not directly comparable to the EGASP dual-genome (category 4) predictions. Because MARS genes are created by overlapping transcript predictions from a number of sources, we were interested to see if these transcript predictions could be statistically combined to produce more accurate consensus gene structures. To directly address this question, we compared the MARS genes to consensus genes produced by GLEAN, a new gene-prediction algorithm that uses dynamic programming to discover gene structures that maximize the probability of several sources of evidence (A Mackey, personal communication). GLEAN was run using the transcript predictions from the six individual pair-wise sets (mouse, rat, chicken, dog, frog, and opossum) as its only input sources of evidence, although the transcript sets cannot be considered independent sources of information and thus represent a non-traditional use of the algorithm likely to reduce its statistical power (A Mackey, personal communication). The GLEAN consensus predictions at the gene and transcript levels were similar to predictions based on either the rat or dog informant only (that is, less sensitive and specific than the mouse or opossum informant, but more sensitive and specific than the chicken or frog informants). For coding exons, the GLEAN consensus predictions are more sensitive than any of the individual informants and less specific than predictions based on mammalian or marsupial informants.

Our analysis of the information in the pair-wise alignments shows that some characteristics of the transcript predictions are a consequence of the alignments themselves. Importantly, the concentration of alignments from the opossum in the coding sequences of the ENCODE regions and the pair-wise predictive accuracy of the opossum informant show that the draft genome sequence of *Monodelphis domestica* is already a valuable tool for dual-genome gene prediction. A more complete or even finished opossum assembly could prove especially powerful for annotating the functional regions in the human genome.

The MARS method is computationally tractable with computational requirements, growing essentially linearly with the number of informant sequences and it can take advantage of additional genome sequences as they become available without extensive reanalysis. Other methods for annotating alternatively spliced transcripts are generally based on information from expressed sequences; thus, the annotations produced are experimentally supported. MARS genes, in general, do not have such support and thus provide a potential pool for experimental validation of novel splice forms [32,33].

Recent reports indicate that alternatively spliced exons have specific sequence features associated with them, such as exonic splicing enhancers (ESEs) and exonic splicing silencers (ESSs) [22]. Moreover, some alternatively spliced exons have conservation patterns unlike constitutively spliced exons [34]. Neither of these observations have been incorporated into the MARS model and doing so could lead to more accurate prediction of alternatively spliced transcripts that are not yet supported in EST databases.

Conclusions

The MARS algorithm is able to predict alternatively spliced transcripts without the use of expressed sequence information, although the number of loci in which multiple predicted transcripts match multiple alternatively spliced transcripts in the GENCODE annotation is relatively small. Based on the current GENCODE annotation, it seems unlikely the majority of alternatively spliced transcripts predicted by MARS are actually produced. However, the results of the EGASP experimental validation of novel predictions show that among the EGASP entries, more MARS predictions were confirmed than for any other method [1]. These results are consistent with the previous reannotation of chromosome 22 in light of additional data that resulted in a significant number of new annotations, including many alternatively spliced transcripts [35]. Finally, the large fraction of incomplete transcripts in the current GENCODE annotation suggests that we are still some distance from finished annotation.

We propose that the selection of other novel alternative transcripts for experimental confirmation may be guided by looking first to those transcripts predicted with identical structure using several informant sequences. In fact, the set of 449 complete transcripts that is common to more than one informant source is approximately three times more specific than the complete set of MARS transcripts.

Materials and methods
Sequences
All predictions were made on the ENCODE regions as mapped to NCBI Build 35 (UCSC id hg17) of the human genome [36] downloaded from the UCSC genome browser [37,38] on 3 June 2004. The human genome was masked for interspersed, but not low-complexity, repeats using RepeatMasker tables provided by UCSC.

Where possible, each ENCODE region was padded on each side with 750,000 base-pairs (bp) of genomic sequence from the corresponding chromosome to ensure that the predictions were made in true genomic context and because genes were expected to extend beyond the boundaries of the ENCODE regions. The size of the sequence context was chosen based on the memory usage of Twinscan. Restricting the input sequences to the exact boundaries of the ENCODE regions results in a small decrease in predictive accuracy of

approximately 1% for all evaluation measures due to incorrectly truncated genes. Informant genome sequences were also downloaded from the UCSC Genome Brower. This set included NCBI Build 33 (UCSC id mm5) of the mouse genome sequence, the canFam1 assembly of the dog genome sequence, the monDom1 assembly of the opossum genome, the rn3 assembly of the rat genome, assembly galGal2 of the chicken genome, and assembly xenTro1 of the frog genome.

Twinscan version and training set
The results in this study use the TwinscanΦ executable [27], which is an updated version of Twinscan 1.1 [6]. Conservation parameters were trained separately for each of the six informant species on a set of 3,072 human RefSeq transcripts from 2,477 loci. Genes in the training set are spread across 112 one megabase fragments of the human genome and selected based on characteristics of the genes on the fragments, including gene density and gene length. These conservation parameters are optimized for accurate whole genome predictive accuracy. The training sequences and annotations are available at [39].

The 13 ENCODE training regions provided in advance of the EGASP submission were used only to determine the optimal size and members of the informant set.

Alignments
All alignments were done with WU-BLAST version 07-14-2004 [40] using a two-stage serial BLAST strategy [41]. First stage BLAST parameters were set at M=1 N=-1 Q=5 R=1 Z=3000000000 Y=3000000000 B=10000 V=100 W=11 X=30 S=30 S2=30 gapS2=30 topcomboN=1. Second stage BLAST alignments used the following more stringent parameters: W=8 X=20 S=15 S2=15 topcomboN=3. For human-chicken and human-frog alignments, Z=1000000000 was used. The seg and dust filters were used for all alignments. BLAST databases were prepared as previously described [6].

Information gain calculation
We calculated the information gain for each of the informant sources using our training set by subtracting the 0th order conditional uncertainty of the annotation given the conservation sequence from the annotation uncertainty as follows:

$$IG_i = H(C_m) - H(C_m|A)$$

where

$$H(C_m) = -Pr(C_m) \times log2\, Pr(C_m)$$

and

$$H(C_m|A) = \begin{aligned}&-Pr(M)\,(Pr(c_m|M)\,log2\,Pr(c_m|M) + Pr(n_m|M)\,log2\,Pr(n_m|M))\\&-Pr(G)\,(Pr(c_m|G)\,log2\,Pr(c_m|G) + Pr(n_m|G)\,log2\,Pr(n_m|G))\\&-Pr(U)\,(Pr(c_m|U)\,log2\,Pr(c_m|U) + Pr(n_m|U)\,log2\,Pr(n_m|U))\end{aligned}$$

Here C_m is defined as a random variable representing whether a given base in the genome should be classified as part of the given portion of the Twinscan conservation model (c_m) or as not a part of the given portion of the model (n_m). We use the maximum likelihood estimate of C_m from our training set. For the conditional uncertainty calculation, we condition the probability of C_m based on whether the corresponding conservation symbol from the given informant sequence is (*M*)atch, (*G*)ap/mismatch, or (*U*)naligned. In this analysis here we use m $\in$ {coding, translation initiation and translation termination signal} portions of the Twinscan conservation model [2].

Evaluation method

All evaluations were performed as described [1] using the GENCODE annotations as a reference.

Availability

MARS source code is available on an open source license. All predictions, training materials, and source code are available at the MARS website [39].

Acknowledgements

The authors would like to thank the organizers of the EGASP workshop, Steve Searle for a helpful discussion about the right way to collapse overlapping transcripts, Aaron Mackey for creating the GLEAN annotations from the informant set transcripts, and Ewan Birney. MRB and the development of the Twinscan code base were supported in part by HG02278 from the National Human Genome Research Institute. Open access publication charges have been paid by the Wellcome Trust.

This article has been published as part of *Genome Biology* Volume 7, Supplement 1, 2006: EGASP '05. The full contents of the supplement are available online at http://genomebiology.com/supplements/7/S1.

References

1. Guigó R, Flicek P, Abril JF, Reymond A, Lagarde J, Denoeud F, Antonarakis S, Ashburner M, Bajic VB, Birney E, *et al.*: **EGASP: The human ENCODE Genome Annotation Assessment Project.** *Genome Biology* 2006, **7**(Suppl 1) :S2.
2. Korf I, Flicek P, Duan D, Brent MR: **Integrating genomic homology into gene structure prediction.** *Bioinformatics* 2001, **17** (Suppl 1):S140-S148.
3. Wiehe T, Gebauer-Jung S, Mitchell-Olds T, Guigó R: **SGP-1: prediction and validation of homologous genes based on sequence alignments.** *Genome Res* 2001, **11**:1574-1583.
4. Batzoglou S, Pachter L, Mesirov J, Berger B, Lander E: **Human and mouse gene structure: comparative analysis and application to exon prediction.** *Genome Res* 2000, **10**:950-958.
5. Waterston RH, Lindblad-Toh K, Birney E, Rogers J, Abril JF, Agarwal P, Agarwala R, Ainscough R, Alexandersson M, An P, *et al.*: **Initial sequencing and comparative analysis of the mouse genome.** *Nature* 2002, **420**:520-562.
6. Flicek P, Keibler E, Hu P, Korf I, Brent MR: **Leveraging the mouse genome for gene prediction in human: from whole-genome shotgun reads to a global synteny map.** *Genome Res* 2003, **13**: 46-54.
7. Parra G, Agarwal P, Abril JF, Wiehe T, Fickett JW, Guigó R: **Comparative gene prediction in human and mouse.** *Genome Res* 2003, **13**:108-117.
8. Alexandersson M, Cawley S, Pachter L: **SLAM: cross-species gene finding and alignment with a generalized pair hidden Markov model.** *Genome Res* 2003, **13**:496-502.
9. Wang M, Buhler J, Brent M: **The effects of evolutionary distance on TWINSCAN, an algorithm for pair-wise comparative gene prediction.** *Cold Spring Harb Symp Quant Biol* 2003, **68**: 125-130.
10. Zhang L, Pavlovic V, Cantor CR, Kasif S: **Human-mouse gene identification by comparative evidence integration and evolutionary analysis.** *Genome Res* 2003, **13**:1190-1202.
11. Gross SS, Brent MR: **Using multiple alignments to improve gene prediction.** *J Comput Biol* 2006, **13**:379-393.
12. Wei C, Lamesch P, Arumugam M, Rosenberg J, Hu P, Vidal M, Brent MR: **Closing in on the C. elegans ORFeome by cloning TWINSCAN predictions.** *Genome Res* 2005, **15**:577-582.
13. Allen JE, Pertea M, Salzberg SL: **Computational gene prediction using multiple sources of evidence.** *Genome Res* 2004, **14**: 142-148.
14. Tenney AE, Brown RH, Vaske C, Lodge JK, Doering TL, Brent MR: **Gene prediction and verification in a compact genome with numerous small introns.** *Genome Res* 2004, **14**:2330-2335.
15. Wu JQ, Shteynberg D, Arumugam M, Gibbs RA, Brent MR: **Identification of rat genes by TWINSCAN gene prediction, RT-PCR, and direct sequencing.** *Genome Res* 2004, **14**:665-671.
16. Kan Z, States D, Gish W: **Selecting for functional alternative splices in ESTs.** *Genome Res* 2002, **12**:1837-1845.
17. Hughes JR, Cheng JF, Ventress N, Prabhakar S, Clark K, Anguita E, Gobbi MD, de Jong P, Rubin E, Higgs DR: **Annotation of cis-regulatory elements by identification, subclassification, and functional assessment of multispecies conserved sequences.** *Proc Natl Acad Sci USA* 2005, **102**:9830-9835.
18. Cáceres JF, Kornblihtt AR: **Alternative splicing: multiple control mechanisms and involvement in human disease.** *Trends Genet* 2002, **18**:186-193.
19. Kan Z, Rouchka EC, Gish WR, States DJ: **Gene structure prediction and alternative splicing analysis using genomically aligned ESTs.** *Genome Res* 2001, **11**:889-900.
20. Sugnet CW, Kent WJ, Ares M, Haussler D: **Transcriptome and genome conservation of alternative splicing events in humans and mice.** *Pac Symp Biocomput* 2004, 66-77.
21. Foissac S, Schiex T: **Integrating alternative splicing detection into gene prediction.** *BMC Bioinformatics* 2005, **6**:25.
22. Ohler U, Shomron N, Burge CB: **Recognition of unknown conserved alternatively spliced exons.** *PLoS Comput Biol* 2005, **1**: 113-122.
23. Yeo GW, Nostrand EV, Holste D, Poggio T, Burge CB: **Identification and analysis of alternative splicing events conserved in human and mouse.** *Proc Natl Acad Sci USA* 2005, **102**:2850-2855.
24. Valenzuela A, Talavera D, Orozco M, de la Cruz X: **Alternative splicing mechanisms for the modulation of protein function: conservation between human and other species.** *J Mol Biol* 2004, **335**:495-502.
25. Pan Q, Bakowski MA, Morris Q, Zhang W, Frey BJ, Hughes TR, Blencowe BJ: **Alternative splicing of conserved exons is frequently species-specific in human and mouse.** *Trends Genet* 2005, **21**:73-77.
26. D'Errico I, Dinardo MM, Capozzi O, Virgilio CD, Gadaleta G: **History of the Tfam gene in primates.** *Gene* 2005, **362**:125-132.
27. Flicek P: **Methods for improving gene prediction with evolutionary conservation.** *PhD thesis*. Washington University, Department of Biomedical Engineering; 2004.
28. ENCODE Project Consortium: **The ENCODE (ENCyclopedia Of DNA Elements) Project.** *Science* 2004, **306**:636-640.
29. **The Encyclopedia of DNA Elements (ENCODE) Project** [http://www.genome.gov/10005107]
30. Pavlovic V, Garg A, Kasif S: **A Bayesian framework for combining gene predictions.** *Bioinformatics* 2002, **18**:19-27.
31. Allen JE, Salzberg SL: **JIGSAW: integration of multiple sources of evidence for gene prediction.** *Bioinformatics* 2005, **21**:3596-3603.
32. Guigó R, Dermitzakis ET, Agarwal P, Ponting CP, Parra G, Reymond A, Abril JF, Keibler E, Lyle R, Ucla C, *et al.*: **Comparison of mouse and human genomes followed by experimental verification yields an estimated 1,019 additional genes.** *Proc Natl Acad Sci USA* 2003, **100**:1140-1145.
33. Eyras E, Reymond A, Castelo R, Bye JM, Camara F, Flicek P, Huckle EJ, Parra G, Shteynberg DD, Wyss C, *et al.*: **Gene finding in the chicken genome.** *BMC Bioinformatics* 2005, **6**:131.
34. Bejerano G, Pheasant M, Makunin I, Stephen S, Kent WJ, Mattick JS, Haussler D: **Ultraconserved elements in the human genome.** *Science* 2004, **304**:1321-1325.
35. Collins JE, Goward ME, Cole CG, Smink LJ, Huckle EJ, Knowles S, Bye JM, Beare DM, Dunham I: **Reevaluating human gene anno-

tation: a second-generation analysis of chromosome 22.** *Genome Res* 2003, **13:**27-36.
36. International Human Genome Sequencing Consortium: **Finishing the euchromatic sequence of the human genome.** *Nature* 2004, **431:**931-945.
37. Kent WJ, Sugnet CW, Furey TS, Roskin KM, Pringle TH, Zahler AM, Haussler D: **The human genome browser at UCSC.** *Genome Res* 2002, **12:**996-1006.
38. **The UCSC Genome Browser** [http://genome.ucsc.edu]
39. **Training Sequences and Annotations** [http://www.ebi.ac.uk/~flicek/MARS/]
40. **WU-BLAST** [http://blast.wustl.edu]
41. Korf I: **Serial BLAST searching.** *Bioinformatics* 2003, **19:**1492-1496.

Research

JIGSAW, GeneZilla, and GlimmerHMM: puzzling out the features of human genes in the ENCODE regions

Jonathan E Allen*†, William H Majoros‡, Mihaela Pertea* and Steven L Salzberg*

Addresses: *Center for Bioinformatics and Computational Biology, University of Maryland, College Park, MD 20742, USA. †Department of Computer Science, John Hopkins University, Baltimore, MD 21218, USA. ‡Institute for Genome Sciences and Policy, Duke University, Durham, NC 27708, USA.

Correspondence: Steven L Salzberg. Email: salzberg@umiacs.umd.edu

Published: 7 August 2006

Genome Biology 2006, **7(Suppl 1)**:S9

The electronic version of this article is the complete one and can be found online at http://genomebiology.com/2006/7/S1/S9

Abstract

Background: Predicting complete protein-coding genes in human DNA remains a significant challenge. Though a number of promising approaches have been investigated, an ideal suite of tools has yet to emerge that can provide near perfect levels of sensitivity and specificity at the level of whole genes. As an incremental step in this direction, it is hoped that controlled gene finding experiments in the ENCODE regions will provide a more accurate view of the relative benefits of different strategies for modeling and predicting gene structures.

Results: Here we describe our general-purpose eukaryotic gene finding pipeline and its major components, as well as the methodological adaptations that we found necessary in accommodating human DNA in our pipeline, noting that a similar level of effort may be necessary by ourselves and others with similar pipelines whenever a new class of genomes is presented to the community for analysis. We also describe a number of controlled experiments involving the differential inclusion of various types of evidence and feature states into our models and the resulting impact these variations have had on predictive accuracy.

Conclusions: While in the case of the non-comparative gene finders we found that adding model states to represent specific biological features did little to enhance predictive accuracy, for our evidence-based 'combiner' program the incorporation of additional evidence tracks tended to produce significant gains in accuracy for most evidence types, suggesting that improved modeling efforts at the hidden Markov model level are of relatively little value. We relate these findings to our current plans for future research.

Background

Predicting complete protein-coding genes in human DNA remains a significant challenge, as the results of the ENCODE Genome Annotation Assessment Project (EGASP) workshop clearly demonstrate. Although much progress has been made of late in the use of increasingly sophisticated models of gene structure, particularly those that utilize homology evidence within a phylogenetic framework (for example, [1,2]), it is clear that there is yet much room for improvement. In the wake of the most recent spate of advances in gene structure modeling, we additionally observe that the sophistication in modeling techniques has to some degree outstripped our ability to ascribe, with high confidence, specific reasons for the difference in performance between competing gene finding systems, particularly those that utilize similar underlying models and/or forms of evidence, but that differ in the particulars of their implementation. Although it is tempting in some cases to ascribe differences in performance to conspicuous differences in the published descriptions of two software systems, it is clear that such reasoning can be highly unreliable when the published descriptions are not complete, when the systems under consideration are highly complex, and when the source code is not available to third parties for detailed comparison. Unfortunately, these conditions hold for most gene finding systems in use today, with few exceptions. An additional complication arises out of the use of different training protocols, which can have a profound effect on the performance of a single system [3], making interpretation of the differences between systems, absent knowledge of precisely how they were trained, very risky indeed. It is clear, however, that accurate interpretation of such differences is essential for progress in the computational science of gene structure modeling.

For these reasons we decided to undertake, in conjunction with our EGASP activities, a series of controlled experiments designed to measure the relative influence of various components in our underlying models. Whereas the high-level EGASP evaluation included in this volume [4] compares disparate systems, each consisting of a complex code base with virtually no shared components between the competing systems, it was our hope that by performing a number of controlled experiments, each within the environment of a single software system, we could help to foster a more fine-grained understanding of the relative merits of different modeling decisions for gene structure prediction. Thus, our hope was to complement the overall EGASP comparison with a smaller-scale (but potentially very valuable) comparison of modeling techniques for human protein-coding genes.

Our efforts can be partitioned into two distinct sets of experiments. The first set involves the inclusion or exclusion of various states in our generalized hidden Markov model (GHMM) gene finder GeneZilla. Starting with a basic state topology for eukaryotic gene structure, we proceeded to incorporate additional states for biological features such as signal peptides and CpG islands, measuring the impact of these modifications on two sets of held-out test genes. We additionally investigated the effect of training set size, as well as the utility of isochore modeling via an external HMM for isochore boundary predictions. We also offer anecdotal observations on the different levels of effort required to achieve similar levels of accuracy in our two GHMM-based gene finders, despite their having nearly identical underlying models and algorithms. The latter observation further bolsters our contention that the differences in performance

between competing systems often cannot be ascribed with any confidence to differences in modeling decisions, due to the many other sources of variation in the training and operation of these complex software systems.

The second set of experiments involved the differential inclusion of various evidence tracks in our comparative and integrative 'combiner' program, JIGSAW, which was found to perform as well as or better than any of the other entries in the GENCODE competition. Because JIGSAW is an integrative program that can combine arbitrary forms of evidence (including the predictions from our other gene finders and sequence analysis programs), our early expectations were that this tool would dominate our submissions to the GENCODE competition, and hence we have concentrated our efforts on this particular tool. Our discussion will therefore focus correspondingly on this most important component of our pipeline.

We give a description of our prediction pipeline and the major components in it, which we have used repeatedly and with much success for the annotation of a number of invertebrate eukaryotic genomes sequenced and/or annotated at The Institute for Genomic Research (TIGR). Because the components that we describe are all released under open-source software licenses, others are thereby enabled to reproduce any of our computational results and to investigate extensions to our methods. In this way, we hope that our efforts will aid others in contributing to the advancement of automated genome annotation techniques.

Results
Accuracy on the ENCODE regions
Results for our *ab initio* predictions seem to place GeneZilla roughly between AUGUSTUS-abinit and GeneMark.hmm in accuracy for this particular test set (for example, Table 5 in [4]). As stated previously, ascribing these differences in accuracy to particular algorithmic and modeling differences between the three systems is difficult at best. In the case of AUGUSTUS and GeneZilla, both systems effectively mimic the earlier program GENSCAN [5] by utilizing nearly identically-structured GHMMs with a generalized Viterbi decoding algorithm. Known differences include the modeling of intron lengths (geometric in GeneZilla and GENSCAN; non-geometric for short introns in AUGUSTUS [6]), the number of isochores modeled (four in GeneZilla and GENSCAN; ten in AUGUSTUS), and the respective training protocols employed in estimating the thousands of parameters required by each of these systems (for example, [3]). We plan to investigate the individual effects of each of these differences within a controlled setting, as in the feature-state experiments described here (see the 'Effects of modeling specific features' section), and to reported these at a later date.

Table I

Results for JIGSAW performance on the 31 ENCODE test regions when utilizing *ab initio* predictions and EST data

	Gene Sn	Gene Sp	Exon Sn	Exon Sp	Nuc Sn	Nuc Sp	Missed Genes	Missed Exons	Inserted Exons
JIGSAW-GeneFinder4	22%	24%	59%	77%	77%	87%	15%	9%	9%
JIGSAW-GeneFinder6	25%	21%	67%	73%	87%	85%	6%	3%	13%
JIGSAW-non-Human-EST	39%	37%	70%	83%	88%	90%	7%	5%	7%
JIGSAW-non-Human-EST+	38%	34%	71%	80%	91%	87%	5%	3%	10%

The percentage of test genes and exons that do not overlap a prediction are listed in the Missed Genes and Missed Exons columns, respectively. The rightmost column shows the percentage of predicted exons inserted into true introns. See text for details. Nuc, nucleotide; Sn, sensitivity; Sp, specificity.

The JIGSAW version designed to recreate the human anno-tation (and submitted to EGASP) is based on the develop-ment of a non-expression-based gene finder. We experimented with input from the four gene finders from UCSC's annota-tion database (GENEID, SGP, TWINSCAN and GENSCAN) plus GeneZilla and GlimmerHMM. Table 1 shows the results of combining the four gene finders downloaded from UCSC (JIGSAW-GeneFinder4) and the addition of GeneZilla and GlimmerHMM (JIGSAW-GeneFinder6). Accuracy is mea-sured on coding regions of the exons for three categories: genes, where the entire gene is correctly predicted from start codon to stop codon including all internal exons; exons, where both splice sites are correct; and the protein coding nucleotide level.

The four gene finders downloaded from UCSC collectively identify 76% of the test exons correctly. Thus, if JIGSAW is provided only the output of these gene finders and if it can always select the correct exon, the theoretical upper bound on its exon sensitivity is 76%. Among the input gene finders, SGP achieves the highest exon sensitivity (61%) and TWINSCAN has the highest exon specificity (73%) with 54% of the exons supported by three or more gene finders. Adding GeneZilla and GlimmerHMM increases the number of correctly identified exons from 76% to 80%, and the additional input supports exons predicted by the other gene finders. With the addition of our gene finders, three or more gene finders support 67% of the exons. Thus, by adding GeneZilla and GlimmerHMM as input, JIGSAW's prediction performance is superior in nearly all categories to the best individual gene finders SGP and TWINSCAN.

Adding expression evidence from non-human sources (non-human RefSeq data and non-human mRNA data) expands the pool of correctly identified exons to 83% of the test set, and shows substantial improvements in prediction accuracy (JIGSAW-non-human EST in Table 1) over the gene-finder-only versions (JIGSAW-GeneFinder4 and JIGSAW-Gene-Finder6). Interestingly, adding the non-expression based evidence sources IsoFinder and PhastCons showed little effect on the gene-finder-only JIGSAW versions. When used

in conjunction with the gene expression evidence, however, sensitivity increased. Adding PhastCons and the IsoFinder track boosted sensitivity at the nucleotide level by 3% with a 3% drop in nucleotide specificity, while also increasing the number of correctly identified exons from 70% to 71%. Surprisingly, 91% of the coding nucleotides are detected using a combination of gene finders, G+C density, sequence conservation, and gene expression evidence from organisms other than human, while maintaining high specificity (87%).

Adding the remaining tracks of expression evidence from human - UniGene, TIGR Gene Index, and mRNAs aligned to the genome with BLAT - expands the pool of correctly identified exons to 87% of the test set. Using just the mRNA alignments and ignoring all other evidence except for the gene finders (JIGSAW-mRNA in Table 2), gives JIGSAW greater specificity, while remaining highly sensitive. This suggests that the human mRNA alignments serve as accurate gene structure predictors, obviating the need to look at other overlapping sources of expression evidence. Incorporating the assembled expressed sequence tags (ESTs) appears to have limited impact, which indicates a high degree of overlap between the ESTs and mRNA alignments (results not shown). Adding the non-human expression sources and the PhastCons and IsoFinder tracks return nucleotide sensitivity to 91% (JIGSAW-All-EST), the same level achieved by the JIGSAW-non-Human-EST+ version shown in Table 1. The use of the human expression evidence improves the percentage of correctly detected exons and genes to 77% and 52%, respectively.

Finally, tracks of evidence derived from curated human genes (KnownGene) and output from the Ensembl automated annotation pipeline were added. Incorporating the KnownGene track along with the six gene finders as input to JIGSAW yields a substantial boost in performance, since the majority of genes in the ENCODE regions overlap KnownGene predictions (JIGSAW-KnownGene). Incorpora-ting the additional evidence sources (JIGSAW-All in Table 2) reduces the number of completely missed genes and exons by 4% and 5%, respectively.

Table 2

Results of applying JIGSAW with all available evidence

	Gene Sn	Gene Sp	Exon Sn	Exon Sp	Nuc Sn	Nuc Sp	Missed Genes	Missed Exons	Inserted Exons
JIGSAW-mRNA	48%	60%	76%	93%	84%	97%	17%	11%	4%
JIGSAW-All-EST	52%	52%	77%	88%	91%	91%	6%	10%	8%
JIGSAW-KnownGene	71%	74%	76%	95%	87%	96%	7%	12%	3%
JIGSAW-All	74%	70%	80%	92%	93%	94%	3%	7%	6%
KnownGene	77%	73%*	78%	82%	89%	94%	13%	10%	4%
JIGSAW-EGASP	73%	66%	81%	89%	95%	92%	4%	6%	8%

* KnownGene predicts multiple transcripts per gene locus with transcript specificity of 47%. The percentage of test genes and exons that do not overlap a prediction are listed in the Missed Genes and Missed Exons columns, respectively. The rightmost column shows the percentage of predicted exons inserted into true introns. See text for details. Nuc, nucleotide; Sn, sensitivity; Sp, specificity.

Performance for JIGSAW using the KnownGene track alone is also listed in Table 2. An important source of Known-Gene's sensitivity is its prediction of multiple isoforms. More than half of the GENCODE genes are annotated with multiple isoforms, but JIGSAW in its current implementation predicts only one isoform per locus. KnownGene averages nearly two predicted transcripts per gene locus, which allows for the possibility of increased sensitivity at the gene level since there is a chance that at least one of the predicted transcripts matches the GENCODE annotation. The drawback, however, is a lower percentage of correctly predicted transcripts compared to JIGSAW; 70% of JIGSAW-All predictions match an annotated transcript, compared to only 47% of KnownGene predictions. Furthermore, the percentage of genes with JIGSAW predictions exactly matching a GENCODE annotation is as high as 74% (Table 2, JIGSAW-All).

JIGSAW output submitted to the EGASP workshop is labeled JIGSAW-EGASP in Table 2 and used input from the TIGR Gene Index, Human mRNAs, UniGene, Non-human RefSeq genes, KnownGene, PhastCons, Ensembl and the six gene finders. Three changes to the inputs were made, which distinguish JIGSAW-EGASP and JIGSAW-All. Two sources were excluded from JIGSAW-EGASP: non-human mRNA alignments and IsoFinder data. The third difference was in the use of RefSeq genes. RefSeq genes were added to the KnownGene track and Ensembl track for use in JIGSAW-EGASP, but excluded from JIGSAW-All. Since RefSeq genes were used for training, they were never used as a separate track of evidence. The difference in input between the two versions was based on changes to the evaluation procedures, pre- and post-EGASP. JIGSAW output submitted to the EGASP workshop was generated without access to GENCODE annotations for 31 of the 44 ENCODE regions and the choice of evidence was based on evaluating performance on a smaller sampling of distinct evidence combinations tested on RefSeq genes and the 13 ENCODE training regions. JIGSAW-All reflects the assessment of JIGSAW accuracy after running additional comparisons of different evidence combinations, evaluating performance on the 31 ENCODE regions using GENCODE annotations.

Our post EGASP-submission JIGSAW performance (JIGSAW-All in Table 2) indicates a modest improvement in gene specificity, but when including input from the KnownGene track, results from different combinations of input show only minor differences in performance. While the addition of several tracks of evidence do not significantly boost performance, it is worth noting that accuracy remains unchanged; thus it appears that we are better off adding more tracks of evidence to JIGSAW, rather than less.

In addition to providing accurate gene structure predictions, an important element of the gene finding problem is detecting more of the 'hard to find' exons. JIGSAW-All-EST (Table 2) identifies 50 exons not identified by KnownGene or Ensembl, which demonstrates the potential benefit of JIGSAW when curated gene information is unavailable. The JIGSAW-non-Human-EST+ version (Table 1) identifies a similar number of novel exons (55), while the final EGASP-submitted version predicts a slightly smaller number (45). Since the EGASP version uses the relatively accurate tracks KnownGene and Ensembl, JIGSAW weighs these evidence sources more heavily, making it less likely that JIGSAW will make predictions without support from these evidence sources. The number of 'novel' identified exons is higher in JIGSAW versions that do not use the curated data as input, which lends support to the idea that JIGSAW-All-EST and JIGSAW-non-Human-EST+ will be useful in identifying novel exons.

Effects of training set size
Results of the training-set-size experiments are shown in Figure 1, which depicts whole-exon accuracy (F score × 100) as a function of the number of training genes (in thousands).

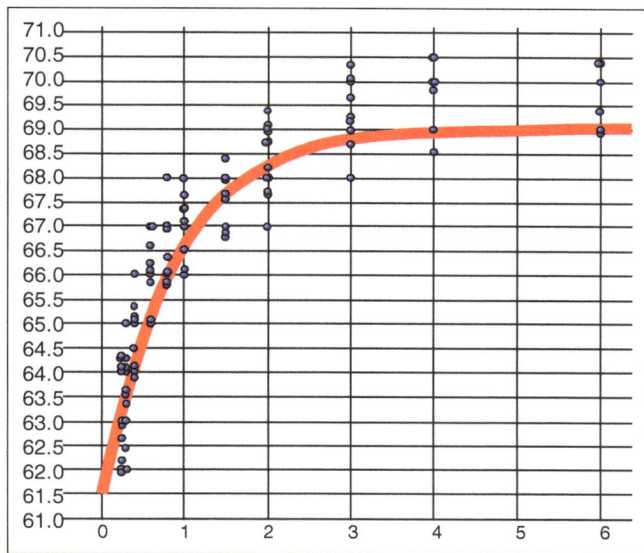

Figure 1
Accuracy as a function of training set size. Percentage of correct exons (*F* score) is shown on the *y*-axis and training set size in thousands is shown on the *x*-axis. Data points (N = 121) are shown in blue; the best fit function of the form $y = a/(1+be^{-cx + d})$ is shown in red; a = 69.01, b = 0.0152, c = 0.0012, d = 2.09. The curve is effectively flat for values of x above 6,000 (not shown). The curve for nucleotide and gene level accuracies and for the second test set are of very similar shape.
$F = 2 \times Sn \times Sp/(Sn + Sp)$.

Table 3

Results of incorporating additional features into GeneZilla evaluated on RefSeq genes

	Nuc	Exon	Gene
polyA	-1%	+1%	+2%
TATA+CAP	+0%	+0%	+1%
TATA (no CAP)	+0%	+0%	+0%
CAP (no TATA)	+0%	+0%	+0%
Branch point	+1%	+1%	+1%
Signal peptide	-1%	-1%	+0%
Intron phase	+1%	+1%	+0%
CpG islands	+0%	+0%	-1%
Isochore switching	+0%	+1%	+1%
UTR (trained on UTR)	-15%	-30%	-7%

Values are changes in nucleotide (Nuc), whole exon (Exon), or whole gene (Gene) accuracy, as measured by *F*-score. Individual features are described in the text.

The trend appears to be effectively flat for sample sizes above 6,000 genes (data not shown). A curve of the form $y = a/(1 + be^{-cx + d})$, fitted to the data via a least squares criterion, is shown superimposed (a = 69.01, b = 0.0152, c = 0.0012, d = 2.09). As can be seen from the figure, increases in sample size improve accuracy very rapidly for small training sets of approximately 250 genes, whereas an asymptote is rapidly approached for samples sizes >3,000 genes. Similar curves were obtained for nucleotide and whole gene level accuracy measures (not shown), supporting roughly the same conclusion regarding the asymptote.

Effects of modeling specific features

The results of the feature-state experiments are summarized in Table 3, where it can be seen that gains from the modeling of additional sequence elements were slight or nonexistent, with some of the additions actually resulting in reduced accuracy. In particular, we found that the polyadenylation signal, branch point, intron phase modeling, and isochore modeling generally improved accuracy by a very small amount, whereas the signal peptide and CpG island states slightly reduced accuracy (though possibly not statistically significantly so). Most surprising was the large decrease in accuracy at all levels, which was observed when the untranslated region (UTR) states were trained on confirmed UTR sequences from GenBank rather than being trained on pooled intergenic sequence.

Discussion

Several factors help to explain JIGSAW's overall strong performance in EGASP. Critical to JIGSAW's success was access to quality cDNA evidence made available through the UCSC genome browser. Inclusion of the KnownGene track, for example, led to a noticeable improvement in predictions at the whole gene level. Equally important was the use of a wide array of evidence sources, including multiple *ab initio* gene finders and non-human expression evidence. The use of a training procedure allowed JIGSAW to conduct its own 'genome annotation assessment project' to compute empirically the most reliable sources of gene structure evidence. Accurate individual evidence sources were identified as well as evidence combinations, where accuracy was dependant on the presence of multiple tracks of evidence. Therefore, gene calls were made in the presence of reliable human cDNAs, but also in the absence of cDNAs when alternative support for a gene was present.

While the EGASP experiment has ably demonstrated the need for further improvements to this community's suite of available computational gene prediction methods, the results of our own study suggest that greater gains in predictive accuracy may be made via advances at the level of integrative evidence-based methods, such as those employed by JIGSAW, than by efforts directed at the improved modeling of individual biological features by *ab initio* HMM-based models. Although such models are clearly necessary for the success of integrative approaches, the impact of expression and homology data on the present study strongly suggests that future efforts may be best spent in improving the fidelity of homology modeling at the higher levels of integrative gene structure modeling. In particular, our successes in utilizing human mRNAs and alignments to curated human proteins suggest that while evolutionary modeling of cross-species conservation may account for a significant portion of the 'low-hanging fruit' that can and should be incorporated into state-of-the-art gene-finding pipelines, improved methods of evaluating similarity to known proteins and mRNAs and of reliably incorporating such evidence within an integrative environment may yet offer significant gains in predictive accuracy. Our own research agenda for the near future includes the application of recent phylogenetic HMM approaches at the level of both *ab initio* and integrative gene finding and, in particular, the application of such approaches within the JIGSAW framework. The fact that JIGSAW was able to perform so well in comparison to the other comparative methods applied within EGASP is an encouraging sign for this line of research.

In contrast to the 'more information is always better' mantra suggested by the JIGSAW results, our experiences in modeling various features within the strict GHMM framework suggest that the higher-fidelity modeling of biological entities within DNA sequence, at least within the probabilistic framework of a GHMM, offers far fewer gains, especially considering the level of effort required in the form of additional software development and testing. Though the precise reasons for this remain somewhat obscure, a number of possible explanations readily present themselves, including the thorny issue of generative versus discriminative modeling for biosequence analysis, which remains somewhat under-characterized in our opinion, though some effort is now being directed at this important issue [3,7,8]. Intuitively, we find it disturbing that the explicit modeling of features of clear biological significance (for example, signal peptides and CpG islands) would seem to provide no advantage in the predictive modeling of protein-coding genes.

Although our own speculations regarding this conundrum point to a basic inadequacy in the HMM modeling formalism for the purpose of optimally parsing gene structures in DNA, work yet remains to be done in order to more rigorously characterize the various modeling paradigms and their applicability to the gene structure modeling problem. In comparing the performance of the individual *ab initio* predictors to that of our integrative program JIGSAW, it is clear that the ability to automatically annotate a single isoform of a gene is much improved from the days of running a single gene finder on a sequence, as shown by the fact that 70% of JIGSAW's predicted gene structures in the ENCODE regions exactly matched the human curation, with 93% of the total protein coding nucleotides correctly detected. We hope in the near future to improve upon these numbers through various enhancements, which we are now in the process of formulating for future investigations.

The perennial question of how much training data is necessary to achieve a certain level of accuracy with an *ab initio* gene finder has been somewhat addressed by the experiments performed within the context of our GHMM-based gene finder. While additional experiments within the contexts of other gene finders remain to be done, our present results suggest that for novel genomes and at the lower end of the sample-size domain, steep gains may be expected for small increases in sample size. The practical significance of this result resides in the way that training data for obscure genomes tends to be produced. For heavily fragmented genomes of obscure organisms, for example, training genes tend to be scarce, and the effort involved in increasing sample sizes may be very laborious. Nevertheless, our results, assuming they generalize to other eukaryotic genomes, suggest that such labor when undertaken with appropriate care may significantly impact the accuracy of the resulting gene finder, thereby justifying the greater effort in developing such training sets.

It is important to note that while both of our GHMM-based gene finders have seen extensive use for genome annotation efforts at TIGR over the past several years, and despite the near equivalence of their state topologies and decoding algorithms, we have often observed that the two programs can produce significantly different accuracy results, with sometimes one or the other program performing better, and no clear trend indicating any overall advantage of either program across all genomes. In contrast, we have often observed that the largest improvements in predictive accuracy have come about through improvements to our training practices [3], as opposed to improvements in the actual GHMM software. The latter observations, which have been further bolstered by our experiences with EGASP, support the notion that gross comparison of predictive accuracy between different software systems may be of limited scientific value in assessing modeling and algorithmic options for gene prediction, and points instead to the need for controlled experiments within the context of an individual software code base, or, more ideally, replicated across several independent software implementations. We believe that the more widespread adoption of such practices could greatly improve computational gene modeling as a rigorous science.

In conclusion, we believe that the more effective integration of multiple forms of evidence (for example, DNA, RNA, and protein), as opposed to higher-fidelity *ab initio* modeling of DNA alone, offers the greatest potential gains for further improvements in human gene prediction. With this in mind, we would suggest that data from other types of experiments, such as protein mass spectrometry, might offer further gains. We have now reached the point where our pipeline predicts roughly three quarters of the genes exactly, missing only 3% of the genes completely. This suggests that further efforts in human gene finding might be more productively applied to refining existing gene annotations than to generating new ones. It is important for the human genome

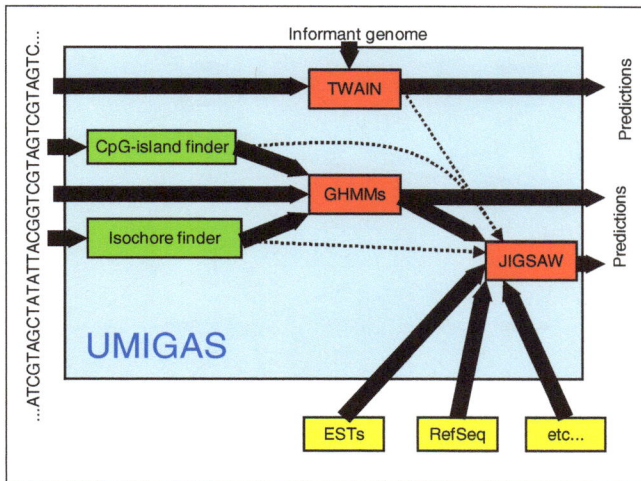

Figure 2
The computational gene finding pipeline UMIAGS (University of Maryland Integrative Analysis of Gene Structure). The raw genomic sequence is shown as an input at left; gene structure predictions are emitted at right. Additional evidence tracks for the combiner program JIGSAW are shown entering from the bottom. See text for details. GHMM, generalized hidden Markov model.

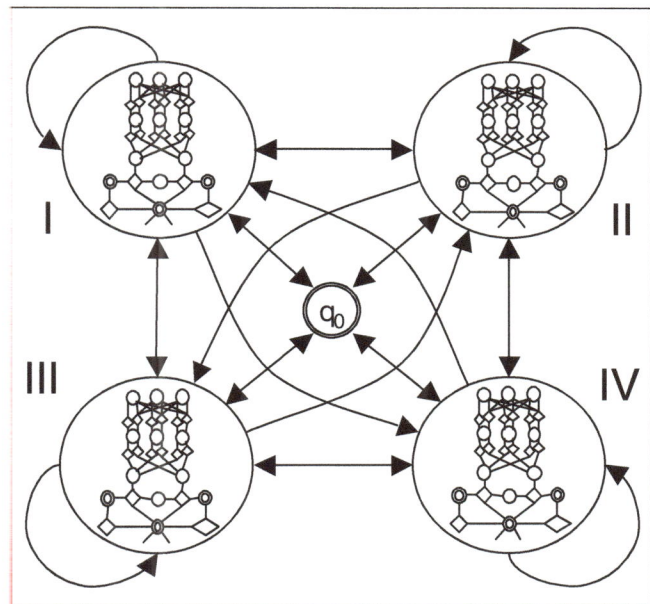

Figure 3
HMM for predicting isochore boundaries. States are shown as large circles, with transitions indicated by directed arrows. Transition probabilities are omitted for clarity. Within each outer state is a GHMM profile. States represent isochores, or discrete ranges of G+C density: I = (0,43%), II = (43-51%), III = (51-57%), and IV = (57-100%).

community to recognize that uncurated *de novo* gene predictions can be highly inaccurate, and this has implications for expression studies and other experiments based on genome annotation. We would also point out that sequencing centers have now completed draft genomes for hundreds of additional species, with many more to come. The data presented in this study makes it clear that in order to predict genes accurately in the countless genomes yet to come, we need both automated gene finders and a steady source of independent evidence such as mRNAs from those species.

Materials and methods
Prediction pipeline
Our prediction pipeline consists of a number of comparative and non-comparative gene finders, as well as several sequence analysis tools, which provide inputs to the other components of the system. The major components are described separately below; here we give a brief overview. The system, tentatively called UMIAGS (University of Maryland Integrative Analysis of Gene Structure) is shown schematically in Figure 2. The gene finders currently in our pipeline are: JIGSAW, GlimmerHMM, GeneZilla, and TWAIN. Because our human gene-finding efforts began only several months ago, not all of these components could be adapted in time for inclusion in the EGASP competition. In particular, our generalized pair hidden Markov model (GPHMM) TWAIN was not included, and is not described further herein, though we hope to adapt it for mammalian gene finding in the near future. The GHMM programs

GlimmerHMM and GeneZilla are described in more detail below, as is the integrative 'combiner' program JIGSAW.

The other two components of our pipeline are the isochore boundary predictor IsoScan and the CpG island predictor Gilligan, which we describe next.

IsoScan
To more accurately model the dependence of GHMM parameter profiles on the local G+C density of a sequence, we constructed a HMM to predict the likely boundaries of isochores. These predictions were then made available to the GHMM gene finders, enabling them to switch parameter profiles during Viterbi decoding at the precise positions of predicted isochore boundaries, without the need for segmenting the input sequence prior to gene finding. The structure of our isochore predictor, called IsoScan, is shown in Figure 3. The states of the HMM, labeled I to IV (not including q_0, which is the silent start/stop state) represent discrete ranges of G+C density: I = (0-43%), II = (43-51%), III = (51-57%), and IV = (57-100%). For the purpose of our GENCODE submissions, we estimated the HMM parameters from the predictions of the IsoFinder program [9] on human chromosome 1. Because the latter program can predict many more than four types of isochores, we coalesced IsoFinder predictions according to the four G+C density ranges given above, and then estimated the emission and transition

probabilities for our IsoScan HMM using maximum likelihood estimates from this data.

Prediction of isochores in IsoScan is accomplished via Viterbi decoding [10]. A post-processing phase allows us to impose a minimum isochore size by identifying predicted isochores smaller than the minimum allowable size and progressively combining them with their neighbors until all remaining isochore segments satisfy size constraints (and such that no two isochores of the same class are adjacent).

Gilligan

Prediction of CpG islands was performed using a modified version of the algorithm given by Larsen *et al.* [11]. Our program Gilligan predicts CpG islands using a sliding window approach. Parameters to the program include: the minimum allowable separation between islands; the size of the sliding window; the minimum allowable island size; the minimum G+C density for an island; and the minimum ratio of observed-to-expected CG dinucleotide counts in predicted islands. These parameters thus impose a set of constraints on predicted CpG islands, which are enforced via an iterative merging process in which islands violating one or more of these constraints are merged with their largest neighbor, until no further merging is required.

Gene finders

GlimmerHMM

The first of our two GHMM-based gene finders is GlimmerHMM, which is depicted in Figure 4. The underlying model is very similar to that of GENSCAN, and features different states for the different forms of exons (initial, internal, final, and single), as well as introns and internal exons of different phases. The signal sensors (that is, fixed-length states such as splice sites and start/stop codons) are implemented using Nth-order weight array matrices (WAM) [5], with N typically set to 2. The variable-length feature states (for example, exons, introns, intergenic regions) are implemented using Nth-order interpolated Markov models (IMM) [12] for $N = 8$. More details about the program can be found in [13,14].

Note that GlimmerHMM was run on the unmasked DNA sequence; we felt this was most appropriate, given that the predictions of the program were to be used as inputs to our integrative gene finder JIGSAW. GlimmerHMM was trained on 6,859 human RefSeq genes; only those training genes not split by an IsoFinder prediction were used. Training protocols roughly followed those used for GeneZilla (see below).

GeneZilla

Our apparatus for the feature-state and training-set-size experiments consisted of the GHMM-based *ab initio* gene finder GeneZilla, previously known as TIGRscan [13]. GeneZilla's basic model topology is similar to that of GlimmerHMM, with the addition of a TATA box state and a

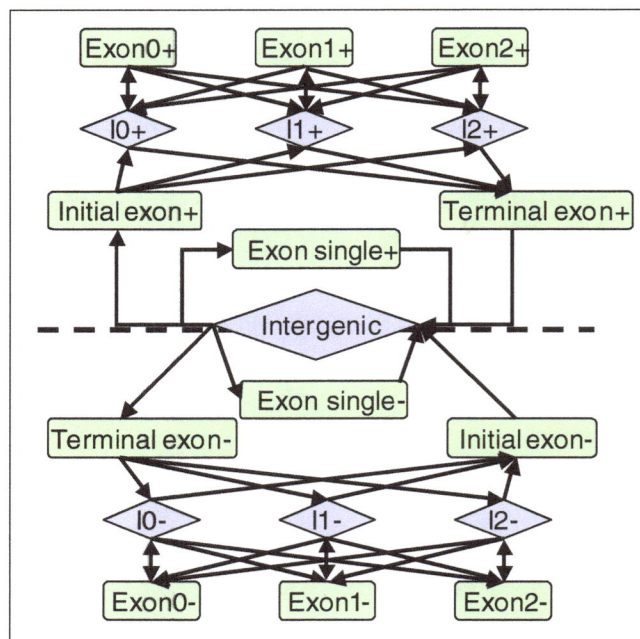

Figure 4
State-transition diagram of the GHMM for GlimmerHMM. The dashed line in the middle separates the positive strand and negative strand portions of the model. Each state in the GHMM is implemented as a separate submodel, such as a weight array matrix or an IMM (interpolated Markov models).

polyadenylation signal state, as well as the UTR states, which they delimit. Modifications were made to the structure of the GHMM to incorporate the following states, as illustrated in Figure 5 (state labels are given in parentheses): CpG islands (CpG); CAP sites (CAP); branch points (b); signal peptides (sigP); phase-specific introns (I_n).

In addition, we investigated the explicit modeling of isochore boundaries, the tying of exon state parameters, the (separate) disabling of the TATA and polyadenylation signal states, and the use of UTR-trained parameters for the UTR models (versus the use of intergenic parameters for those states).

The base gene finder (not including the above added states) was trained on 8,259 human RefSeq [15] genes rendered non-redundant via BLASTN [16], so that no two genes were more than 80% identical over 80% of the gene length at the nucleotide level. Genes known to have multiple isoforms were also removed prior to training, since GeneZilla currently predicts only one form for each putative gene. For the experiments addressing the effect of sample size, training sets of 250 to 16,000 RefSeq genes were randomly selected from our full set of 17,477 nonredundant RefSeq transcripts.

A fixed-length state ('CpG') was used to represent the 5' end of a predicted CpG island, where predictions were produced

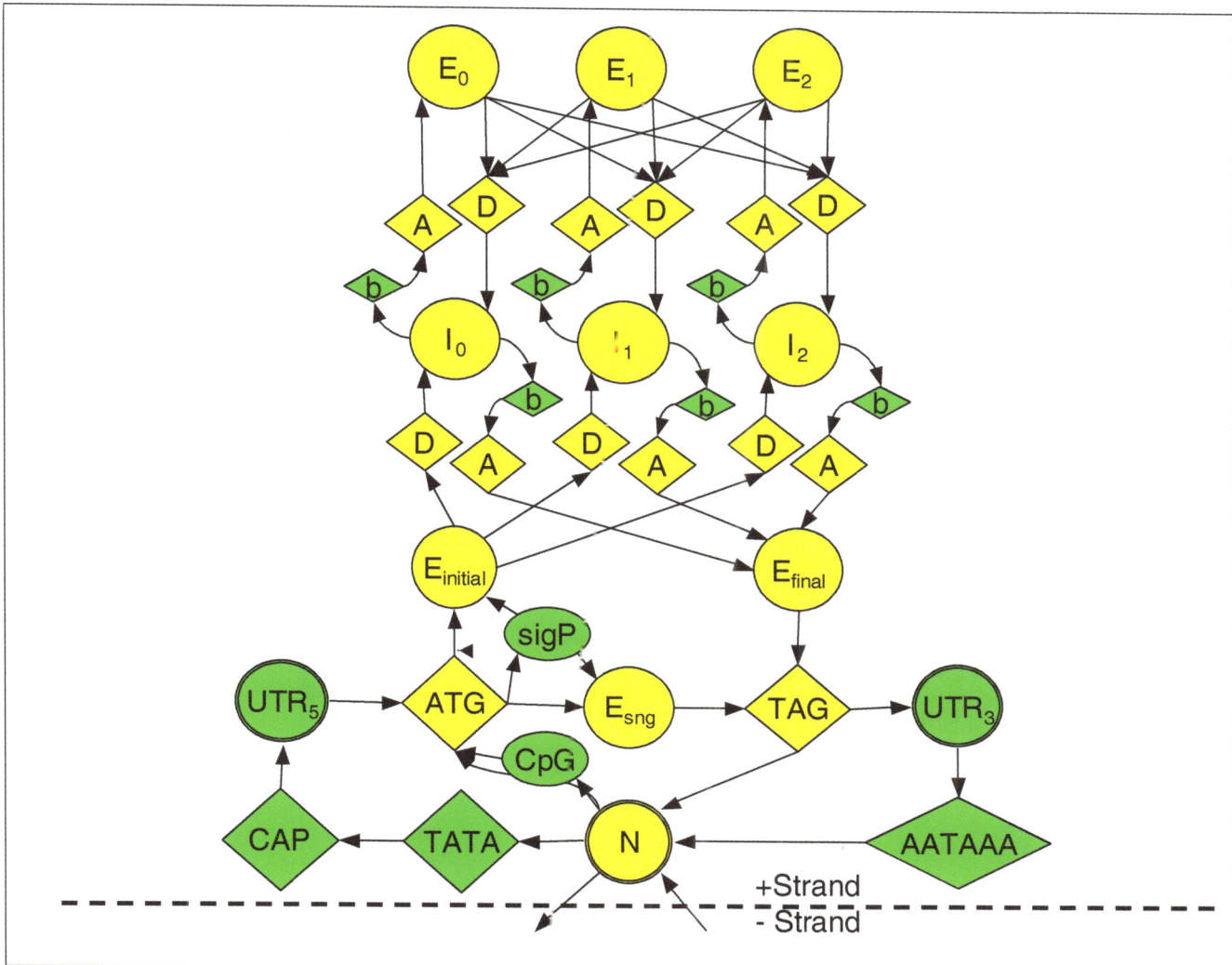

Figure 5
State-transition diagram of the GHMM-based gene finder GeneZilla. Green states were differentially included for the feature-state experiments. Reverse-strand states have been omitted for brevity. A, acceptor site; AATAAA, polyadenylation signal (including ATTAAA); ATG, start codon; b, branch point; CAP, cap site; CpG, CpG island; D, donor site; E, exon; I, intron; N, intergenic; sigP, signal peptide; TATA, TATA box; TAG, stop codon (including TAA and TGA); UTR, untranslated region.

via the program Gilligan (see above). The decision to explicitly model only the 5' end of a CpG island in the GHMM was based on our observation that predicted CpG islands often overlapped the 5' region of a coding sequence (CDS; data not shown). Because Viterbi decoding algorithms generally do not allow for the prediction of overlapping features, we instead opted to model the 5' end of each CpG island (for each strand) as an upstream element of a putative gene on the same strand.

The polyadenylation signal state ('polyA') was implemented by a 16 base-pair (bp) 2nd order WAM trained on 10,046 examples labeled as 'polyA_signal' features in human GenBank entries. (All GenBank entries were extracted in

April 2005). Two consensus sequences were allowed for this signal: AATAAA and ATTAAA. Only one isochore was modeled for this feature because the range of G+C densities for the example sequences were mostly <43%. Because the WAM was trained via simple maximum likelihood and is, therefore, not guaranteed to provide optimal discrimination power for the gene finder as a whole [3], we incorporated two additional parameters related to this state and explored a broad range of values for these parameters in an attempt to discover a maximally discriminative parameterization. The additional parameters were $R_{3'}$, a multiplicative factor that adjusts the existing $L_{3'}$ (mean 3' UTR length) parameter; and O_{poly} ('poly-A optimism'), another multiplicative factor that is applied to the (pre-logarithm) WAM score. Larger window sizes were investigated for the WAM but were found to provide no advantage over the 16 bp window, so all further experiments utilized a configuration similar to that described in [5].

The promoter state ('TATA') was implemented using a model very similar to the one used in GENSCAN, consisting of a TATA-box followed by a CAP site with a variable 14 to 20 bp 'spacer' region between. Difficulty in obtaining reliable CAP site features from GenBank compelled us to use the existing CAP model from TRANSFAC 3.2 [17], a weight matrix (WMM) trained from 303 putative CAP sites. The spacer region was modeled using simple 0th order intergenic nucleotide frequencies.

The TATA-box WMM was trained on 548 examples extracted from human 'TATA_signal' elements in GenBank. These sequences were filtered to include only those having one of the following consensuses, based on patterns observed in a previously published TATA-box model [18]: TATA, CATA, GATA, AATA, TAAA, TATT, TATG. Although a wider range of degeneracy may be present in functional TATA-box elements, the linear-time performance of the GHMM decoding algorithm requires that the number of potential matching sites be relatively small, and this is most readily accommodated by employing a limited consensus list [19]. Weight array matrices of up to 5th order were also investigated, though preliminary experiments showed no advantage to using the latter.

As with the polyA state, two additional parameters related to the promoter state were incorporated and tuned so as to maximize accuracy: $R_{5'}$, a multiplicative factor for the mean 5' UTR length; and O_{prom} ('promoter optimism'), which is multiplied by the promoter model score. Note that the tuning of these extra parameters was performed on the first of two test sets; to avoid undesirable *post hoc* effects as a result of 'peeking' at the test set, our final results were measured on a second, unseen, test set (described below).

Putative signal peptide sequences S were evaluated by the signal peptide model M_{sp} via:

$$P(S \mid M_{sp}) = \prod_{\substack{codons \\ c\,in\,S}} P(amino(c) \mid M_{sp})\, P(c \mid amino(c))$$

where $amino(c)$ is the amino acid encoded by codon c. $P(amino(c) \mid M_{sp})$ was estimated by observing frequencies of amino acids in the set of training signal peptides; $P(c \mid amino(c))$ was estimated by observing the codon usage statistics of the training genes. Training data for this state consisted of 1,048 'sig_peptide' features extracted from human GenBank entries.

The test sets for the feature-state experiments consisted of 458 and 481 individual human genes selected randomly from the set of all nonredundant RefSeq genes available at

the beginning of the study, with a margin of 1,000 bp retained before and after the CDS portion of each gene when segmenting the sequence for input to the gene finder. This was done because we wished to test the ability of the gene finder to accurately model the structure of genes, rather than to assess the false positive rate for entire genes. However, for experiments targeting the utility of the polyA, promoter, and UTR states, a margin of 50 kb was instead used, since most UTRs in the training set were seen to be shorter than 50 kb in length. Under these latter conditions the test sets each comprised 62 Mb of sequence, or roughly 2% of the genome. Likewise, for the isochore-switching experiments we selected margin sizes so that each test chunk was ≥300 kb in length, as per the commonly accepted definition of isochores [20]. Note that because these experiments were performed in part to help us prepare for the EGASP submissions, we were unable to perform the tests on the final EGASP annotations, which had not yet been released; hence, these experiments were not limited to the ENCODE regions.

The 5' and 3' UTR states were trained on 18,432 and 19,977 untranslated regions, respectively, extracted from GenBank. These states were also retrained from scratch using pooled intergenic sequences, and the differences in accuracy resulting from this change were recorded.

The remaining parameters of the GHMM were initially trained via maximum likelihood estimation from the 8,259 RefSeq training genes, and then a handful of the parameters (including transition probabilities, WMM and WAM sizes, WAM and Markov chain orders, and mean intron and intergenic lengths) were tuned by hand so as to maximize accuracy on the first of the two test sets. Results are reported only on the second, unseen test set.

Note that GeneZilla, like GlimmerHMM, was run on unmasked sequence; for this reason, direct comparisons with other GHMM-based gene finders in the EGASP exercise are not appropriate for those programs that were applied to masked sequence.

JIGSAW

JIGSAW predictions are based on the set of available gene structure evidence aligned to the genome. An overview of the method is given here to highlight key aspects of the prediction strategy; further details are described in [21]. A graphical model similar to the GeneZilla and GlimmerHMM is used to model protein coding gene structure. A state q is an element of the gene structure label set taking one of six values: single exon, internal exon, initial exon, terminal exon, intron or intergenic. Gene prediction involves parsing the sequence S into non-overlapping intervals $t = (t_0, t_1, ... , t_n)$, where each interval $t_i = (b_i, e_i, q_i)$ aligns state q_i to the subsequence $S[b_i, e_i]$ from position b_i to e_i inclusive. Input to JIGSAW is the genomic sequence S and a parameter E denoting the evidence aligned to S. An example sequence parse is shown in Figure 6.

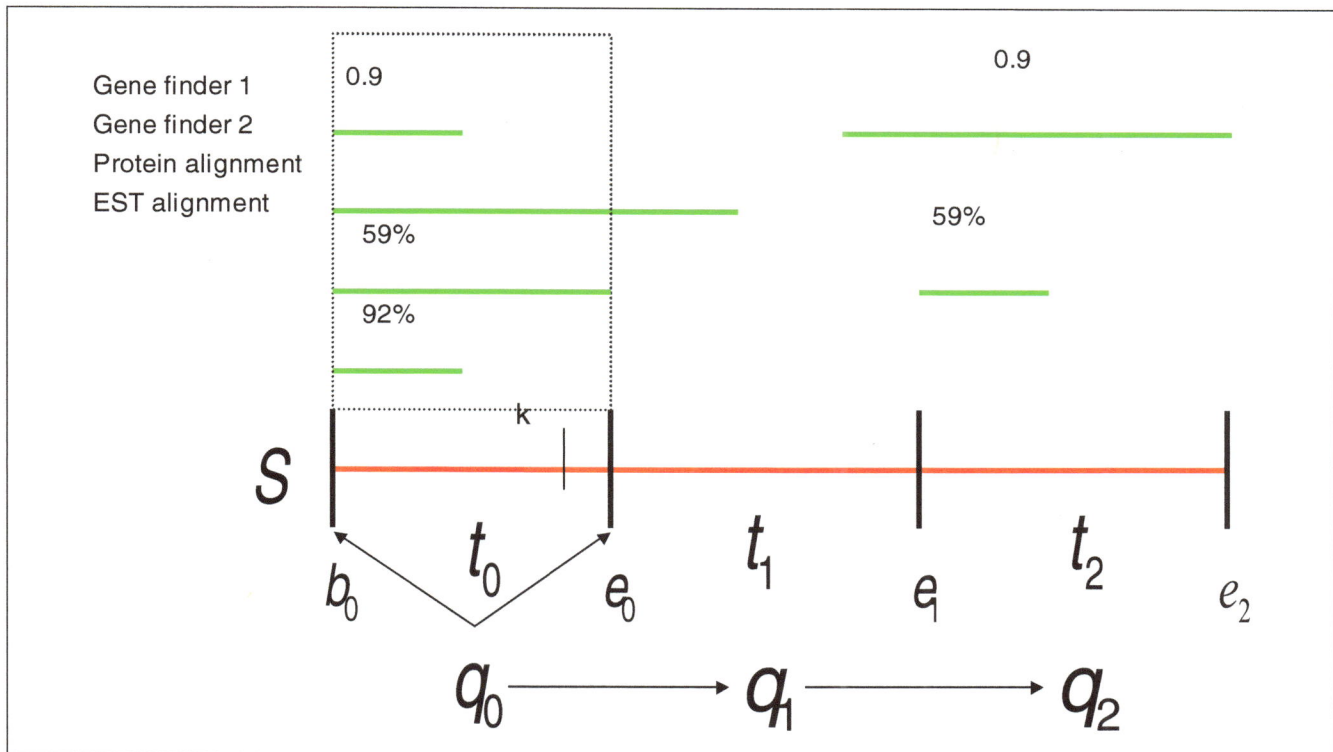

Figure 6
Parsing sequence S into three non-overlapping intervals t_0, t_1 and t_2 with the state assignments q_1, q_2 and q_3, respectively. Position k marks an index in S. The dashed box highlights the evidence overlapping the first interval from position b_0 to e_0.

The input parameter E refers to the collection of gene finder, protein, and EST evidence. A conditional probability $P(t \mid S, E)$ is computed, which assumes that the probability of aligning q_i is dependent only on the previous state q_{i-1} along with the sequence and evidence overlapping the interval from b_i to e_i. The probability of a parse is:

$$P(t \mid S, E) = P(q_0 \mid S[b_0, e_0], E) \cdot P(q_0) \prod_{i=1}^{n} P(q_i \mid S[b_i, e_i], E) \cdot P(q_i \mid q_{i-1})$$

A dynamic programming algorithm is used to find the most probable parse of the sequence. The evidence parameter E is defined by feature vectors, which record each evidence source's predictions at each nucleotide in the sequence. Six distinct feature vectors record each predicted occurrence of the following six gene features at position k in the sequence: start codon, v_{start}^k; stop codon, v_{stop}^k; donor site, v_{donor}^k; acceptor site, $v_{acceptor}^k$; coding interval, v_{coding}^k; intron interval, v_{intron}^k.

Each entry in a feature vector corresponds to a specific evidence source. Using the evidence listed in Figure 6, ordered from top to bottom, the coding feature vector at position k in this example is $v_{coding}^k = (0,1,0.59,0)$ since Gene Finder 2 and the protein alignment overlap position k. Probabilities are estimated to reflect the likelihood of each

feature type occurring in position k given the gene feature type's matching feature vector - $P(type \mid S, v_{type}^k)$. The probability of aligning state q_i to the sequence is the product of probabilities of each gene feature occurring from b_i to e_i consistent with q_i. For example, if state q_i is a single exon this means that b_i is the beginning of a start codon, e_i is the end of a stop codon, with a protein-coding interval from b_i to e_i. Therefore, the scoring function computes the probability of a start codon at b_i, the probability of a stop codon at e_i, the probability of a coding interval from b_i to e_i and the probability that no conflicting gene features occur. At each position k in the sequence the product of six probabilities for the six gene feature types (start, stop, acceptor, donor, coding, and intron) is computed,

$$\prod_{type} h(type \mid S, v_{type}^k),$$

where h is a function that returns the probability of the occurrence of $type$, if $type$ is consistent with q_i, and $1 - P(type \mid S, v_{type}^k)$ otherwise. The probability of an intergenic sequence is computed as the probability of no gene features occurring in the sequence.

Figure 7 illustrates the training procedure used to obtain probability models conditioned on the sequence and the

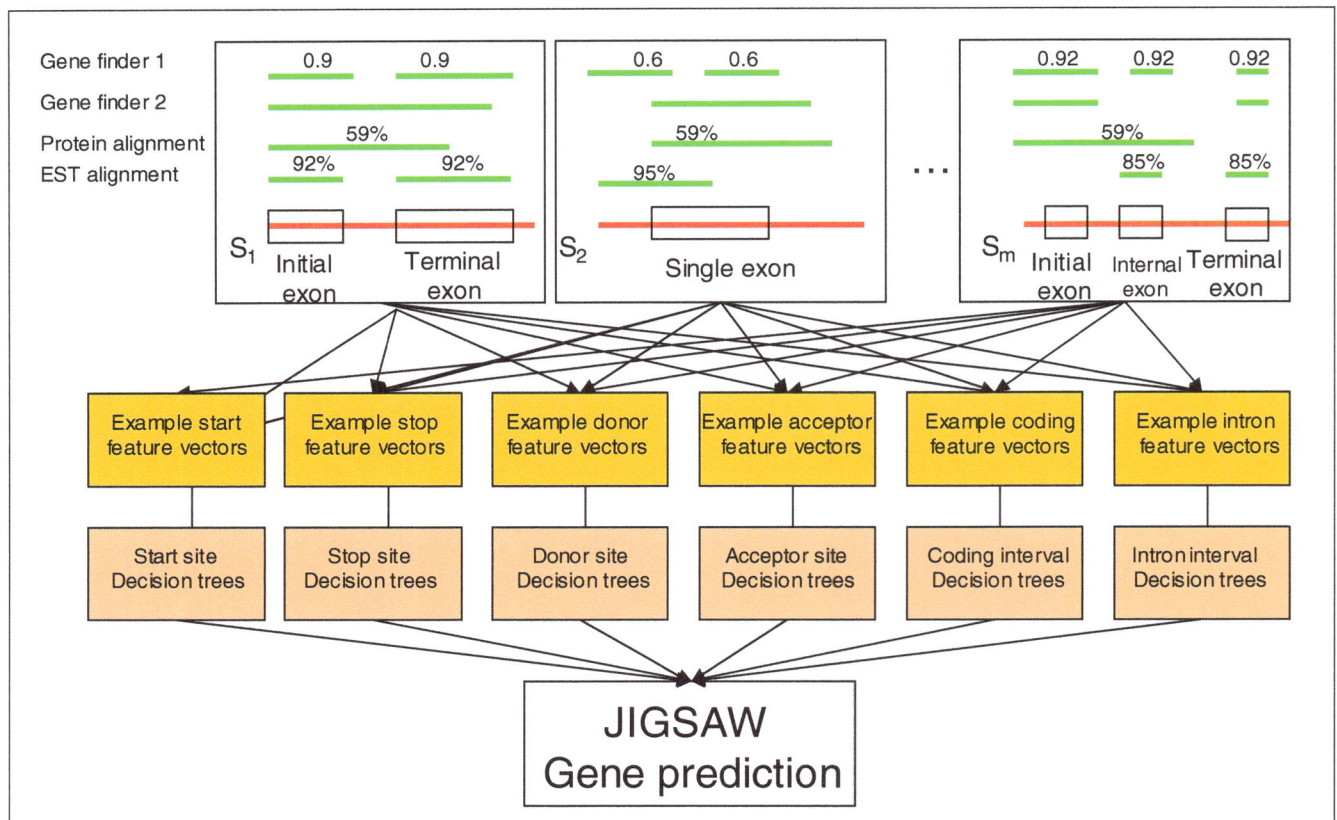

Figure 7
Training procedures for building JIGSAW prediction models. Feature vectors are collected from *m* examples and separated according to each of the six gene feature types. Decision trees are induced for each of the separated training sets, and their output is combined during the prediction procedure.

evidence. The statistics for the feature vectors observed in the training set are collected to estimate $P(type \mid S, v^k_{type})$. These statistics reflect the accuracy of each observed combination of evidence in predicting each gene feature type. Using the coding feature vector at position k from Figure 6 as an example - $v^k_{coding} = (0,1,0.59,0)$ - the training procedure checks the percentage of times the observed feature vector (0,1,0.59,0) correctly predicts a protein coding nucleotide. This percentage is taken to be the probability of coding given the observed feature vector. To handle both boolean predictions and continuous values (such as percent similarity values from alignments), a decision tree [22] is induced to group accurate and inaccurate feature vectors into distinct groups. The average probability of the feature vectors grouped together by the decision tree is taken as the final probability value.

Data preparation

To train our gene finders for the EGASP exercise, we downloaded from the NCBI the complete set of human RefSeq genes available at the beginning of our study. This comprised a total of 26,941 transcripts belonging to 22,487 genes, all having canonical start and stop codons. Because

the programs in our pipeline each predict at most one isoform per locus, we discarded any RefSeq gene having more than one isoform in the downloaded set, thereby reducing our set to 19,838 genes. We further reduced this set by eliminating overlapping genes (based on a comparison of their genome coordinates) and those found by BLASTN to be at least 80% identical over 80% of their length. The final set, which we call R_{NR}, consisted of 17,477 transcripts.

We then took a random sample of 8,308 genes from R_{NR} to use as training data for the final versions of our GHMM gene finders (but note that the sample-size experiments, described below, use larger subsets of R_{NR}). This training set we refer to as R_T. From the set of unused genes $R_U = R_{NR}-R_T$, we then took two random samples to produce test sets T_1 (458 genes) and T_2 (481 genes), with $T_1 Ç[ED]T_2 = \emptyset$, as described previously.

Evaluation of evidence tracks

To evaluate the utility of various evidence tracks in JIGSAW, we performed a series of experiments in which individual tracks were progressively added to the gene finder's set of available inputs. For each experiment, JIGSAW was retrained using 1,024 RefSeq genes excluded from GlimmerHMM and GeneZilla training. Prediction accuracy was evaluated on the 31 ENCODE regions using the GENCODE annotations with JIGSAW running on unmasked sequence. GENCODE data were used to evaluate both JIGSAW's ability to recreate the human annotation and the program's performance in the absence of reliable human cDNA evidence.

Input to JIGSAW was taken from the UCSC gene structure annotation database (build hg17; [23]) plus three auxiliary sources: IsoFinder, GeneZilla, and GlimmerHMM. Evidence used from the UCSC genome browser included: UniGene [15] and TIGR Gene Index [24] (assembled human ESTs); human mRNA, non-human RefSeq and non-human mRNA (BLAT alignments; [25]); KnownGene [23], Ensembl [26] (Curated sources); GENEID [27], SGP [28], GENSCAN [5], and TWINSCAN [29] (GeneFinders); PhastCons [1] (Cross-species conserved elements).

Evaluation of training data quantity

As a final experiment, we addressed the perennial question of how much training data would be sufficient to achieve near-optimal performance for an *ab initio* gene finder. Although we are often asked this question by prospective users of our gene finders, we know of very few studies addressing this most practical issue. Training sets of between 250 and 16,000 genes were randomly sampled from the set R_T and used to retrain GeneZilla from scratch. The gene finder was then evaluated on test set T_2 and its exon-level accuracy (that is, percentage of perfectly predicted exons) was scored using the F measure:

$$F = 2 \times Sn \times Sp/(Sn + Sp)$$

where Sn is sensitivity and Sp is specificity. A total of 121 (training and test) runs were performed with sample sizes chosen uniformly at random within the above specified range.

Acknowledgements

This work was supported by grants R01-LM06845 and R01-LM007938 to SLS from the National Institutes of Health. JIGSAW, GeneZilla, and GlimmerHMM were developed by JEA, WHM, and MP, respectively. Processing of the training data was performed by MP, WHM, and JEA. All design, development, and experimental activities were undertaken by JEA, MP, and WHM, under the direction of SLS. The manuscript was written jointly by all authors.

This article has been published as part of *Genome Biology* Volume 7, Supplement 1, 2006: EGASP '05. The full contents of the supplement are available online at http://genomebiology.com/supplements/7/S1.

References

1. Siepel A, Haussler D: **Combining phylogenetic and hidden Markov models in biosequence analysis.** In *Proceedings of the Seventh Annual International Conference on Computational Molecular Biology (RECOMB 2003).* April 10-13. Berlin Germany. 2003:277-286.
2. Pedersen JS, Hein J: **Gene finding with a hidden Markov model of gene structure and evolution.** *Bioinformatics* 2003, 19:219-227.
3. Majoros WH, Salzberg SL: **An empirical analysis of training protocols for probabilistic gene finders.** *BMC Bioinformatics* 2004, 5:206.
4. Guigo R, Flicek P, Abril JF, Reymond A, Lagarde J, Denoeud F, Antonarakis S, Ashburner M, Bajic VB, Birney E, et al.: **EGASP: The human ENCODE genome annotation assessment project.** *Genome Biology* 2006, 7(Suppl 1):S2.
5. Burge C, Karlin S: **Prediction of complete gene structures in human genomic DNA.** *J Mol Biol* 1997, 268:78-94.
6. Stanke M, Waack S: **Gene prediction with a hidden Markov model and a new intron submodel.** *Bioinformatics* 2003, 19:II215-II225.
7. Jaakkola T, Haussler D: **Exploiting generative models in discriminative classifiers.** In *Advances in Neural Information Processing Systems (NIPS'11).* Edited by Kearns M, Solla S, Cone DA. MIT press; Cambridge MA. 1998:487-493.
8. Raina R, Shen Y, Ng AY, McCallum A: **Classification with hybrid generative/discriminative models** [http://www.cs.stanford.edu/~rajatr/nips03.ps]
9. Oliver JL, Carpena P, Hackenberg M, Bernaola-Galvan P: **IsoFinder: computational prediction of isochores in genome sequences.** *Nucleic Acids Res* 2004, 32:W287-92.
10. Viterbi AJ: **Error bounds for convolutional codes and an asymptotically optimal decoding algorithm.** *IEEE Trans on Inf Proc* 1967, 13:260-269.
11. Larsen F, Gundersen G, Lopez R, Prydz H: **CpG islands as gene markers in the human genome.** *Genomics* 1992, 13:1095-1107.
12. Salzberg SL, Pertea M, Delcher AL, Gardner MJ, Tettelin H: **Interpolated Markov models for eukaryotic gene finding.** *Genomics* 1999, 59:24-31.
13. Majoros WH, Pertea M, Salzberg SL: **TIGRscan and GlimmerHMM: two open-source *ab initio* eukaryotic gene finders.** *Bioinformatics* 2004, 20:2878-2879.
14. **GlimmerHMM** [http://www.cbcb.umd.edu/software/glimmerhmm/]
15. Wheeler DL, Church DM, Federjen S, Lash AE, Madden TL, Pontius JU, Schuler GD, Schriml LM, Sequeira E, Tatusova TA, Wagner L: **Database resources of the National Center for Biotechnology.** *Nucleic Acids Res* 2003, 31:28-33.
16. Altschul SF, Gish W, Miller W, Myers EW, Lipman DJ: **Basic local alignment search tool.** *J Mol Biol* 1990, 215:403-410.
17. Wingender E, Kel AE, Kel OV, Karas H, Heinemeyer T, Dietze P, Knuppel R, Romaschenko AG, Kolchanov NA: **TRANSFAC, TRRD and COMPEL: Towards a federated database system on transcriptional regulation.** *Nucleic Acids Res* 1997, 25:265-268.
18. Lodish H, Berk A, Zipursky LS, Matsudaira P, Baltimore D, Darnell J: *Molecular Cell Biology.* 4th edition. WH Freeman; New York, NY. 2000.
19. Majoros WH, Pertea M, Delcher AL, Salzberg SL: **Efficient decoding algorithms for generalized hidden Markov model gene finders.** *BMC Bioinformatics* 2005, 6:16.
20. Bernardi G: **Isochores and the evolutionary genomics of vertebrates.** *Gene* 2000, 241:3-17.
21. Allen JE, Salzberg SL: **JIGSAW: integration of multiple sources of evidence for gene prediction.** *Bioinformatics* 2005, 21:3596-3603.
22. Murthy SK, Kasif S, Salzberg SL: **A system for induction of oblique decision trees.** *J Artif Intell Res* 1994, 2:1-32.
23. Karolchik D, Baertsch R, Diekhans M, Furey TS, Hinrichs A, Lu YT, Roskin KM, Schwartz M, Sugnet CW, Thomas DJ, et al.: **The UCSC genome browser database.** *Nucleic Acids Res* 2003, 31:51-54.
24. Lee Y, Tsai J, Sunkara S, Karamycheva S, Pertea G, Sultana R, Antonescu V, Chan A. Cheung F, Quackenbush J: **The TIGR gene indices: clustering and assembling EST and known genes and integration with eukaryotic genomes.** *Nucleic Acids Res* 2005, 33:D71-D74.
25. Kent WJ: **BLAT - the BLAST-like alignment tool.** *Genome Res* 2002, 12:656-664.
26. Curwen V, Eyras E, Andrews TD, Mongin E, Searle SM, Clamp M: **The Ensembl automatic gene annotation system.** *Genome Res* 2004, 14:942-950.
27. Guigo R, Knudsen S, Drake N, Smith T: **Prediction of gene structure.** *J Mol Biol* 1992, 226:141-157.
28. Parra G, Agarwal P, Abril JF, Wiehe T, Fickett JW, Guigo R: **Comparative gene prediction in human and mouse.** *Genome Res* 2003, 13:108-117.
29. Korf I, Flicek P, Duan D, Brent MR: **Integrating genomic homology into gene structure prediction.** *Bioinformatics* 2001, 17:S140-S148.

Research

Automatic annotation of eukaryotic genes, pseudogenes and promoters

Victor Solovyev*, Peter Kosarev[†], Igor Seledsov[†] and Denis Vorobyev[†]

Addresses: *Department of Computer Science, Royal Holloway, University of London, Egham, Surrey TW20 0EX, UK. [†]Softberry Inc., Radio Circle, Mount Kisco, NY10549, USA.

Correspondence: Victor Solovyev. Email: victor@cs.rhul.ac.uk

Published: 7 August 2006

Genome Biology 2006, 7(Suppl 1):S10

The electronic version of this article is the complete one and can be found online at http://genomebiology.com/2006/7/S1/S10

Abstract

Background: The ENCODE gene prediction workshop (EGASP) has been organized to evaluate how well state-of-the-art automatic gene finding methods are able to reproduce the manual and experimental gene annotation of the human genome. We have used Softberry gene finding software to predict genes, pseudogenes and promoters in 44 selected ENCODE sequences representing approximately 1% (30 Mb) of the human genome. Predictions of gene finding programs were evaluated in terms of their ability to reproduce the ENCODE-HAVANA annotation.

Results: The Fgenesh++ gene prediction pipeline can identify 91% of coding nucleotides with a specificity of 90%. Our automatic pseudogene finder (PSF program) found 90% of the manually annotated pseudogenes and some new ones. The Fprom promoter prediction program identifies 80% of TATA promoters sequences with one false positive prediction per 2,000 base-pairs (bp) and 50% of TATA-less promoters with one false positive prediction per 650 bp. It can be used to identify transcription start sites upstream of annotated coding parts of genes found by gene prediction software.

Conclusions: We review our software and underlying methods for identifying these three important structural and functional genome components and discuss the accuracy of predictions, recent advances and open problems in annotating genomic sequences. We have demonstrated that our methods can be effectively used for initial automatic annotation of the eukaryotic genome.

Background

The successful completion of the Human Genome Project has demonstrated that large-scale sequencing projects can generate high-quality data at a reasonable cost. In addition to the human genome, researchers have already sequenced the genomes of a number of important model organisms that are commonly used as test beds in studying human biology. These are chimpanzee, mouse, rat, two puffer fish, two fruit flies, two sea squirts, two roundworms, and baker's yeast.

Currently, sequencing centers are close to completing working drafts of the genomes of chicken, dog, honey bee, sea urchin and a set of four fungi, and variety of other genomes are currently in the sequencing pipelines [1].

Many new genomes lack such rich experimental information as the human genome and, therefore, their initial computational annotation is even more important as a starting point for further research to uncover their biology. The more comprehensive and accurate are such computational analyses, the less time-consuming and costly experimental work will have to be done to determine all functional elements in new genomes. Using computational predictions, the scientific community can get at least partial knowledge of a majority of real genes, because gene finding programs usually correctly predict most exons of each gene.

The National Human Genome Research Institute (NHGRI) has initiated the ENCODE project to discover all human genome functional elements [2]. Its pilot phase is focused on performance evaluation of different techniques of genome annotation, including computational analysis, on a specified 30 Mb of human genome sequence. The 2005 ENCODE gene prediction workshop (E-GASP '05) [3] was organized to evaluate how well automatic annotation methods are able to reproduce manual annotations.

This paper describes computational methods for identifying three important structural and functional genome components: genes, pseudogenes and promoters. We used Softberry gene finding software to predict genes, pseudogenes and promoters in 44 ENCODE sequences. We review the performance of our software and underlying methods for identifying these three important structural and functional genome components, and discuss the accuracy of predictions, recent advances and open problems in annotating genomic sequences.

Results and discussion
Running Fgenesh++ on ENCODE sequences
Two sets of ENCODE sequences were prepared to run on the gene prediction pipeline: 44 original ENCODE sequences, and 44 ENCODE sequences with repeats masked by N. Files with coordinates of repeats were downloaded from UCSC web pages devoted to ENCODE project [4]. Low complexity regions and simple repeats were not masked. All three steps of the pipeline were run to annotate ENCODE sequences.

Step 1: mapping known mRNAs and selecting good mappings
A set of known human mRNA sequences was prepared from RefSeq. Only RefSeq records with an accession prefix NM_ and a status key REVIEWED, that is, those corresponding to curated and reviewed RefSeq mRNA records, were taken into account. Known mRNAs were mapped by Est_map to 44 ENCODE sequences, and good mappings were automatically selected by the pipeline. Areas corresponding to mapped mRNAs were masked to exclude them from subsequent gene prediction steps.

Step 2: mapping known proteins by Prot_map followed by protein homology-based gene prediction by Fgenesh+
In this step, genes are predicted based on homology to known proteins - as a rule, it improves quality of predicted gene models. The NR (non-redundant) database of protein

sequences was used as a source of known proteins. First, gene models were predicted using a combination of Prot_map and Fgenesh+: Prot_map maps the NR database to genomic sequences, and Fgenesh+ predicts more refined gene models in regions corresponding to mapped proteins. Then, predicted gene models were additionally filtered by a script that analyses blast2 alignment between predicted proteins and protein homologs. Only reliable models that have a blast score >100 and coverage >80% for both proteins and homologs were selected.

Step 3: ab initio gene prediction
In this step, special scripts prepared sequence fragments that contained no gene models from steps 1 and 2. Then gene models in these sequence fragments were predicted *ab initio* by Fgenesh. Finally, gene predictions were converted from the Fgenesh-like output format into GTF format, which is required for submission of results to E-GASP '05.

Results of Fgenesh++ application to ENCODE sequences
While doing calculations for EGASP, we annotated ENCODE regions of the hg16 version (NCBI build 34). HAVANA annotation, against which results were compared by EGASP, was done on the hg17 version. Four ENCODE sequences were changed upon transition from hg16 to hg17: ENm006, ENm014, ENr131, ENr211. We re-annotated these four sequences (after the EGASP deadline), and the results presented here include this correction.

When calculating the prediction accuracy, only coding sequence (CDS) blocks, from Softberry predictions as well as from the HAVANA annotation, were taken into account. We used the HAVANA annotation file '44regions_coding.gff', the version of 7 June 2005, which describes 1,078 transcripts with CDS containing 673,501 nucleotides (the HAVANA annotation was taken into account only within the range of ENCODE sequences). The accuracy results are presented in Table 1. At the nucleotide level we estimated sensitivity (Sn) as the percentage of true coding bases that were correctly predicted as coding, and specificity (Sp) as the percentage of bases predicted to be in coding regions that were actually coding. We observed Sn = 0.9 and Sp = 0.8 at the nucleotide level. To measure accuracy at the CDS level, a non-redundant set of CDS was considered. Sensitivity (Sn) at the CDS level is the number of CDS predicted correctly divided by the number of known CDS, and specificity (Sp) is the number of predicted CDS that are correct divided by the number of all predicted CDS. When calculating the accuracies, CDS orientation is checked for known and predicted CDS. We observed Sn = 0.78 and Sp = 0.74 at the CDS level. More than 50% of predicted coding bases were predicted with the help of homologous proteins from the NR database, and approximately 35% were predicted with the help of mRNAs from RefSeq.

Performance at the level of exact prediction of all CDS in a gene is presented in Table 2. We can see that all CDSs were

Table 1

Accuracy of coding exon prediction by Fgenesh++ pipeline

	All genes, Sn/Sp (%)	mRNA supported, Sp (%)	Protein supported, Sp (%)	Ab initio, Sp (%)
Nucleotide level	93.00/79.54	94.19	86.64	13.98
CDS EXACT	78.42/74.18	90.20	84.28	7.85
CDS 1EDGE	91.55/78.51	95.00	88.77	10.47
CDS OVERLAP	92.31/78.85	95.53	89.02	10.74

CDS is considered to be predicted correctly if: both CDS coordinates are predicted correctly (CDS EXACT); at least one CDS edge is predicted correctly (CDS 1EDGE);or predicted CDS overlaps with known CDS (CDS OVERLAP).

Table 2

Exact prediction of all CDS in a gene

	All genes	mRNA supported	Protein supported	Ab initio
No. of nucleotides predicted (%)	787,505 (100%)	274,889 (34.91%)	417,202 (52.98%)	104,003 (13.21%)
No. of transcripts predicted (%)	820 (100%)	314 (38.29%)	298 (36.34%)	208 (25.37%)
No. of SoftBerry transcripts identical to HAVANA transcripts (%)*	346 of 820 (42%)	191 of 314 (61%)	154 of 298 (52%)	1 of 208 (0.48%)

*Transcripts identical means that their CDS parts (including protein coding exons and coding parts of 5' and 3' exons) are identical; the percentage of SoftBerry transcripts is relative to the number of Softberry transcripts predicted in the corresponding category: mRNA supported, protein supported or *ab initio*.

predicted exactly for 61% and 52% of genes computed with mRNA and protein support, respectively. It is interesting to note that the Sp for *ab initio* predictions, which comprise approximately 13% of all predicted nucleotides, is very low. If we exclude *ab initio* predictions and calculate an accuracy only for mRNA and protein supported predictions, the specificity rises up to 89.5% at the nucleotide level with just a slight decrease in sensitivity (Table 3). On the other hand, if we run just *ab initio* predictions for 44 ENCODE sequences, we have Sn = 0.88 and Sp = 0.74 at the nucleotide and CDS levels, respectively (Table 3). That is significantly higher than the values for *ab initio* predictions in Table 1. It might indicate that regions having neither known mRNAs nor homology to known proteins can contain genes that are missed in the HAVANA annotation. Another interesting observation is that *ab initio* gene finding demonstrates a good performance at the nucleotide level (Sn = 0.88, Sp = 0.74), while it is relatively weak at the level of exact CDS prediction, compared to mRNA- or protein-supported predictions. *Ab initio* predictions seem to usually contain one or several errors in a set of gene CDSs, as well as tend to split one gene into two or merge neighbor genes more often.

We did not use expressed sequence tag (EST) information [5] in the generation of our predictions that resulted in the smaller number of predicted alternative transcripts compared with the HAVANA annotation. EST data also can be used for extension of terminal coding exons to their 5' or

Table 3

Performance data for annotating 44 ENCODE sequences by either mRNA and protein supported or *ab initio* predictions

	mRNA + protein supported, Sn/Sp (%)	Ab initio, Sn/Sp (%)
Nucleotide level	91.14/89.54	88.44/74.46
CDS EXACT	77.19/86.48	67.54/64.22
CDS OVERLAP	90.60/91.4	85.00/71.71
SoftBerry transcripts identical to HAVANA transcripts*	56.37% (of 612)	14.75% (of 590)

*Transcripts identical means that their CDS parts (including protein coding exons and coding parts of 5' and 3' exons) are identical; the percentage of SoftBerry transcripts is relative to the number of Softberry transcripts predicted in the corresponding category: mRNA supported, protein supported or *ab initio*.

3' non-coding parts. Including EST data as well as inter-genome similarity data can further improve the annotation quality of our gene prediction pipeline.

Prediction of pseudogenes

We used Softberry gene PSF (pseudogene finding) to identify pseudogenes in 44 ENCODE sequences. This program, described in Materials and methods, recognizes pseudogene sequences using some characteristics of genome alignment regions with their parent proteins. Examples of two types of

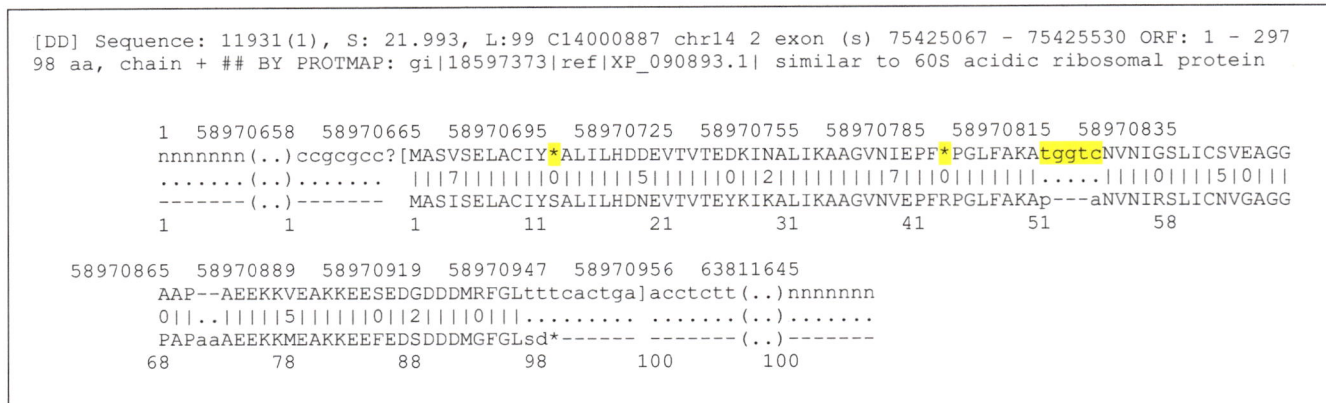

```
[DD] Sequence: 11931(1), S: 21.993, L:99 C14000887 chr14 2 exon (s) 75425067 - 75425530 ORF: 1 - 297
98 aa, chain + ## BY PROTMAP: gi|18597373|ref|XP_090893.1| similar to 60S acidic ribosomal protein


        1  58970658  58970665  58970695  58970725  58970755  58970785  58970815  58970835
        nnnnnnn(..)ccgcgcc?[MASVSELACIY*ALILHDDEVTVTEDKINALIKAAGVNIEPF*PGLFAKAtggtcNVNIGSLICSVEAGG
        .......(..).......  |||7||||||||||||||||5|||||||0||2||||||||||||7|||0|||||.....||||0||||5|0|||
        -------(..)-------  MASISELACIYSALILHDNEVTVTEYKIKALIKAAGVNVEPFRPGLFAKAp---aNVNIRSLICNVGAGG
        1          1       1          11        21        31        41        51        58

58970865  58970889  58970919  58970947  58970956  63811645
AAP--AEEKKVEAKKEESEDGDDDMRFGLtttcactga]acctctt(..)nnnnnnn
0||..||||||5|||||||0||2||||0|||.........(..).......
PAPaaAEEKKMEAKKEEFEDSDDDMGFGLsd*------ -------(..)-------
68        78        88        98        100       100
```

Figure 1

Example of a processed pseudogene. Alignment versus protein encoded by the parent gene. Identity, 83.7%; coverage of protein sequence, 93.9%; number of internal stop codons, 2; number of frameshifts, 1; K_a/K_s, 0.484.

pseudogenes, processed and non-processed, and their characteristics are presented in Figures 1 and 2.

We presented to EGASP two sets of pseudogenes found in ENCODE sequences (hg16 release). Four ENCODE sequences were changed upon transition from hg16 to hg17 (ENm006, ENm014, ENr131, ENr211) and the results presented here exclude them. One set, which we called 'reliable set', contained 56 processed pseudogenes, 93% of which almost completely overlap with 52 of 145 HAVANA pseudogenes. Overall, 80 (59%) of 135 pseudogenes from two sets overlapped 82 (57%) of 145 HAVANA pseudogenes.

We improved our PSF automatic pseudogene predictor and reran it. As a result, we found 181 potential pseudogenes, 118 of which had a significant overlap with the annotated 145 HAVANA pseudogenes. Of these 118 pseudogenes, 68 (58%) had only one exon and could be classified as processed pseudogenes: 58 had the parent gene with more than one exon and 7 others had polyA tail. Of the 118 pseudogenes, 106 (90%) had one or more defects in their open reading frames (ORFs). Among the remaining 12, there are 4 pseudogenes with a single exon (while their parents have 4 or more exons), 4 contain both polyA signal and polyA tract, 4 have only a polyA tract, and 2 have only high Ka/Ks ratios (0.59 and 1.04).

PSF did not find 27 HAVANA annotated pseudogenes. Three of them were not reported because they are located in introns of larger pseudogenes (AC006326.4-001, AC006326.2-001 and AL162151.3-001). The other 10 represent fragments of some human proteins and are missing stop codons or frameshifts. We did not include pseudogenes corresponding to fragments of proteins in our pseudogene set. The remaining 14 HAVANA pseudogenes were not found, probably because of some limitation of our

program and the processed datasets. Some of them might have parent genes that were absent from our initial protein set compiled by the Fgenesh++ gene prediction pipeline. Some of the 63 pseudogenes that have been predicted by PSF but were absent from the HAVANA set might have appeared because of imperfect predictions by the pipeline, which produced frameshifts when a pseudogene candidate and its parent gene were aligned. However, some of these 'over-predicted' pseudogenes might be actual pseudogenes missed by the HAVANA annotators (see Figure 3 for such an example).

To summarize, the PSF pseudogene prediction program found 81% of annotated pseudogenes. Its quality can further be improved by improving the quality of parent gene-protein sets.

Pol-II promoter recognition

Since each eukaryotic polymerase II promoter has a unique selection and arrangement of regulatory elements, which provide unique instructions for gene expression, the computational identification of promoters in genomic DNA is an extremely difficult problem [6]. This task is two-fold: finding the exact position of a transcription start site within a long upstream region of a typical eukaryotic gene; and avoiding false positive predictions within exon and intron sequences. To resolve the second problem, some authors of promoter finding software include special procedures for recognition of coding parts of gene blocks inside promoter prediction programs [7,8]. However, gene prediction software such as Genscan [9] or Fgenesh [6,10] provides much better accuracy in the identification of coding exons and introns than any such procedures. We think that the best promoter identification strategy is to combine prediction of all gene components in one program. While trying to create such a program, we decided to use some

```
[RD] Sequence: 35522(1), S: 50.463, L:423 C7000711 chr7 3 exon (s) 51197888 - 51195897 ORF: 1 - 1269
422 aa, chain - ## BY PROTMAP: gi|27481026|ref|XP_209794.1| similar to hypothetical protein DKFZp43


   1 63659329  63659336  63659366  63659385  63659392  63659422  63659452  63659472
   nnnnnnn(..)tacagtc?[PTSASQQILHAQcatctac(..,gtggaccPQAKLPTFQQLLHTQLPPASGLFRPatggggcSFLTTAFP
   .......(..).......  |2||||50|||||.......(..,.......||5|0|2|50|022||0||||||||.......||||||||
   -------(..)-----mg  PASASQRTLHAQlala---(..,----slrpPQSKAPAFRPLRQAQLLPASGLFRP------sSFLTTAFP
   1          1        3         13         19         23         33         43         48

 63659498  63659528  63659558  63659588  63659618  63659648  63659678  63659708  63659738
   GPVFPFRRPLRAQNLLKSASPDPLAPSGRSLRAQLFFLVGSPGPIPASQQPLWTQCLPISWRPWSAHSFLKPSSPGPGQASRWPLQDELL
   ||7|||5|||5|||||||0||0||||||||05||||2022||||0||||||||||||||||||||||||||||||||||||||||||6||
   GPIFPFQRPLQAQNLLKLASPGPLAPSGRPLQAQLFLPAASPGPTPASQQPLWTQCLPISWRPWSAHSFLKPSSPGPGQASRWPLQDQLL
   57         67         77         87         97         107        117        127        137

 63659768  63659798  63659828  63659858  63659888  63659907  63659952  63659971  63660001
   PSDGISRPQMVSGRWAPPRQGWASRRLPQAQVVLKSGSPGPA3QQ]gtaagca(..)tttgtag[APNFLQPSSEGPPPASWWPVQF*HW
   ||||7|||||||||||||02|||||||00|||||||2|||||||| .......(..)      ||||||||2||||||0|||||000|
   PSDGVSRPQMVSGRWAPPRPAWASRRLPQAQVVLKSASPGPA3QQ -------(..)------ APNFLQPSSSGPPPASRWPVQAQLW
   147        157        167        177        187        192        192        197        207

 63660031  63660061  63660089  63660119  63660147  63662724  63662731  63662748  63662766
   LENSLCRPRPCLPgGPLQAQLLPPRRPPGAKSLPASQQPgc]gtgcggc(..)tctccag[gPDSGccgactccagVPTTSLDSAPAQLP
   ||||||||||0||.||||||0||5|||||||||5||.. .......(..).......|||||.......5|00||||||||||||
   LENSLCRPRSCLP-GPLQAQLSPPQRPPGAKSLPASRQP-- -------(..)------ aPDSG----------LPIRSLDSAPAQLP
   217        227        236        246        255        255        255        260        264

 63662796  63662826  63662856  63662884  63662914  63662944  63662974  63663004  63663034
   AALVGPQLP*AKLPRPSSGLAVASPGSAPgALR*HLQAPNGL3SVGSSRPSLGLPAASAGPNRPEVSLSRLSSSLPAASAGPSRPQVGLE
   ||||||||||0||||||||||2|||||||||||0||||||||||||||||||||||||||2|||0||2|||||||0||||||||
   AALVGPQLPEAKLPRPSSGLTVASPGSAP-ALRRHLQAPNGL3SVGSSRPSLGLPAASAGPNRPEVGLSRPSSGLPAASAGLSRPQVGLE
   274        284        294        303        313        323        333        343        353

 63663064  63663094  63663124  63663154  63663184  63663214  63663244  63811645
   VGLEEQQVGLPGPSSVLSTASPGAKLPRVSLSRPSSSCLPVA3FSPAQLMALGGLRRPCF*]cttttgg(..)nnnnnnn
   |||||0||||||||||||2|||||||||||||||||||||||||2|||||||2|0||0|| .......(..).......
   VGLEELQVGLPGPSSVLSAASPGAKLPRVSLSRPSSSCLPVA3FGPAQLMALGSLPRPRF* -------(..)------
   363        373        383        393        403        413        423        424
```

Figure 2

Example of a pseudogene that has not been processed. Alignment versus protein encoded by the parent gene. Identity, 86.4%; coverage of protein sequence, 97.6%; number of internal stop codons, 3; number of frameshifts, 4; K_a/K_s, 0.594.

---

intermediate variant that includes the following steps: computation of gene annotation using a gene prediction pipeline, and promoter prediction within 5' regions upstream of the annotated coding regions of predicted genes.

We extracted 5' regions (upstream from the first CDS) from predicted genes and ran Fprom on these sequences. For each region, we selected one predicted promoter closest to the CDS and presented it in our results. There are no data on the exact location of transcription start sites for most of genes. But 5' ends of 'full length' mRNAs from Refseq could, on average, be considered pretty close to actual transcription start sites, whereas their 3' ends are often incomplete. With this in mind, we estimated the accuracy of promoter prediction on 251 genes derived from known Refseq mRNAs with >40 bp in their 5' non-coding sequence. Promoters were predicted for 90% (226) of them. Among them, there were 95 TATA+ and 131 TATA- promoters. Figure 4 shows how close predicted promoters are to starts of corresponding mRNAs.

In Figure 5, we see a sharp peak showing that a substantial fraction of predicted promoters is as close as several bases to mRNA starts. The ability to find many promoters with such precision is a remarkable characteristic of our program that distinguishes it from many promoter finding programs that usually assign promoter within a 200 to 1,000 bp range around actual Transcription Start Site (TSS). We should take into account the occurrence of multiple transcription start sites, especially in genes with TATA-less promoters; therefore, some scattering of the predictions around the annotated TSS should not be unusual. Some predictions that deviate significantly from known 5' ends of mRNAs could belong to alternative promoters, which are not unusual for human genes. The histogram (Figure 5) might serve as an approximate criterion of program quality and can be used to compare results produced by different approaches. The ideal test should be done with experimentally verified transcription start sites, but accounting for multiple TSSs will present complications even in such a setting.

```
[DD] Sequence: 622(1), S: 27.323, L:153 C6000781 chr6 6 exon (s) 840966 - 845318 ORF: 1 - 459  152 aa,
chain + ## gi|6755368|ref|NP_035426.1| ribosomal protein S18 [Mus musculus] gi|11968182|ref ## 152


1     151509    151516    151546    151576    151606    151636    151664    151694    151724
caaannn(..)tcctgct?[MSLVIPEKFQRILRILNSNINGQQKIGFAITAIKDVG*QYTHaVLRKADVDLTKWAGELTEDEMERVMTIM
.......(..).......  ||||||||||2||7||5||5|55||2||||||||0||05|2|.||||||7||||0||||||||5|||5|||
-------(..)-------  MSLVIPEKFQHILRVLNTNIDGRRKIAFAITAIKGVGRRYAHvVLRKADIDLTKRAGELTEDEVERVITIM
1       1          1         11        21        31        41        51        61        71


   151754    151784    151814    151844    151874    151904    151934    151964
QNPCQYKIPDWFLNRRKDVKDGKYSQVLASGLDKKLRADVERLKKIQAHRGPHHFWGLRVRGQHTKTTGHHGCTMGGSKKK*]gtctgca(..)aaaataa
||0|||||||||||||5|||||||||||||5||2|||0|5|||||||5|||02||||||||||||||||||22|0|5|0|||||  .......(..).......
QNPRQYKIPDWFLNRQKDVKDGKYSQVLANGLDNKLREDLERLKKIRAHRGLRHFWGLRVRGQHTKTTGRRGRTVGVSKKK*  -------(..)-------
         81        91       101       111       121       131       141       151
```

Figure 3
Pseudogene in ENm004 sequence, absent from HAVANA annotation. The alignment has a stop codon close to position 151636.

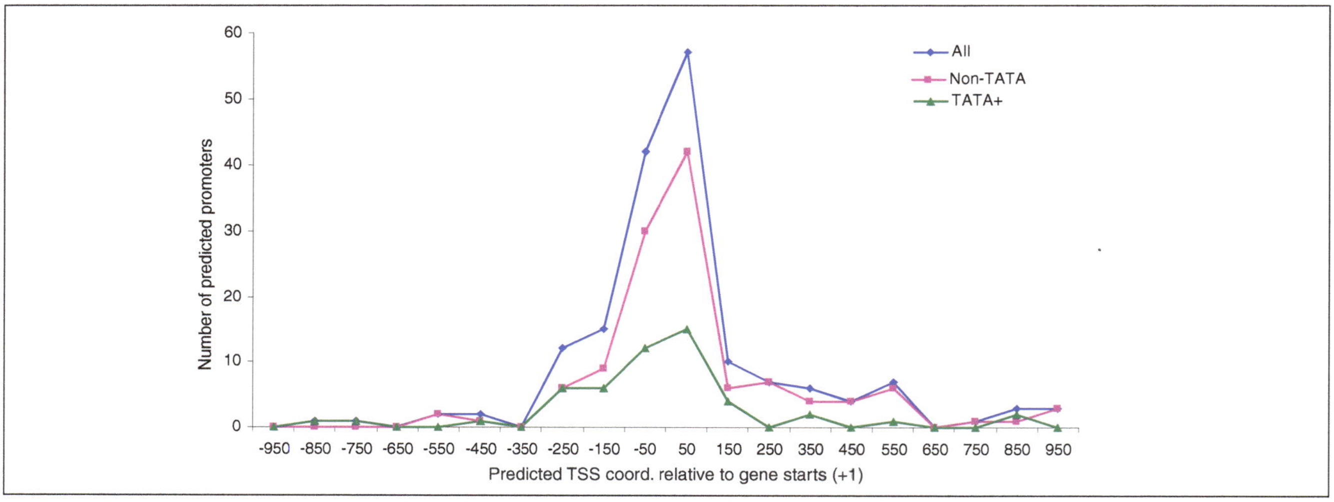

Figure 4
A distribution of predicted TSS relative to the start of mRNA sequences. Figures on the x-axis are centers of 100 bp intervals, for example, mark 50 corresponds to [+1,+100] interval.

Conclusions

In this paper we present an implementation of three computational pipelines (Fgenesh++, PSF and Fprom) for automatic identification of protein coding genes, pseudo-genes and promoters in eukaryotic genomes. These pipelines, applied to analysis of 44 selected ENCODE sequences, demonstrated an ability to reproduce, to a significant extent, the manual ENCODE-HAVANA annotation. Fgenesh++ gene prediction pipeline can identify 91% of coding nucleotides with a specificity of 90%. The automatic pseudogene finder (PSF program) found 90% of manually annotated pseudogenes and some new ones. Fprom promoter prediction program identifies 80% of TATA promoter sequences with one false positive prediction per 2,000 base pairs (bp), and 50% of TATA-less promoters with one false positive per 650 bp. It can be used to identify transcription start sites upstream of annotated coding parts of genes found by gene prediction software. Thus, the pipelines could be used for easy and fast production of reasonably accurate first pass annotation of a new genome. The described software and its components can be run on computers with Unix operation systems, as well as with Windows as part of the Molquest program package.

Materials and methods
Fgenesh++ gene identification pipeline
About 41% of sequenced human DNA consists of different kinds of repeats. Only approximately 3% of the genome sequence contains protein coding exon sequences. Gene sizes can be as large as hundreds of megabases in verte-brates, especially in primates. The average size of an exon is

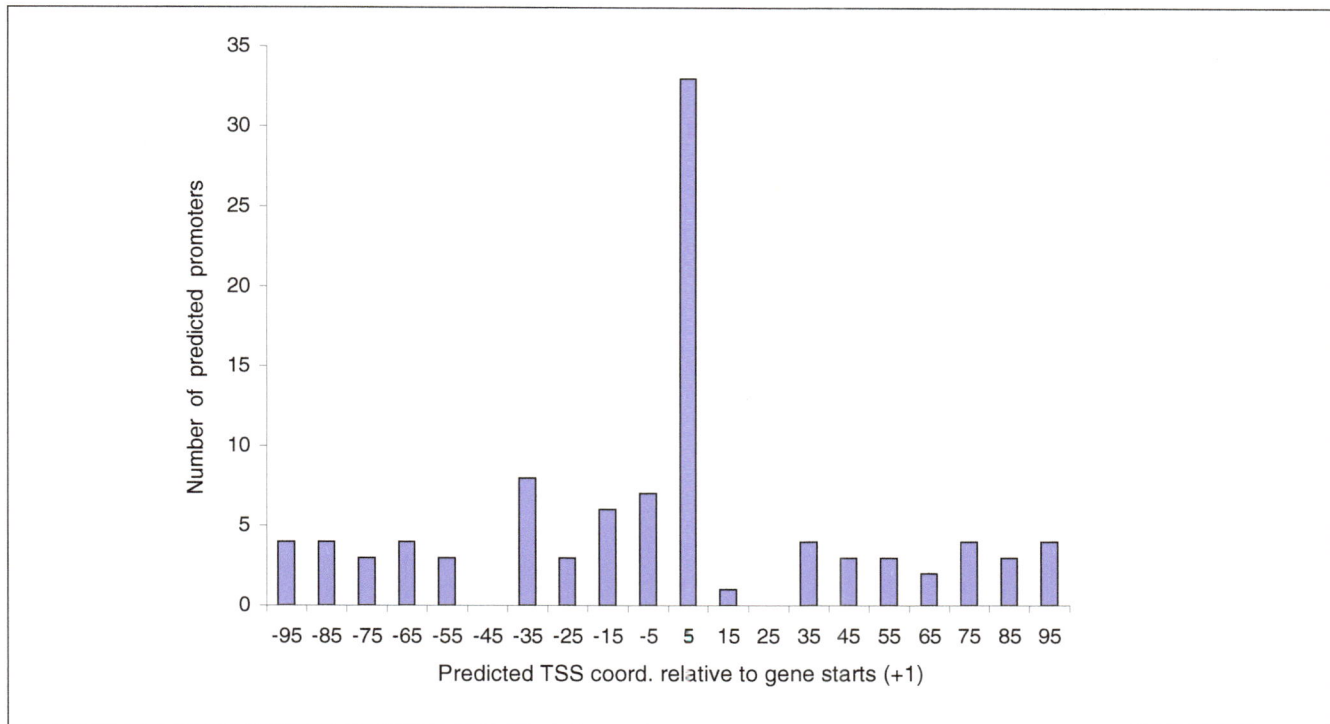

Figure 5
A distribution of predicted TSS near the start of mRNA sequences. Figures on the x-axis are centers of 10 bp intervals, for example, mark 5 corresponds to [+1,+10] interval.

about 190 bp, which is close to the DNA length associated with a nucleosome particle. Human exons are significantly smaller than genes. There are many exons as short as several bases. Moreover, the same DNA sequences may code several different proteins due to alternative promoters or terminators and alternative splicing. These processes make computational gene finding a rather nontrivial task.

Hidden Markov model based eukaryotic gene identification
Exons, introns, 5' and 3' UTRs regions are different components (states) of gene structure that occupy k non-overlapping subsequences of a sequence. There are 35 states in a eukaryotic gene model, considering direct and reverse chains as possible gene location. A gene structure can be considered as an ordered set of state/sub-sequence pairs, $\phi = \{(q_1,x_1),(q_2,x_2), \ldots ,(q_k,x_k)\}$, called a parse. A parse ϕ is considered a predicted gene structure if probability $P(X, \phi)$ of generating X according to ϕ is maximal over all possible parses, or when a score is optimal in some meaningful sense. This probability can be computed using statistical parameters describing a particular state and generated from a training set of known gene structures and sequences. Successive states of this hidden Markov model (HMM) are generated according to the Markov process with inclusion of explicit state duration density. A simple technique based on a dynamic programming method for finding an optimal

parse, or the best sequence of states, is the Viterbi algorithm, which requires $o(N^2D^2L)$ calculations, where N is the number of states, D is the longest duration and L is the sequence length [11]. A helpful technique to reduce the number of states and simplify computations by modeling non-coding state length with a geometrical distribution to predict multiple genes was initially implemented in the Genscan algorithm [9]. Several other successful HMM-based gene finding programs, such as HMMgene [12], a variant of Genie [13] and GeneMark [14], and Fgenesh [6,10] have been developed. Fgenesh (Find GENES using Hmm) is currently one of the most accurate and the fastest program. The run time of Fgenesh is practically linear, and the current version has no practical limit on length of analyzed sequence. Predicting genes in 34.5 Mb of human chromosome 22 sequence takes about 1.5 minutes with a EV6 Dec-alpha processor and is even faster on modern Linux computers.

An *ab initio* gene prediction program such as Fgenesh predicts about 93% of all coding exon bases and exactly predicts about 80% of human exons when applied to single gene sequences (Table 4). Analysis of multi-gene, long genomic sequences is a more complicated task. A program can erroneously join neighboring genes or split a gene into two or more. To improve automatic annotation accuracy, we

Table 4

Speed of gene prediction programs

	Fgenesh+	Prot_map	GeneWise
88 sequences of genes <20 kb	1 minute	1 minute	90 minutes
8 sequences of genes >400 kb	1 minute	1 minute	1,200 minutes

All values are approximate.

developed a pipeline Fgenesh++, which can take into account available supporting data such as mRNA or homologous protein sequences.

Components of the Fgenesh++ gene prediction pipeline

Fgenesh++ is a pipeline for automatic prediction of genes in eukaryotic genomes without human modification of results. It uses the following sequence analysis software.

Fgenesh

Fgenesh is a HMM-based *ab initio* gene prediction program.

Fgenesh+

Fgenesh+ is a gene prediction program that uses homologous protein sequence to improve performance.

Est_map

Est_map is a program for mapping known mRNAs/ESTs to a genome, producing genome alignment with splice site identification.

Prot_map

Prot_map is a program for mapping a protein database to genomic sequence.

Est_map

Est_map can map a set of mRNAs/ESTs to a chromosome sequence. For example, 11,000 full-length mRNA sequences from a NCBI reference set were mapped to a 52 Mb unmasked Y chromosome fragment in approximately 20 minutes. Est_map takes into account statistical features of splice sites for more accurate mapping.

Prot_map

The Prot_map program maps a set of protein sequences to a genomic sequence, producing gene structures and corresponding alignments of coding exons with similar or identical protein queries. Prot_map uses a genomic sequence and a set of protein sequences as its input data, and reconstructs gene structure based on protein identity or homology, in contrast to a set of unordered alignment fragments generated by Blast [15]. The program is very fast (Table 4), produces gene structures with similar accuracy to those of the relatively slow GeneWise program [16] and does not require knowledge of protein genomic location. The accuracy of gene reconstruction can further be significantly improved using the Fgenesh+ program on the output of Prot_map, that is, a fragment of genomic sequence and the protein sequence mapped to it.

Comparison of accuracy of gene prediction by *ab initio* Fgenesh and gene prediction with protein support by Fgenesh+ or GeneWise and Prot_map was performed on a large set of human genes with homologous proteins from mouse or *Drosophila*. We can see that Fgenesh+ shows the best performance with mouse proteins (Table 5). With *Drosophila* proteins, *ab initio* prediction by Fgenesh works better than GeneWise for all ranges of similarity, and Fgenesh+ is the best predictor if similarity is higher than 60% (Table 6).

Table 5

Accuracy of human gene prediction using similar mouse proteins

	Sn ex	Sno ex	Sp ex	Sn nuc	Sp nuc	CC	%CG
Similarity of mouse protein >90% in 921 sequences							
Fgenesh	86.2	91.7	88.6	93.9	93.4	0.9334	34
Genwise	93.9	97.6	95.9	99.0	99.6	0.9926	66
Fgenesh+	97.3	98.9	98.0	99.1	99.6	0.9936	81
Prot_map	95.9	98.3	96.9	99.1	99.5	0.9924	73
80% < similarity of mouse protein <90% in 1,441 sequences							
Fgenesh	85.8	92.1	87.7	94.0	93.4	0.9334	30
Genewise	92.6	98.0	94.1	98.9	99.5	0.9912	58
Fgenesh+	96.8	99.0	97.2	99.1	99.5	0.9929	77
Prot_map	93.9	98.5	94.1	98.9	99.3	0.9898	60

Sn_ex, sensitivity on the exon level (exact exon predictions); Sno_ex, sensitivity with exon overlap; Sp_ex, specificity on the exon level; Sn_nuc, sensitivity, nucleotides; Sp_nuc, specificity, nucleotides; CC, correlation coefficient; %CG, percent of genes predicted completely correctly (no missing and no extra exons, and all exon boundaries are predicted exactly correctly).

Table 6

Accuracy of gene prediction using similar *Drosophila* proteins

	Sn ex	Sno ex	Sp ex	Sn nuc	Sp nuc	CC	%CG
Similarity of *Drosophila* protein >80% in 66 sequences							
Fgenesh	90.5	93.8	95.1	97.9	96.9	0.950	55
Genewise	79.3	83.9	86.8	97.3	99.5	0.985	23
Fgenesh+	95.1	97.8	97.0	98.9	99.5	0.9914	70
Prot_map	86.4	95.3	88.1	97.6	99.0	0.982	41
60% < similarity of *Drosophila* protein <80% in 290 sequences							
Fgenesh	88.6	93.1	90.8	94.9	93.8	0.941	34
Genewise	76.3	91.8	82.9	92.8	99.4	0.959	7
Fgenesh+	89.2	94.4	92.7	95.5	98.5	0.968	44
Prot_map	75.1	92.5	74.9	91.4	97.5	0.941	10

Besides the programs listed above, the Fgenesh++ package also includes files with gene finding parameters for specific genomes, configuration files for programs and a number of Perl scripts. In addition, the Fgenesh++ package uses the following public software and data: BLAST executables blastall and bl2seq [15], the NCBI NR database (non-redundant protein database) formatted for BLAST, and the NCBI RefSeq database [17].

Fgenesh++ requires genome sequences and, optionally, the same sequences with repeats masked by N. Sequences can be either complete chromosomes or their fragments, such as scaffolds, contigs, and so on. When preparing repeat-masked sequences, we recommend not masking low complexity regions and simple repeats, as they can be parts of coding sequences.

Three main steps of the Fgenesh++ pipeline
There are three main steps in running the pipeline: step 1 involves mapping known mRNAs/cDNAs (for example, from RefSeq) to genomic sequences; step 2 involves the prediction of genes based on homology to known proteins (for example, from NR); and step 3 involves *ab initio* gene prediction in regions having neither mapped mRNAs nor genes predicted based on protein homology.

A user can skip some steps while running the pipeline. For example, to take a first very cursory look at gene models, a user can skip the first two steps and go right to *ab initio* gene predictions. Generally, step 1 (mapping known mRNAs) can be skipped in the following cases: if there is no representative collection of known mRNAs for a query genome, that is, RefSeq does not contain enough entries and the user does not have their own collection; and if genomic sequences are fragmented, so that individual mRNAs are likely to be broken among several genomic fragments. The output of the

pipeline consists of predicted gene structures and corresponding proteins. It also indicates whether particular gene structure was assigned based on mRNA mapping, protein homology, or *ab initio* gene prediction.

The pseudogene annotation program (PSF)
Our method of searching for pseudogenes can work with two types of initial information available. One type contains exon-intron structures of annotated genes and their protein sequences for a genome under analysis. To get such information, we can execute a gene finding pipeline, such as Fgenesh++. In this case, we run Prot_map program with a set of protein sequences to find possible significant genome-protein alignments that do not correspond to a location of a gene for mapped protein. Another type of initial data can be a set of known proteins for a given organism. Having such data, we can restore gene structure of a given protein using the Prot_map program. For each mapped protein, we can select the best scoring mapping and the computed exon-intron structure as the 'parent' gene structure of this protein. If the alignment of a protein with its own parent has obvious internal stop codons or frameshifts, this locus could be included in the list of potential pseudogenes, but we need to keep in mind more trivial explanations, such as sequencing errors. Such loci cannot be analyzed on the basis of their Ka/Ks or checked for intron losses. In any case, for each of two cases we have a set of protein sequences, their parent gene structures, and protein-genome alignments for further analysis to identify pseudogenes.

Selecting potential pseudogenes
Using genome-protein alignments generated by the Prot_map program, the PSF program produces a list of alignments possessing the following properties for each protein. First, the identity in blocks of alignment exceeds a certain value. Second, a substantial portion of protein

sequence is included in the alignment. Third, the genomic location of alignment differs from that of parent gene. And fourth, at least one of four events is observed: damage to an ORF - there is one or more frameshifts or internal stop codons; a single exon with a close poly-A site – the poly-A site is too close to a 3' end of an alignment, while the carboxyl terminus of the protein sequence is aligned to the last amino acid, and a single exon covers 95% of protein sequence; loss of introns - protein coverage by alignment is at least 95%, and the number of exons is fewer than in the parent gene by a certain number; or the protein sequence is not preserved - the ratio of non-synonymous to synonymous replacements exceeds a certain threshold (Ka/Ks > 0.5). Ka/Ks is calculated relative to the parent gene by the method presented by Nei and Gojobori [18].

Selecting a reliable part of alignment
The procedures described apply to a so-called reliable part of alignment. The necessity of introducing this concept appears due to imperfections in aligning a protein against a chromosome sequence. There are complex cases where accurate alignment cannot be produced, such as very short (1 to 3 bp) exons separated by a large intron, or because of some errors in the protein or genome draft sequence that prevent perfect alignment. For instance, if a protein as a whole is well aligned to a chromosome, but about 20 amino acids on its 5' end cannot be aligned in one continuous block, Prot_map will most likely try to align these 20 amino acids by scattering them along several short blocks. Most likely, these blocks will not have any relation to a gene or a pseudogene. Therefore, when searching for pseudogenes, we remove short insignificant trailer blocks. The rest of the alignment is considered its reliable part. To find the reliable part of an alignment, we evaluate the quality of the alignment blocks (exons). For each exon found by Prot_map, we calculate the number of aligned amino acids (M), the number of non-aligned amino acids (AI) and nucleotides (NI) within an exon, and the number of aligned amino acids (AO) and nucleotides (NO) located outside of the exon region to the left and to the right side of an exon. Also, we compute the 'correctness' of splice site conserved dinucleotides (SSC) that flank an exon. If an exon is an amino or carboxy-terminal one, we also compute the 'correctness' of corresponding start or stop codons. The length of an intron (IL) that separates an exon from its nearest exon in the direction of the longest mapped exon is also computed. The empirical 'quality' measure is defined by the following formula:

$$Q = M - P_{AI}(AI) - P_{NI}(NI) - P_{AO}(AO) - P_{NO}(NO) + B_{SSC}(SSC) - P_{IL}(IL)$$

where P_{AI}, P_{NI}, P_{AO} and P_{NO} are the penalties for the internal and external unaligned amino acids and nucleotides, B_{SSC} is a bonus for the correctness of splice sites or start/stop codons, and P_{IL} is the penalty for high intron length. The reliable part of the alignment consists of a set of neighboring alignment exons that each have Q > 5.

Table 7

Characteristics of promoter sequences used by Fprom for identification of TATA+ promoters

Characteristics	D^2 for TATA+ promoters
Hexaplets in region [-200, -45]	3.1
Hexaplets in region [1, 40]	4.0
TATA box score in region [-45, -25]	2.3
TATA box average score in region [-45, -25]	2.2
Triplets in region [-200, -45]	2.2
Triplets in region [0, 40]	2.9
Position triplet matrix in region [-50, +30]	7.0
Protein-induced deformability	2.9
CpG content	3.0
Similarity in region [-200, -100]	1.0
Motif density in region [-200, -100]	4.5
Protein-DNA-twist	0.3
Motif density in region [-100, -1] (reverse chain)	2.3
Total Mahalonobis distance	14.8
Number of promoters/non-promoters	366/18600

D^2 is the Mahalonobis distance [26] showing the strength of characteristics to separate promoter from non-promoter test set sequences.

After Prot_map mapping, many loci on a chromosome include alignments with more than one protein. In such cases, we choose only one most reliable alignment, based on a sum of included exon's qualities.

FPROM Pol-II promoter recognition program
The gene annotation pipeline was described above. Here we present our promoter recognition program Fprom (find promoter), which is based on further development of an algorithm realized earlier in the TSSW/TSSG programs [6,19]. It was assumed that TATA+ and TATA- promoters have very different sequence features, so these groups were analyzed separately. Potential TATA+ promoter sequences were selected according to the score value of a Bucher TATA box weight matrix [20], with the threshold close to the minimal score value for the TATA+ promoters in the learning set. Selected significant characteristics of the TATA+ promoter group found by discriminant analysis are presented in Table 7.

For each position on a given sequence, the Fprom program evaluates the occurrence of TSS using two linear discriminant functions (separate for TATA+ and TATA- promoters) with characteristics computed in the [-200, +50] region around a given position. If it finds a TATA-box (using a TATA-box weight matrix) in the region, then it computes the value of Linear Discriminant Function (LDF) for TATA+

Table 8

Fprom predictions on some GenBank entries with experimentally verified TSS

Gene	GenBank accession number	Length (bp)	True TSS	Predicted TSS	Number of false positives
CXCR4	AJ224869	8,747	2,632	2,632	7
HOX3D	X61755	4,968	2,280	2,223	7
DAF	M64356	2,003	733	795	2
GJB1	L47127	950	404	415	2
ID4	AF030295	1,473	1,066	1,095	2
C inhibitor	M68516	15,571	2,200	2,049	2
MBD1	AJ132338	2,951	1,964	1,891	1
Id −3	X73428	2,481	665	637	4

GenBank entries are from [27].

promoters, otherwise the value of LDF for TATA-less promoters. Only one prediction with the highest LDF score and that is greater then a certain threshold is selected within any 300 bp region.

Examples of Fprom predictions are presented in Table 8. The distances between true TSSs and correctly predicted ones varied from matching exactly to 151 bp. It should be noted that experimental mapping of TSSs has the estimated precision of ± 5 bp [20].

Testing Fprom on a control set of 366 TATA and 650 TATA-less promoter sequences demonstrated that the program identified 80% of TATA promoter sequences, with one false positive prediction per 2,000 bp, and 50% of TATA-less promoters, with one false positive prediction per 650 bp. The prediction algorithm described above uses the propensities of each Transcription Factor (TF) binding site [21] independently, not taking into account their mutual orientation and positioning. At the same time, it is well known that transcription regulation is a highly cooperative process, involving simultaneous binding of several transcription factors to their corresponding sites. In future algorithms, we should analyze patterns of regulatory sequences where mutual orientation and location of individual regulatory elements are necessary requirements for their function.

Prediction of genes, ORFs, promoters, and splice sites using the methods described above is available via the web. Fgenesh (ab initio gene finding program with parameters for 27 organisms), Fgenesh-M (program for prediction of alternative spliced gene variants), Fgenesh+ (gene prediction based on protein homology), Fgenesh_c (gene prediction with EST support), and Fgenesh2 (gene prediction with support of second, homologous genome sequence) can be found at [22]. Prot_map and Est_map (mapping protein or mRNA/EST, correspondingly, to a genome with exon-intron gene structure reconstruction) is available at [23]. Finding

promoter sequences and transcription start sites by Fprom can be executed at [24]. Pseudogene finding software (PSF) is available as a part of Windows-based Molquest package [25] that includes more than a hundred sophisticated sequence analysis programs, including several pipelines and complex visualization components for computational work with biomedical data.

Acknowledgements

We are very grateful to Dr Asaf Salamov for his active involvement in development of initial versions of the gene prediction pipeline and gene/promoter finding programs and Oleg Fokin for participation in the development of alignments procedures used by the Fgenesh+ and Prot_map programs.

This article has been published as part of Genome Biology Volume 7, Supplement 1, 2006: EGASP '05. The full contents of the supplement are available online at http://genomebiology.com/supplements/7/S1.

References

1. Collins FS, Green ED, Guttmacher AE, Guyer MS: **A vision for the future of genomic research.** Nature 2003, **422:**835-847.
2. The ENCODE Project Consortium: **The ENCODE (ENCyclopedia Of DNA Elements) Project.** Science 2004, **306:**636-639.
3. Guigo R, Reese MG: **EGASP collaboration through competition to find human genes.** Nat Methods 2005, **2:**575-577.
4. **ENCODE Project** [http://genome.ucsc.edu/ENCODE/]
5. Boguski MS, Lowe TM, Tolstoshev CM: **dbEST - database for "expressed sequence tags".** Nat Genet 1993, **4:**332-333.
6. Solovyev VV: **Finding genes by computer: probabilistic and discriminative approaches.** In Current Topics in Computational Biology. Edited by Jiang T, Smith T, Xu Y, Zhang M. Massachusetts: The MIT Press; 2002:365-401.
7. Scherf M, Klingenhoff A, Frech K, Quandt K, Schneider R, Grote K, Frisch M, Gailus-Durner V, Seidel A, Brack-Werner R, Werner T: **First Pass Annotation of promoters of human chromosome 22.** Genome Res 2001, **11:**333-340.
8. Bajic VB, Seah SH, Chong A, Zhang G, Koh JLY, Brusic V: **Dragon promoter Finder: recognition of vertebrate RNA polymerase II promoters.** Bioinformatics 2002, **18:**198-199.
9. Burge C, Karlin S: **Prediction of complete gene structures in human genomic DNA.** J Mol Biol 1997, **268:**78-94.
10. Salamov A, Solovyev V: **Ab initio gene finding in Drosophila genomic DNA.** Genome Res 2000, **10:**516-522.
11. Rabiner L, Juang B: Fundamentals of Speech Recognition. New Jersey: Prentice Hall; 1993.

12. Krogh A, Mian, IS, Haussler D: **A hidden Markov model that finds genes in E. coli DNA.** *Nucleic Acids Res* 1994, **22:**4768-4778.

13. Kulp D, Haussler D, Rees M, Eeckman F: **A generalized Hidden Markov Model for the recognition of human genes in DNA.** In *Proceedings of the Fourth International Conference on Intelligent Systems for Molecular Biology: June 12-15, 1996; St. Louis.* Edited by States D, Agarwal P, Gaasterland T, Hunter L, Smith R: Menlo Park, CA: AAAI Press; 1996:134-142.

14. Lukashin AV, Borodovsky M: **GeneMark.hmm: new solutions for gene finding.** *Nucleic Acids Res* 1998, **26:**1107-1115.

15. Altschul SF, Madden TL, Schaffer AA, Zhang J, Zhang Z, Miller W, Lipman DJ: **Gapped BLAST and PSI-BLAST: a new generation of protein database search programs.** *Nucleic Acids Res* 1997, **25:**3389-3402.

16. Birney E, Durbin R: **Using GeneWise in the Drosophila annotation experiment.** *Genome Res* 2000, **10:**547-548.

17. Pruitt KD, Tatusova T, Maglott DR: **NCBI Reference Sequence (RefSeq): a curated non-redundant sequence database of genomes, transcripts and proteins.** *Nucleic Acids Res* 2005, **33:**D501-D504.

18. Nei M, Gojobori T: **Simple methods for estimating the numbers of synonymous and non-synonymous nucleotide substitutions.** *Mol Biol Evol* 1986, **3:**418-426.

19. Solovyev VV, Salamov AA: **The Gene-Finder computer tools for analysis of human and model organisms genome sequences.** In *Proceedings of the Fifth International Conference on Intelligent Systems for Molecular Biology: June 21-25; Halkidiki, Greece.* Edited by Rawling C, Clark D, Altman R, Hunter L, Lengauer T, Wodak S: Menlo Park, CA: AAAI Press; 1997:294-302.

20. Perier CR, Praz V, Junier T, Bonnard C, Bucher P: **The eukaryotic promoter database (EPD).** *Nucleic Acids Res* 2000, **28:**302-303.

21. Ghosh D: **Status of the transcription factors database (TFD).** *Nucleic Acids Res* 1993, **21:**3117-3118.

22. **SoftBerry: Gene Finding** [http://sun1.softberry.com/berry.phtml?topic=index&group=programs&subgroup=gfind]

23. **SoftBerry: Sequences Alignment, Alignment with Genome, Alignment Genomes** [http://sun1.softberry.com/berry.phtml?topic=index&group=programs&subgroup=scanh]

24. **SoftBerry: Search for Promoters/functional Motifs** [http://sun1.softberry.com/berry.phtml?topic=index&group=programs&subgroup=promoter]

25. MolQuest – Bioinformatics Toolbox for analysis of biomedical data. [http://www.molquest.com/molquest.phtml]

26. Afifi AA, Azen SP: *Statistical Analysis. A Computer Oriented Approach.* New York: Academic Press; 1979.

27. Benson DA, Boguski MS, Lipman DJ, Ostell J, Ouellette BF, Rapp BA, Wheeler DL: **GenBank.** *Nucleic Acids Res* 1999, **27:**12-17.

AUGUSTUS at EGASP: using EST, protein and genomic alignments for improved gene prediction in the human genome

Mario Stanke, Ana Tzvetkova and Burkhard Morgenstern

Address: Institut für Mikrobiologie und Genetik, Universität Göttingen, Goldschmidtstraße, 37077 Göttingen, Germany.

Correspondence: Mario Stanke. Email: mstanke@gwdg.de

Published: 7 August 2006

Genome **Biology** 2006, **7(Suppl 1)**:S11

The electronic version of this article is the complete one and can be found online at http://genomebiology.com/2006/7/S1/S11

Abstract

Background: A large number of gene prediction programs for the human genome exist. These annotation tools use a variety of methods and data sources. In the recent ENCODE genome annotation assessment project (EGASP), some of the most commonly used and recently developed gene-prediction programs were systematically evaluated and compared on test data from the human genome. AUGUSTUS was among the tools that were tested in this project.

Results: AUGUSTUS can be used as an *ab initio* program, that is, as a program that uses only one single genomic sequence as input information. In addition, it is able to combine information from the genomic sequence under study with external hints from various sources of information. For EGASP, we used genomic sequence alignments as well as alignments to expressed sequence tags (ESTs) and protein sequences as additional sources of information. Within the category of *ab initio* programs AUGUSTUS predicted significantly more genes correctly than any other *ab initio* program. At the same time it predicted the smallest number of false positive genes and the smallest number of false positive exons among all *ab initio* programs. The accuracy of AUGUSTUS could be further improved when additional extrinsic data, such as alignments to EST, protein and/or genomic sequences, was taken into account.

Conclusions: AUGUSTUS turned out to be the most accurate *ab initio* gene finder among the tested tools. Moreover it is very flexible because it can take information from several sources simultaneously into consideration.

Background

With an increasing number of completely or partially sequenced genomes, computational prediction of protein-coding genes has become one of the most active fields of research in bioinformatics. This task is particularly challenging for eukaryotes, where protein-coding exons are usually separated by non-coding introns of varying length. Previous studies have shown that the accuracy of the currently available tools for gene finding in human is not satisfactory [1].

AUGUSTUS is a method for gene finding in eukaryotes [2]. The original version of the program used intrinsic information only, that is, information contained in the genomic sequence that is to be annotated. A recent extension of the program is also able to integrate extrinsic information from arbitrary sources for improved prediction accuracy [3].

At the ENCODE genome annotation assessment project (EGASP) workshop that took place in May 2005 in Cambridge, UK, some of the currently used methods for gene prediction were systematically evaluated and compared, including some of the most widely used gene-finding tools [4,5]. AUGUSTUS was among the methods that were evaluated at this workshop.

Results and discussion

Ab initio gene prediction is an important tool for the task of finding new genes for which sufficient evidence from transcribed sequences is not available. It is particularly important in genome projects of species where a large fraction of the genes cannot be constructed using expressed sequence tag (EST) evidence. *Ab initio* gene prediction is typically one of the first annotation steps in eukaryotic genome projects. For the test set of EGASP, the predictions of five *ab initio* single genome programs were evaluated by the organizers of the workshop [6]. Besides AUGUSTUS, these programs were GENSCAN [7], GeneID [8], GeneMark.hmm [9] and Genezilla [10]. All programs predicted only the coding parts of the genes and only one transcript per gene. On the gene and transcript level AUGUSTUS outperformed all the other programs with respect to both sensitivity and specificity. AUGUSTUS achieved a gene level sensitivity of 24.3% and a gene level specificity of 17.2%; for about one-quarter of the genes it predicted one splice variant exactly as annotated and 17.2% of the genes predicted by AUGUSTUS are correct according to the annotation. The second most sensitive program, Genezilla, had a gene level sensitivity of only 19.6% and also had the disadvantage that it predicted many more false positive genes. Only 8.8% of the genes predicted by Genezilla were correct. The second most specific program on the gene level after AUGUSTUS was GENSCAN. But even GENSCAN had a gene level specificity of only 10.1% and a gene level sensitivity of only 15.5%.

On the base and exon level the situation was less clear. Taking the mean between sensitivity and specificity, however, AUGUSTUS also had the best values on the base and exon level, very closely followed by GeneID. Apparently, GeneID is as good as AUGUSTUS at finding exons but is less successful at chaining the exons to genes: compared to AUGUSTUS, it correctly predicted less than half the number of genes. Also, the fraction of predicted genes that are correct is about half the number predicted by AUGUSTUS.

In the category of genome-genome comparisons, the predictions of eight programs were evaluated with respect to their ability to predict the coding regions of genes. The program NSCAN [11], which used mouse, rat and chicken as informant genomes, clearly performed best. AUGUSTUS performed second best with respect to the average of sensitivity and specificity at the base level and also with respect to the average of sensitivity and specificity at the exon level. The program MARS, which also uses multiple informant genomes, performed second best at the gene level. According to the average of sensitivity and specificity, AUGUSTUS was

the most accurate comparative gene prediction method at the base, exon and gene levels that is based on just one informant species (mouse in the case of AUGUSTUS).

The use of expression data as a source of information improved the accuracy of AUGUSTUS dramatically. For example, the gene level sensitivity increased to 47.6% and the gene level specificity increased to 37%. However, many programs could reconstruct the genes much better than AUGUSTUS.

What went right?

AUGUSTUS turned out to be the most accurate program among the participating programs and the University of California Santa Cruz (UCSC) hosted programs when no other data than the human genome was used or just one other informative genome was used. Furthermore, the method of incorporating hints makes it a flexible program that can use external information from various sources. The hints are collected by independent programs and stored in a standard file format. AUGUSTUS can use hints from one source alone or use hints from several sources at the same time. This is particularly important for species where one source of hints is not yet available, for example, because a closely related species is not yet sequenced or not enough ESTs are available.

What went wrong?

Compared to newly sequenced genomes, human genes are, on average, extremely well supported by experimental data. For most of the genes in the EGASP test set there were full length mRNA, ESTs or protein alignments supporting one or more splice variants. When the ESTs, mRNA and protein data are available, the task of gene prediction consists more of reconstructing the (alternative) transcripts from the available evidence than of predicting new genes. This explains why AUGUSTUS compared favorably with the other programs in the absence of extrinsic evidence but was outperformed by some of them when ESTs, mRNA and protein data were available. Our method of finding evidence using EST and protein BLAST alignments is more geared towards weak evidence, for example, evidence from other species. In the presence of a large number of human ESTs and even full length cDNAs, spliced alignment should be preferred over BLAST alignments. This is particularly so because, in contrast to BLAST, spliced alignment methods assume the presence of long gaps corresponding to introns and such methods are likely to be more precise at inferring intron boundaries. Another disadvantage of our program that makes hints from protein alignments is the fact that it treats alignments with human sequences the same as alignments with sequences from other species.

Conclusions

For genomes with extensive high quality expression data we should generate hints for AUGUSTUS using spliced alignments, such as from BLAT [12], instead of BLAST alignments. Also, we will work to improve comparative gene prediction using multiple species and more sensible methods for extracting information about the location of splice sites from multiple species genomic alignments. Furthermore, we are currently extending the model to the untranslated regions of genes.

Materials and methods
Hints to AUGUSTUS from extrinsic evidence

Evidence about the location of exons, introns and biological signals of a given input DNA sequence s can be retrieved in various ways, such as by comparing s to genomic sequences of other species or by comparing s to ESTs or proteins from a database. We refer to this as 'extrinsic' evidence as it is derived from sources other than the sequence s itself. In contrast, 'intrinsic' evidence is evidence derived from the sequence s itself, such as a long open reading frame or the occurrence of typical splice site patterns.

The model underlying the program AUGUSTUS has been extended to a model that we call AUGUSTUS+. Both the original model and the extended model are implemented in the same program AUGUSTUS. There is only one version of the program but two different models, depending on whether extrinsic evidence is given as input or not. AUGUSTUS+ incorporates certain pieces of extrinsic evidence, which we call 'hints', as input and balances it with the intrinsic evidence to produce a most likely gene structure that takes both the intrinsic and extrinsic evidence into account.

As our method of incorporating hints has been described in [3,13], we here only describe the practical effects of our model, an extension to it that allows the formulation of hints about introns and its application in EGASP 2005.

Each hint is a piece of information of one of the following types: *start*, the position of a translation start site; *stop*, the position of a translation stop; *ass*, the position of an acceptor splice site; *dss*, the position of a donor splice site; *exonpart*, the interval that is part of an exon; or *exon*, the interval that is exactly an exon.

For the first four types, the hint specifies the sequence position of the biological signal and a strand. *Exonpart* and *exon* hints specify a range of sequence positions, a strand and a reading frame. Each hint is also assigned a grade from a small discrete set of grades that may depend on the type of the hint and the sources of available extrinsic information.

The grade makes it possible to distinguish hints with different degrees of reliability. For example, both alignments with ESTs and protein sequences yield *dss* hints. However, it turns out that those *dss* hints we derive from protein alignments coincide on a training set more often with true donor splice sites. Giving all *dss* hints from proteins one grade and all *dss* hints from ESTs another grade allows us to distinguish their reliability. Another typical application would be to map a score of an alignment to a grade of the hint derived from the alignment; for example, by introducing three grades for a low, medium and a large score. The set of grades is an abstract set; grades are not numbers. The parameters measuring the reliability of the hints and its dependency on the grade are estimated on a training set with known annotation.

The model underlying the program AUGUSTUS is a so called generalized hidden Markov model (GHMM). HMMs and GHMMs for gene prediction typically define a probability for each pair (φ,s) of a sequence s and a gene structure φ. Here, the term 'gene structure' refers to a parse of the input sequence into exons, introns and intergenic regions. By contrast, the model AUGUSTUS+ defines a probability distribution on the set of all triples (φ,s,h), where h is a set of hints. Such hints can come from arbitrary sources of additional information that are available to the user, for example, alignments to expressed sequences or any kind of expert information. The distribution is such that the marginal distribution of (φ,s) is the same as in the *ab initio* AUGUSTUS model not incorporating hints. Given a sequence s and a set of hints h, AUGUSTUS searches the most likely gene structure $\hat\varphi$ given s and h, that is, a gene structure satisfying:

$$\hat\varphi = \mathrm{argmax}_\varphi p(\varphi \mid s,h)$$

As in standard HMM theory, this is equivalent to searching a gene structure that satisfies:

$$\hat\varphi = \mathrm{argmax}_\varphi p(\varphi,s,h)$$

When s and h are given, we refer to $p(\varphi,s,h)$ as the likelihood of the gene structure φ. For the decision which gene structure has the highest likelihood - and is therefore predicted - only the likelihood of the gene structures relative to each other is relevant, not their absolute value. The introducton of hints changes the relative likelihood of gene structures. We observe two effects: the 'bonus effect', where the introduction of a hint increases the likelihood of gene structures that are compatible with the hint relative to gene structures that are not compatible with the hint (we say the compatible gene structures are 'upvalued'); and the 'malus effect', where exons and signals that are not supported by hints become less likely than in the *ab initio* model. For example, suppose we have searched for extrinsic evidence about genes in a sequence region and have found no hints. Then the posterior probability of a gene in this region in the above model is smaller than its posterior probabilty in the *ab initio* model. Unsuccessful searches for hints tend to result in a prediction with fewer exons or genes. 'No information' is also infor-

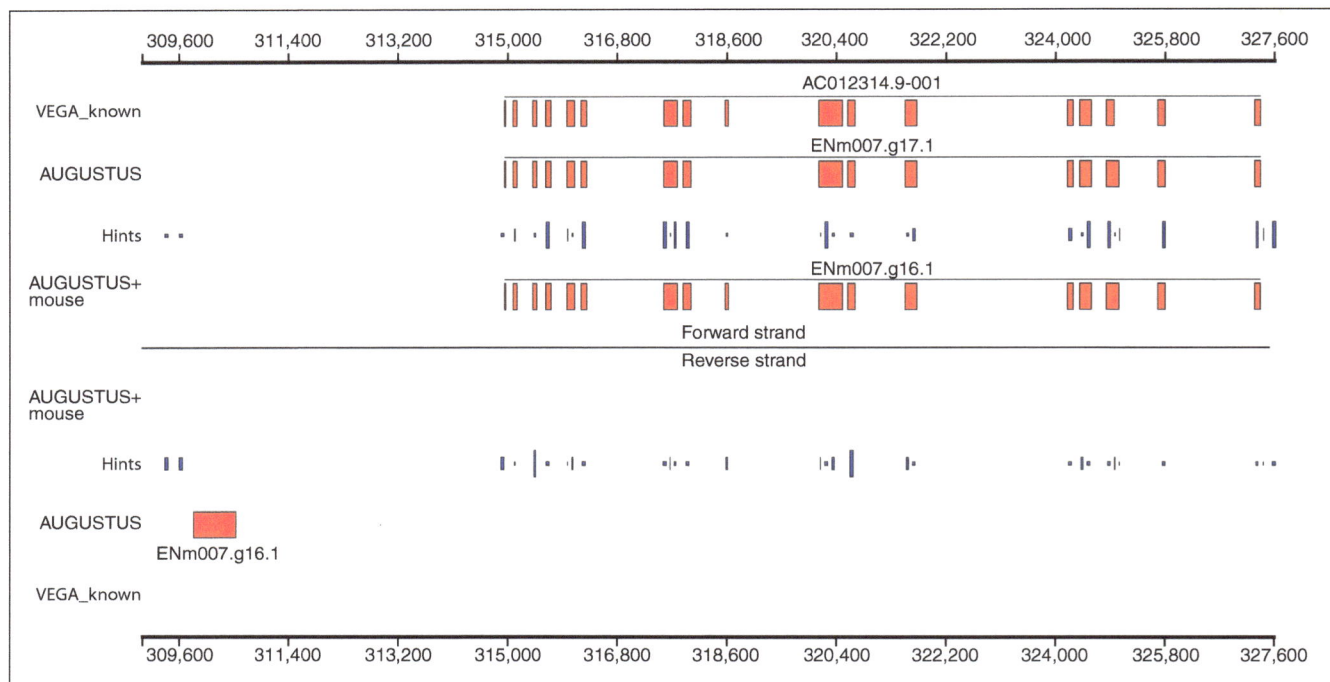

Figure 1
Annotation of the protein coding regions of a part of the human ENCODE region ENm007. The line labeled 'VEGA_Known' shows one known gene on the forward strand. The *ab initio* program AUGUSTUS (labeled 'AUGUSTUS') predicts this gene almost correctly but completely misses the 9th exon annotated around position 318,600. Furthermore, as an *ab initio* program, AUGUSTUS predicts a false positive gene on the reverse strand around position 310,000. The lines labeled 'hints' show the hints derived from a comparison to the mouse genome. The height of the rectangles depends on their estimated reliability. The hints indicate the presence of an exon where AUGUSTUS missed the annotated exon. Also, there are no hints about coding regions where AUGUSTUS predicted a gene on the reverse strand. When the given hints are used by AUGUSTUS (labeled 'AUGUSTUS+mouse'), the missed exon is correctly predicted and the false positive gene is not predicted anymore. The former is a consequence of the bonus effect and the latter a consequence of the malus effect. Note that the hint about the exon around position 318,600 was helpful, although that exon is more likely to be on the reverse strand according to the hints alone. This plot has been obtained using gff2ps [28].

mation. Both effects are illustrated using an example in Figure 1.

Hints about introns

Suppose the input sequence *s* can be well aligned to an EST or a protein sequence, in such a way that a segment $s[a,b]$ of *s* extending from position *a* to position *b* is aligned to a large gap in the other sequence. Furthermore, suppose that the splice site dinucleotide consensus occurs at the boundaries *a* and *b* of that segment. Then it is reasonable to assume that this segment is likely to be an intron. We would like to be able to formulate this as 'intron hint', which states that there is a likely intron extending exactly from *a* to *b*. The presence of such a hint should upvalue every gene structure that is compatible with the hint and has an intron going from *a* to *b*.

This problem turns out to be tricky. If HMM based gene prediction programs modeled complete introns as an emission from one state only, the time to compute the commonly used Viterbi recursion for intron states would be proportional to the maximum allowed intron length. As introns can be hundreds of kilobases long, for performance

reasons programs do not model the intron as a complete emission but model introns piecewise using states that emit just one base at a time [7,9,14,15] or a bounded number of bases [2,16]. Therefore, the Viterbi algorithm does not allow a gene structure to be upvalued based on a complete long intron.

The Viterbi algorithm does allow, for example, the probability of the gene structures to be upvalued by a constant factor for each base of an intron that overlaps the interval from *a* to *b*. However, such a positionwise bonus would give a bonus to a predicted intron that just overlaps with $[a,b]$ and has different splice sites and also would depend too strongly on the length of the intron hint [13,17].

Using hints to the splice sites is not solving the problem, either, because then gene structures that have splice sites at *a* and *b* but have an exon in this range are upvalued although they are not compatible with the hint and contradict the alignment. Of course, it is easily possible to force a program to predict an intron exactly from *a* to *b*, but this does not account for the fact that such hints can be wrong.

We present here a heuristic approach that allows hints about introns to be incorporated into a HMM. The idea is that the possibility of emitting a complete intron in one step exceptionally arises when the intron is exactly as given by a hint. This way, gene structures that exactly obey an intron hint can be upvalued arbitrarily and the overall additional computational cost is proportional only to the number of intron hints.

To illustrate the concept, just consider the forward strand of the DNA sequence and assume that the HMM has one state *acc* that models the first exonic base downstream of an acceptor splice site and one state *don* that models the first exonic base upstream of a donor splice site. Let q be a state of the GHMM and let i be a position in the sequence s. If $q = acc$ or $i = b + 1$ is not the exon base following the intron hint, then we use the normal Viterbi-recursion:

$$\gamma_{q,i} = \max_{q'} t_{q',q}\, \gamma_{q',i-1}\, p(s_i \mid q)$$

In the maximum q' ranges over all states, $t_{q',q}$ denotes the probability of the transition from state q' to state q and $p(s_i \mid q)$ the probability of the ith base of s under the model of state q. Without the adjustment below, the Viterbi-variables $\gamma_{q,i}$ have the usual meaning [18]. In those cases where $q = acc$ and where $i = b + 1$, we use a different recursion formula and take into account, as an additional alternative, that the intron is emitted in one step.

$$\gamma_{acc,b+1} = \max \begin{cases} \max\limits_{q'} \gamma_{q',b} \cdot t_{q',q} \cdot p(s_{b+1} \mid acc), \\ \gamma_{don,a-1} \cdot p(s[a,b] \mid intr.) \cdot r \cdot p(s_{b+1} \mid acc) \end{cases}$$

Here, $p(s[a,b] \mid intr.)$ is the probability of the intron sequence of the hint under the intron model and includes the probability of the length. It is identical to the probability of the intron when modeled piecewise using a sequence of emissions and transitions. r is a bonus factor that upvalues the probability of gene structures that are compatible with the intron hint; r could depend on the reliability of the hint. In particular, this method can be used to enforce some introns; however, here we chose a fixed $r = 10$. The effect of the above method is that the probability of gene structures that are completely compatible with the hint are upvalued by a factor of r relative to all other gene structures.

Predictions of AUGUSTUS on the ENCODE regions

In each of the categories of the EGASP workshop, we used the same program, AUGUSTUS, to predict the genes. The difference lies in the set of hints that are given to AUGUSTUS. In each case the hints were generated automatically using the available information in the respective category. The hints are given to AUGUSTUS in the form of a file in GFF format. In none of the categories was human intervention necessary. All predictions were made on the repeat masked sequence.

Ab initio single genome

The program AUGUSTUS has been described in [2]. We here only summarize it briefly and state what is additionally relevant when running the human version on large sequences. When run as an *ab initio* single genome gene finder, AUGUSTUS takes as input a DNA sequence s only and proceeds internally, that is, hidden from the user, as follows. It cuts s into non-overlapping pieces of length ≤ 200 kb, such that the cutting points are likely to be in the intergenic region. These cutting points are chosen using preliminary predictions of the model. For each such piece, the GC content is computed and a parameter set out of 10 possible GC content dependent sets is chosen. Then the most likely gene structure for each piece is searched using the Viterbi algorithm and the results are mapped back to the original sequence.

Currently, AUGUSTUS just predicts the coding sequence (CDS) and not the untranslated regions. In EGASP, AUGUSTUS predicted just one transcript per gene. However, after the workshop it has been extended to be able to predict multiple transcripts per gene [19]. The human version of AUGUSTUS was trained on a training set with 1,286 genes retrieved in 2002 from GenBank. This training set is available from the AUGUSTUS web server [20]. The running time for the 21.9 Mb of the EGASP test regions was 4 hours on a single processor PC of 2.4 Ghz. Everything stated above also applies to the following three sections when hints are used as additional input to AUGUSTUS; in particular, the hints do not slow down the program significantly.

Dual genome based

In this section we describe how we predicted the genes in a human input sequence s using the mouse genomic sequence as additional information. The method is based on the observation that functional regions tend to be more conserved between human and mouse than nonfunctional regions. Conversely, high conservation of a segment pair at the amino acid level is (weak) evidence that this segment is coding in both species.

The application flow of our method is as follows (see also [21]). First, we parse the precomputed UCSC BLASTZ [22] alignments of human with mouse to obtain large (up to 100 kb) alignable human/mouse sequence pairs. Second, for each such sequence pair we use CHAOS [23] to find alignment anchor points. Third, between the anchor points we use DIALIGN [24] to find fragments conserved on the peptide level. Fourth, we process the DIALIGN fragments and make a set of hints h from them. Fifth, AUGUSTUS predicts genes on s using the hints h.

The alignments we use in step 1 were downloaded from [25]. In step 2 we break the alignment problem down into alignment problems of smaller size by anchoring the alignment at

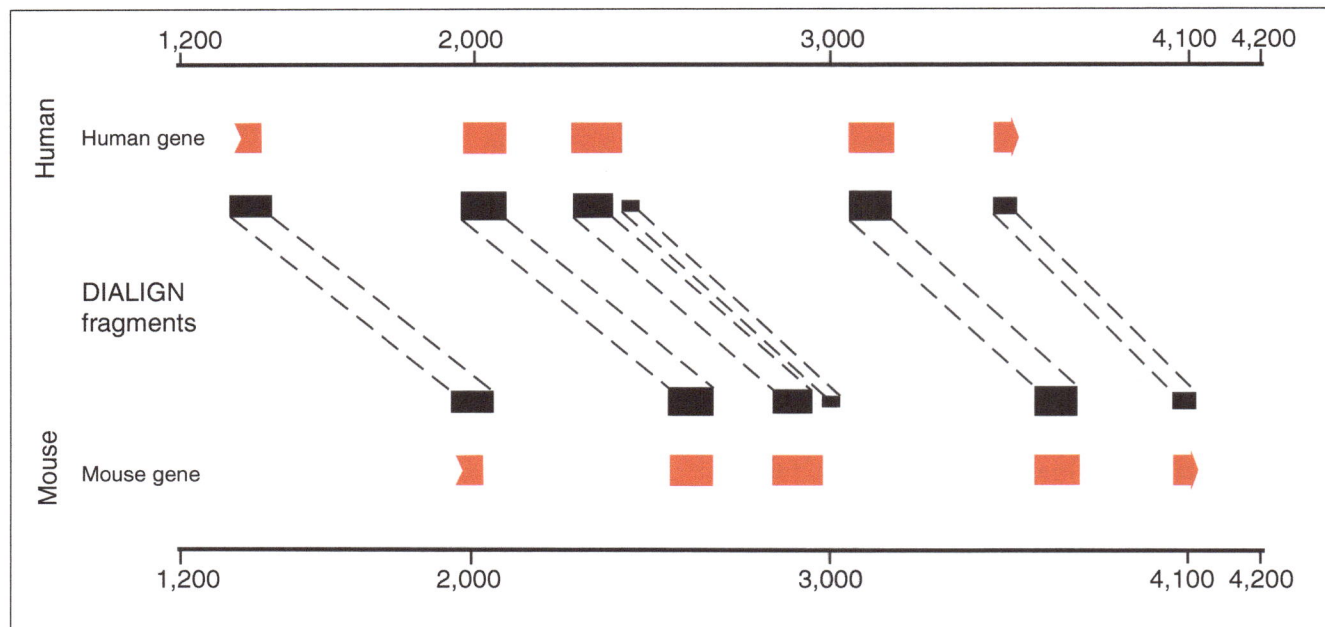

Figure 2
A syntenic human-mouse sequence pair and its DIALIGN alignment. Each sequence contains one gene with five exons (only CDS shown). The fragments are segment pairs with high similarity at the protein level.

pairs of positions that are outstandingly similar, so-called anchor points. As the running time of DIALIGN is superlinear, this reduces the running time of the subsequent runs of DIALIGN.

DIALIGN (step 3) is an alignment program that we use in this context to find sequence pairs in human and mouse that have significantly high similarity at the amino acid level and are likely to correspond to coding exons. DIALIGN uses the term 'fragment' to denote a gap free local pairwise alignment. The weight of a fragment f when aligning two sequences of lengths λ_1, λ_2 is defined as:

$$weight(f) = -\log p,$$

where p is the probability that two random sequences of lengths λ_1, λ_2 contain a fragment of the same length as f and with at least the BLOSUM62 score of f. DIALIGN then searches a chain of non-overlapping fragments, where the sum of the weights is maximal. The regions between the fragments remain unaligned. In the following step only the fragments of the optimal chain with a weight above the threshold of 20 were considered. Figure 2 shows an example of the DIALIGN alignment of a human-mouse sequence pair containing orthologous genes.

In particular, each such DIALIGN fragment defines a weight and in the human sequence an interval of sequence

positions, a strand and a reading frame. In step 4, for each DIALIGN fragment several hints of type *exonpart* are generated for the human sequence. We now describe the hints that are generated for each single fragment. The interval of each of the hints is the interval of the fragment interval minus 33 base-pairs on each side. This cutoff accounts for the fact that, typically, some part of the introns flanking an exon are also conserved (Figure 2). We generate one hint for each reading frame and strand combination, that is, six hints in total per DIALIGN fragment. It is true that a DIALIGN hint specifies the strand and the reading frame but this information sometimes is wrong, although the fragment does indeed correspond to an exon (see the example in Figure 1). The reason for this is that an exon pair with very high sequence similarity will usually have high similarity at the amino acid level in any reading frame or on either strand. Nevertheless, for those fragments for which the hints interval indeed fell completely into a coding exon, the strand specified by DIALIGN was correct 72% of the time and the reading frame specified by DIALIGN was correct 61% of the time (estimated on a subset of the ENCODE training regions). Thus, the strand and reading frame given by DIALIGN contains useful information as it is much more often correct than guessing would be. However, it is not correct often enough for AUGUSTUS to be able to rely on it.

The bonus a gene structure gets when a coding region fully contains the hint interval depends only on the DIALIGN weight, whether the strand is as given by DIALIGN and whether the reading frame is as given by DIALIGN (D). For the weight we distinguish only two cases, a weight of at least 45 or a weight below 45. The set of grades contains $2 \times 2 \times 2 = 8$ elements:

{weight $\geq$ 45, weight < 45}
$\times$ {strand as D, strand not as D}
$\times$ {frame as D, frame not as D}

The 8 parameters for the bonus as well as the one parameter for the malus have been computed using the 13 ENCODE training regions. The computation of the bonus of each grade is based on the count of the hints of that grade that is compatible with the annotated gene structure. The 8 relative bonuses range between 2.3 and 85.1, the latter for hints with weight $\geq$ 45 and both strand and score as given by Dialign. Therefore, the likelihood of a gene structure that has an exon that is supported by a DIALIGN fragment with weight greater than 45 on the same strand and with matching reading frame is upvalued by a factor of 85.1 relative to other gene structures. The computation of the malus is based on how often annotated exons are not supported by any hint. Here, the malus is 0.951, which means, in particular, that an exon of length λ that is not supported by any DIALIGN fragment is punished by the factor 0.951^{λ}. For details on the parameter estimation see [3,13].

The running time of the first four steps of the automatic pipeline described above is dominated by the running time of DIALIGN. For the 31 test ENCODE regions encompassing 21.9 Mb, these steps took about 3 hours on a single CPU. An example of the constructed hints is shown in Figure 1.

EST and protein based
To make use of evidence derivable from ESTs and protein sequences, we automatically generated hints about the gene structure in the input sequence s using an EST and a protein database. For the EST database we used *est_human* from the NCBI. For the protein database we used the NCBI *nr* database. The application flow in this category of predictions is as follows. First, we search for local alignments of s to ESTs using WUBLASTN [26] (parameters -Q 15 -R 15 -B 250 -V 250) and to protein sequences using WUBLASTX (parameters -B 250 -V 250). Second, a program called AGRIPPA [27] parses the BLAST alignments of step 1, starts new WUBLASTX searches and generates a set of hints h. Third, AUGUSTUS predicts genes on s using the hints h.

The EST alignments are used to generate hints of types *exon*, *intron*, *exonpart*, *dss* and *ass*. The protein alignments are used to generate hints of all seven types. In addition, AGRIPPA generates hints using a combined EST and protein search in the following way. After the EST database has been used to partially reconstruct the mRNA, each presumable part τ of an mRNA sequence is searched against the protein database. The idea behind this is that parts of τ that are aligned to an amino acid sequence are relatively likely to be coding. Thus, this search is a means of separating noncoding exons from coding exons. For details on the generation of hints from transcribed data see [3,13].

The reliability of the hints depends on whether they were derived from ESTs, proteins or from a combined search. For example, hints to donor splice sites were much more reliable when they came from protein alignments than when they were from EST alignments. When an identical hint is derived both by EST and by protein alignment, we keep only the hint from the more reliable source. We introduce three grades for hints according to their source of information. One grade is assigned to all hints from ESTs, one grade is assigned to all hints from proteins and one grade is assigned to hints from a combined EST-protein search. None of the hints depend on the BLAST e-value. We treated each entry in the protein database equally, no matter if the species was human or not. Also, we treat each entry in the EST database equally. We again estimated the parameters for the hints on the 13 training regions. For details we refer to [3,13].

In the above pipeline, by far the most time consuming step is the blast runs, particularly the WUBLASTX run from step 1 against the protein database, which requires many computing resources. Blasting the 31 test sequences against the *nr* database took about 50 CPU days. However, when the BLAST results have been precomputed and the BLAST output is given to AGRIPPA as input it takes time in the order of minutes to generate the hints.

EST, protein and dual genome based
The evidence about the gene structure coming from genome to genome comparisons extends and partially complements the evidence from the similarity to transcribed sequences. In order to incorporate both kinds of information we simply take as a set of hints h the union of the two sets of hints described above, that is, we concatenate the GFF file containing the hints from ESTs and proteins and the GFF file containing the hints from comparisons to the mouse genome. The set of possible grades for each type of hint also encompasses the union of the two sets of possible grades of the above two categories. For example, for hints of type *exonpart* there are now $8 + 1 + 1 + 1 = 11$ grades possible in order to distinguish between the reliabilities of the 8 grades of hints from DIALIGN and of hints from protein alignments, from EST alignments and from combined EST-protein searches. The parameters for this new configuration have again been estimated on the 13 training regions.

Acknowledgement
This article has been published as part of *Genome Biology* Volume 7, Supplement 1, 2006: EGASP '05. The full contents of the supplement are available online at http://genomebiology.com/supplements/7/S1.

References

1. Guigó R, Agarwal P, Abril J, Burset M, Fickett J: **An assessment of gene prediction accuracy in large DNA sequences.** *Genome Res* 2000, **10:**1631-1642.
2. Stanke M, Waack S: **Gene prediction with a hidden Markov model and new intron submodel.** *Bioinformatics* 2003, **19**(Suppl 2):ii215-ii225.
3. Stanke M, Schöffmann O, Morgenstern B, Waack S: **Gene prediction in eukaryotes with a generalized hidden Markov model that uses hints from external sources.** *BMC Bioinformatics* 2006, **7:**62.
4. Abbott A: **Competition boosts bid to find human genes.** *Nature* 2005, **435:**134.
5. Guigó R, Reese M: **EGASP: collaboration through competition to find human genes.** *Nat Methods* 2005, **2:**575-577.
6. Guigó R, Flicek P, Abril JF, Reymond A, Lagarde J, Denoeud F, Antonarakis S, Ashburner M, Bajic VB, Birney E, *et al.* **EGASP: The ENCODE Genome Annotation Assessment Project.** *Genome Biology* 2006, **7**(Suppl 1):S2
7. Burge C, Karlin S: **Prediction of complete gene structures in human genomic DNA.** *J Mol Biol* 1997, **268:**78-94.
8. Parra G, Enrique B, Guigó R: **GeneID in Drosophila.** *Genome Res* 2000, **10:**511-515.
9. Lukashin AV, Borodovsky M: **GeneMark.hmm: new solutions for gene finding.** *Nucleic Acids Res* 1998, **26:**1107-1115.
10. Majoros WH, Pertea M, Salzberg SL: **TIGRscan and GlimmerHMM: two open-source ab initio eukaryotic gene finders.** *Bioinformatics* 2004, **20:**2878-2879.
11. Gross SS, Brent MR: **Using multiple alignments to improve gene prediction.** In *Proceedings of the Ninth Annual International Conference on Research in Computational Molecular Biology (RECOMB 2005):* May 14-18 2005. Cambridge, Ma, USA. Springer, Berlin. 2005:374-388.
12. Kent W: **BLAT - the BLAST-like alignment tool.** *Genome Res* 2002, **12:**656-664.
13. Stanke M: **Gene prediction with a hidden Markov model.** *PhD thesis.* Universität Göttingen, Institute for Numerical and Applied Mathematics [http://webdoc.sub.gwdg.de/diss/2004/stanke/]; 2003.
14. Kulp D, Haussler D, Reese M, Eeckman F: **A generalized hidden Markov model for the recognition of human genes in DNA.** *Proc Int Conf Intell Syst Mol Biol* 1996, **4:**134-142.
15. Korf I, Flicek P, Duan D, Brent MR: **Integrating genomic homology into gene structure prediction.** *Bioinformatics* 2001, **17** (Suppl 1):S140-S148.
16. Brejova B, Brown DG, Li M, Vinar T: **ExonHunter: a comprehensive approach to gene finding.** *Bioinformatics* 2005, **21**(Suppl 1): i57-i65.
17. Krogh A: **Using database matches with HMMGene for automated gene detection in Drosophila.** *Genome Res* 2000, **10:**523-528.
18. Durbin R, Eddy SR, Krogh A, Mitchison G: *Biological Sequence Analysis.* Cambridge, UK: Cambridge University Press; 1998.
19. Stanke M, Keller O, Gundez I, Hayes A, Waack S, Morgenstern B: **AUGUSTUS: ab initio prediction of alternative transcripts.** *Nucleic Acids Res* 2006, **34**(**Web Server Issue**):W435-W439.
20. **AUGUSTUS Web Server** [http://augustus.gobics.de]
21. Tzvetkova A: **Improving gene prediction in human using alignments with mouse genome sequences.** *Bachelor's thesis.* Universität Göttingen, Institute of Microbiology and Genetics; 2005.
22. Schwartz S, Kent WJ, Smit A, Zhang Z, Baertsch R, Hardison RC, Haussler D, Miller W: **Human-mouse alignments with BLASTZ.** *Genome Res* 2003, **13:**103-107.
23. Brudno M, Chapman M, Göttgens B, Batzoglou S, Morgenstern B: **Fast and sensitive multiple alignment of large genomic sequences.** *BMC Bioinformatics* 2003, **4:**66.
24. Morgenstern B: **DIALIGN 2: improvement of the segment-to-segment approach to multiple sequence alignment.** *Bioinformatics* 1999, **15:**211-218.
25. **University of California Santa Cruz BLASTZ Alignments** [http://hgdownload.cse.ucsc.edu/goldenPath/hg17/vsMm5/axtTight/]
26. **WUBLAST** [http://blast.wustl.edu]
27. Schöffmann O: **Gewinnung extrinsischer Informationen zur Genvorhersage und Einbindung in ein Hidden Markov Modell.** *Diplomarbeit.* Universität Göttingen, Institute for Numerical and Applied Mathematics; 2003.
28. Abril J, Guigó R: **gff2ps: visualizing genomic annotations.** *Bioinformatics* 2000, **16:**743-744.

Research

AceView: a comprehensive cDNA-supported gene and transcripts annotation

Danielle Thierry-Mieg and Jean Thierry-Mieg

Address: National Center for Biotechnology Information, National Library of Medicine, NIH, Bethesda, MD 20894, USA.

Correspondence: Danielle Thierry-Mieg. Email: mieg@ncbi.nlm.nih.gov

Published: 7 August 2006

Genome Biology 2006, **7(Suppl 1)**:S12

The electronic version of this article is the complete one and can be found online at http://genomebiology.com/2006/7/S1/S12

Abstract

Background: Regions covering one percent of the genome, selected by ENCODE for extensive analysis, were annotated by the HAVANA/Gencode group with high quality transcripts, thus defining a benchmark. The ENCODE Genome Annotation Assessment Project (EGASP) competition aimed at reproducing Gencode and finding new genes. The organizers evaluated the protein predictions in depth. We present a complementary analysis of the mRNAs, including alternative transcript variants.

Results: We evaluate 25 gene tracks from the University of California Santa Cruz (UCSC) genome browser. We either distinguish or collapse the alternative splice variants, and compare the genomic coordinates of exons, introns and nucleotides. Whole mRNA models, seen as chains of introns, are sorted to find the best matching pairs, and compared so that each mRNA is used only once. At the mRNA level, AceView is by far the closest to Gencode: the vast majority of transcripts of the two methods, including alternative variants, are identical. At the protein level, however, due to a lack of experimental data, our predictions differ: Gencode annotates proteins in only 41% of the mRNAs whereas AceView does so in virtually all. We describe the driving principles of AceView, and how, by performing hand-supervised automatic annotation, we solve the combinatorial splicing problem and summarize all of GenBank, dbEST and RefSeq into a genome-wide non-redundant but comprehensive cDNA-supported transcriptome. AceView accuracy is now validated by Gencode.

Conclusions: Relative to a consensus mRNA catalog constructed from all evidence-based annotations, Gencode and AceView have 81% and 84% sensitivity, and 74% and 73% specificity, respectively. This close agreement validates a richer view of the human transcriptome, with three to five times more transcripts than in UCSC Known Genes (sensitivity 28%), RefSeq (sensitivity 21%) or Ensembl (sensitivity 19%).

Background

Annotating the genes, transcripts and proteins of the human genome is a significant challenge. How many genes will ultimately be identified, what mechanisms control transcription, alternative splicing, the stability of the transcripts, translatability, what role do non-coding genes play and are there identifiable signals encoded in the genome sequence that control these events are all questions that need to be resolved so that we can hope to annotate the human genome faithfully. To address this type of question, the ENCODE project [1], launched by the National Human Genome Research Institute, encourages a concentration of international efforts and expertise on 1% of the human genome, in 44 carefully selected regions taken as representative of the whole genome, in the hope that mature annotation techniques will be developed, validated, and further applied to the entire genome.

The UCSC genome browser [2] provides fast and open access to a highly configurable view of a wealth of sequence-based genome annotations. The evidence-based or predicted gene tracks are an open repository for genome-wide annotations of the genes, and most tracks are well documented. All the data can easily be retrieved in a uniform format. The submission process is also simple and friendly, and there are no signs of limits to the amount of data that can be displayed and distributed by this group: the UCSC genome browser was naturally selected as the official repository for sequence-related data for the ENCODE project [3].

The Human and Vertebrate Analysis and Annotation (HAVANA) teams are expert at manual gene annotation [4]. They "require that all annotated gene structures (transcripts) are supported by transcriptional evidence, either from cDNA, expressed sequence tag (EST) or protein sequences, and as such not all annotated transcripts are necessarily complete". They typically bring to the curator, in a specialized Acedb-based display, a combination of evidence from alignment of mRNAs, ESTs and proteins, from human and other vertebrates. Curators hand select the best supported transcript models, and occasionally experimentally extend or confirm a model, using reverse transcription polymerase chain reaction and/or rapid amplification of cDNA ends. In this way, the Sanger Institute group carefully annotated the 44 ENCODE regions. Their gene models on these regions are called Gencode. They identify five times more variants than RefSeq, yet all their transcripts should be considered experimentally validated.

The ENCODE gene annotation assessment project (EGASP) [5,6] launched a competition among gene-predicting programs to try to best reproduce the Gencode annotations, taken as a reference, and/or to predict novel transcripts; the most promising novel genes would eventually be validated by RT-PCR. The Gencode solutions for 13 training regions were released at the end of 2004, and interested parties were asked to annotate the remaining 31 test regions before the solutions were unveiled in May 2005. Sixteen teams contributed complete mRNA or protein models; AceView was one of them.

The AceView program [7], developed at NCBI, provides a strictly cDNA-supported view of the human transcriptome and the genes by summarizing all quality-filtered human cDNA data from GenBank, dbEST and the RefSeq. The nematode version (also known as WormGenes) is even more

evolved and heavily hand curated: it uses over 280,000 cDNA sequencing traces, provided by the Kohara laboratory (Y Kohara, T Shin-i, Y Suzuki, S Sugano, D Thierry-Mieg and J Thierry-Mieg, personal communication) and the worm community, that we hand edit and use as a training set to handle automatically EST sequence basecall errors. AceView was written from scratch and guided over the years by visual expert evaluation and users' reports; it uses heuristics to closely reproduce manual curation in an automatic way. Annotation is a difficult and dynamic problem, and we do not claim to have a final solution, yet we hope to stimulate experiments and accelerate discovery. Our results are frequently updated as new cDNA sequences are submitted to the nucleotide databases, and they have been publicly available at NCBI since 2000 [7]. AceView, previously called Acembly, is also displayed as one of the UCSC gene tracks and as a DAS track on the Ensembl browser.

We submitted to EGASP a lightly hand edited version of the public AceView, with the note: "AceView: All mRNAs and cDNAs available in GenBank, excluding NMs, were co-aligned on the Gencode regions. The results were then examined and filtered to resemble HAVANA. The very restrictive view of HAVANA on CDS was not reproduced, due to a lack of experimental data." Our special treatment consisted of eliminating single exon genes, unless they had a Pfam annotation (as an unfortunate consequence, we lost a number of olfactory receptors correctly represented in the standard AceView) and discarding the RefSeqs, to avoid second-hand annotation. We also removed several recent retroposon-type pseudogenes that had escaped our standard filters and about 50 cDNA clones aligning with non-standard introns (not GT/AG or GC/AG). Then, after the workshop, we modified our program to automatically perform some of this extra filtering: the current genome-wide public AceView, dated August 2005, benefits from these Gencode-driven improvements and rates even slightly better than the EGASP version. Numbers quoted in the text correspond to the genome-wide version, because it is more relevant to the AceView users, but the closely similar AceView/EGASP performances are displayed graphically and in numbers in Additional data files 2-4.

In the main EGASP paper, Guigo *et al.* [6] thoroughly analyze the novel gene predictions and all regions annotated as protein coding by Gencode; they also present some basic comparisons of the mRNAs. Their preferred mode is to project on the genome the features to be compared, for instance nucleotides or exons, and to count each element only once per gene, in a way flattening the alternative transcript variants. However, as Gencode indicates, human gene transcription and splicing patterns are complex, yet not combinatorial, and the exon-intron chaining cannot be rendered in projection. We therefore undertook a complementary analysis of the same data, but focused on the complete mRNAs, irrespective of whether or not they have an annotated coding sequence (CDS). We took special interest in comparing the alternative variants across the tracks, using the complete chains of introns as signatures. We wrote a standalone program, UCSCtrackCompare, to compare the models of the various tracks to the Gencode validated transcripts (October 2005 freeze).

In general agreement with Table 6 from [6] for the projected view, but much more clearly when we look separately at the alternative variants, we show that at the mRNA level, AceView transcripts are by far the closest match to Gencode transcripts: all nucleotides used in spliced variants are common, except for 8% specific to Gencode or 12% specific to AceView. All introns are common, except for 10% specific to Gencode or 14% specific to AceView. In contrast, due to a lack of large scale protein sequence evidence, we have chosen different strategies to annotate the proteins: Gencode proteins (of which 31% are partial) are annotated in only 41% of the transcripts. It does not necessarily mean that HAVANA predicts that the remaining 59% transcripts are non-coding, but this is definitely what the Guigo *et al.* analysis [6] assumed, and this biased their CDS analysis. AceView, like all other methods except Gencode, conservatively chose to annotate the best predicted CDS in nearly all transcripts. Notice also that the Gencode hand annotation of transcripts is available at this high quality only in the ENCODE regions: their annotation of chromosome 20, for example, is far less comprehensive (Thierry-Mieg and Thierry-Mieg, unpublished) [4,7]; it may be too time consuming to annotate a whole chromosome manually at this depth while the cDNA data are accumulating so fast. In contrast, because AceView is automatically generated, it can provide for the whole genome a regularly updated annotation of the intron-exon structure of the genes and their alternative transcript variants that, as we show here, is of a quality comparable to the manual Gencode annotation. If deemed desirable, AceView mRNAs could easily be re-annotated with parsimonious Gencode-like CDSs.

The excellent agreement in gene structure between Gencode and AceView provides a cross validation of both annotations. The cDNA-supported consensus transcriptome includes close to five times more transcripts than RefSeq, and three times more than UCSC 'Known Genes', a track summarizing the human protein-coding quality-filtered data from RefSeq, GenBank mRNAs, and UniProt.

Results and discussion
Comparison of gene models using the UCSCtrackCompare program
To compare the performance of all programs at reproducing the rich Gencode annotation, we wrote a standalone program, UCSCtrackCompare, available in source form in Additional data file 1. The program compares, in chosen genomic regions, a UCSC track or a collection of tracks taken as reference (usually Gencode) against any number of selected tracks. It produces, in about 10 minutes, support for the analysis presented here; for a more detailed comparison of transcripts see Additional data file 2 and for coding regions see Additional data file 3. The direct outputs of UCSCtrackCompare were pasted in an Excel document (Additional data file 4). For the analyses of coding regions comparable to [6], our results almost exactly match those obtained by Guigo *et al.*, and our results lie within 0.2% of theirs (Tables 4 and 5 in [6]). However, there are uneven discrepancies that cannot be rationalized, but that may reach up to 8% in sensitivity and 13% in specificity, between our mRNA comparisons (Table 6 in [6]; see Additional data file 2.1). Methods that show an advantage in [6] include Ensembl, Exogean and Pairagon, and methods that show a disadvantage include AceView, ECgene, SGP2 and eight others. Yet, the general ordering of the methods is consistent across the two evaluations.

On 14 December 2005, we downloaded from UCSC all tracks with gene models in the 31 ENCODE test regions (see Additional data file 2.1). We selected for comparison to Gencode (October 2005 freeze) 14 non-redundant EGASP tracks released before the solutions, and 10 genome-wide tracks, distinguished on all diagrams by addition of an asterisk in front of the track name. Nine tracks, labeled P in front of their names, predict protein-only models by using *ab initio* methods, often integrating evolutionary sequence conservation; all others also use, or only use, mRNA and/or EST evidence. Statistics of the 25 tracks and hints on their inputs are summarized in Table 1.

Since the UCSC files give directly the exons of all tracks in a uniform chromosome-based coordinate system, the comparison of coordinates is straightforward and easy to duplicate (Additional data file 1). The choices offered in our program are either to decompose the models into their elements, exons, introns and nucleotides, or to consider them in their entirety; and then either to count each element with given genomic coordinates only once, thereby providing a rationalized projected measure of the unique elements (as was done in [6]), or to count each element as many times as it occurs in the alternative variants, providing a quantitative appraisal of the biological complexity of the gene. The two measures are complementary. Another option is to compare only the part of the models annotated as protein-coding. When applied to identical transcripts, this option allows the preferred hypotheses on choice of CDS and Start codon to be reverse engineered (Additional data file 2.5).

Comparison of introns, exons and nucleotides in whole models
Consistently, when we compare whole mRNA models (rather than CDSs) to the Gencode reference, AceView fares remarkably well, better than any other track. This is true in the projected mode, and even more striking in the quanti-

Table 1

Statistics of the 25 selected tracks, arranged in the order of the UCSC genome browser

UCSC track	Model with introns	Model with introns and CDS	Single exon model (some clipped)	Unique introns in mRNA	All introns in mRNA	Input or method
HAVANA Gencode (Sanger, UK) known + putative	1,691	**649**	70	3,618	9,693	MEP,CA,H
EGASP model submissions						
AceView (NCBI, US)	1,630	1,460	24	3,530	9,597	ME,(H)
UP Dogfish (Sanger, UK)	204	204	15	1,679	1,679	CA
Exogean (ENS, France)	554	538	2	2,855	6,178	MEP,CA
UP ExonHunter (U Waterloo, Canada)	807	807	220	3,237	3,237	MEP,CA
Fgenesh (U London, UK)	462	458	97	2,610	3,241	P,CA
UP Geneid (IMIM, Spain)	267	267	51	1,905	1,905	A
UP GeneMark (Georgia IT, US)	551	551	81	2,185	2,185	A
UP Jigsaw (TIGR, US)	259	259	67	2,168	2,168	MEP,CA
PairagonAny (Wash U, US)	471	437	38	2,300	3,470	MEP?,CA
UP SGP2 (IMIM, Spain)	552	552	159	2,645	2,645	P,CA
P Twinscan-MARS (Wash U,US)	547	547	108	2,501	4,943	CA
UP Augustus Any (U Göttingen, Germany)	312	316	87	2,291	2,291	MEP,CA
UP GeneZilla (TIGR, US)	477	477	179	2,758	2,758	A
UP Saga (UC Berkeley, US)	331	331	47	1,737	1,737	CA
UCSC gene tracks						
*Known Gene (UCSC)	501	477	53	2,264	4,427	MP
*P CCDS	201	201	14	1,296	1,508	MP,H
*RefSeq (NCBI, US)	342	325	41	2,082	2,922	M(E)P,H
*MGC	323	310	19	1,400	2,101	M
*Ensembl (EBI, UK)	427	418	58	2,429	3,548	MEP,CA
*AceView (Aug 2005 NCBI)	1,792	1,627	902	3,812	9,792	ME, (H)
*ECgene (Korea)	3,851	3,551	2,569	3,942	30,660	ME,C
*U NscanEst (Wash U, US)	282	252	27	2,292	2,292	ME,CA
*UP GenScan (MIT, US)	395	395	59	3,042	3,042	A

The number of models, with or without introns (after clipping at region boundaries), the number of spliced coding models, and the number of unique and multiply used introns are given over the 31 ENCODE test regions. Coded information has been added in front of the track name: asterisks distinguish standard gene tracks, available genome-wide, from an ENCODE only track; a U track predicts a unique model per gene; P predicts protein coding regions only. According to their documentation, the programs use different input or methods: M, E, P stand for human mRNA, EST, protein sequences or alignments, respectively; C stands for for conservation, or use of cDNA or protein evidence from other species; A stands for *ab initio* prediction; H stands for Hand curation; and parenthesized letters stand for minimal use of the particular type. Notice the low proportion of Gencode mRNA models with an annotated CDS (in bold).

tative mode, where alternative variants are counted separately (see details in Additional data files 2 and 4).

Nucleotides provide a global appraisal of the transcribed regions: AceView and Gencode spliced transcripts cover almost exactly the same nucleotides in the genome (92% sensitivity, 88% specificity; Additional data file 2.2). The structural precision of the models is best defined by the exact position of intron-exon boundaries. As shown in Figure 1a, most of the unique Gencode introns are used in AceView and few are added (sensitivity 90%, specificity 86%). Only AceView and ECgene (86.5%) detect more than 70% of the introns: the other EGASP tracks probably do not succeed in aligning with sufficient precision all the available ESTs and mRNAs. For reference, on 14 December 2005, RefSeq sees 56% of the Gencode validated introns and the

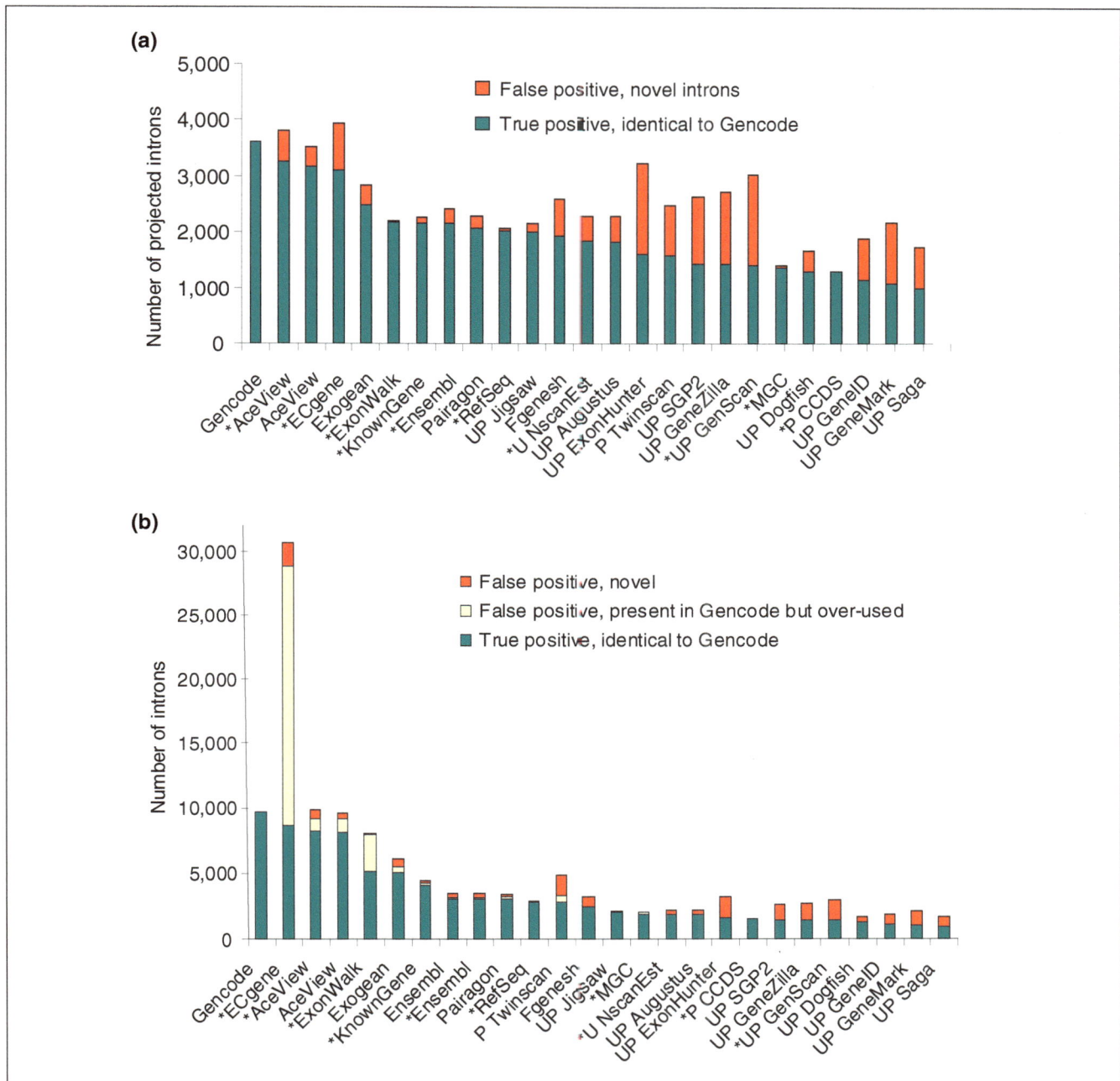

Figure 1
Comparison of introns between the Gencode reference and the 24 tracks, ordered by decreasing sensitivity, over the 31 test regions. Gencode validates 3,618 unique introns and a total of 9,693 introns in its alternative transcripts. **(a)** Projected measure: each intron is counted only once per method. Introns with the same coordinates as Gencode introns are shown in green and novel introns in red. The Gencode introns missed in each track (false negative) correspond to the distance between the 'true positive' bar and the Gencode reference, but are not explicitly represented. **(b)** Quantitative measure: all alternative variants are counted separately. Introns identical to Gencode introns, but over-used relative to Gencode are counted (in yellow) separately from novel introns that are not known to Gencode.

Mammalian Gene Collection (MGC) 38%. As expected, the tracks relying mostly on *ab initio* predictions detect less Gencode introns, but more novel intron candidates, usually unique to each program (Additional data file 2.4); ExonHunter and Genscan are the most creative.

If all introns from alternative variants are counted separately (Figure 1b), Gencode uses close to three times the number of unique introns. AceView does too, but remarkably in this expansion, most of the introns remain exactly the same as in Gencode (sensitivity 85%, specificity 84%). In contrast, the specificity of ECgene drops to 28%, because of their drastic combinatorial use of the Gencode introns. This over-use feature is apparent to a lesser extent in the ExonWalk track, possibly because of excessive use of partial cDNAs in their step 3 (for documentation, see [3]).

The comparison of exons (Additional data file 2.3) can also assess the intron-exon boundaries of the models, but it is dominated by the effects of terminal exons, which represent 20% to 42% of all exons (depending on the method), and are often partial in transcript models. Furthermore, in contrast to intron boundaries, which are accurate and can in principle be verified by PCR or microarray experiments, the boundaries of terminal exons cannot be defined precisely even in a truly complete transcript because, biologically, the first (capped) base and the polyA addition site fluctuate *in vivo* [8] (D Thierry-Mieg and J Thierry-Mieg unpublished observations). Indeed, when we compare exons between Gencode and any other track, both sensitivity and specificity drop because of the terminal exons (Additional data file 2.3).

Another advantage of comparing spliced models through introns is that methods that predict only coding regions (labeled P(name) in all figures and tables) are less disadvantaged in intron than in exon mode, because most introns are located in the coding regions. This is true for example for 92% (2,075/2,264) of the unique introns in the UCSC 'Known Gene' track. We therefore chose to perform intron-based comparisons of whole spliced transcripts. Single exon genes and transcripts will be discussed separately below.

N to N comparison of entire mRNAs across methods proposing alternative transcript models

On average, Gencode genes with introns have 4.5 transcripts per gene; each transcript has 5.7 introns, but in projection only contributes 2.1 unique introns. If we limit this to coding transcripts, coding genes have on average 2.6 annotated CDSs, and each CDS has 8 introns, but in projection only contributes 3.7 unique introns (Additional data file 4, SummaryStats). Just a few of the possible combinations of introns correspond to supported models: this is called the combinatorial splicing problem. To properly compare performance across the tracks, we need to evaluate how the introns and exons are chained in the models. Let us consider each model as a non-separable chain of introns, the set of coordinates of its intron boundaries on the genome provides a precise signature. As illustrated in Figure 2a, we define the one-to-one best matching Gencode-to-track-X model pair by comparing all pairs and scoring intron boundaries: each boundary counts +1 if it is shared by the two models or -1 if it is unique to either of the two. In this way, identical models always score best. All pairs of models with one intron boundary in common are listed and sorted, and the two models from the best rating pair are flagged as 'best match', irrespective of their score. We then look recursively for the next best rating pair where none of the two models are

flagged, and flag them. The remaining unflagged models in the list become 'additional variants in Gencode genes'. Finally, the models that do not occur in the list are orphans and belong to new genes, specific of method X, or to missed genes, present only in the Gencode reference. Some of the 'best matches' are actually 'identical to Gencode' in terms of intron-exon structure.

Figure 2b shows the results of this analysis (see also Additional data file 4, complete models). More than 70% of the Gencode transcripts have an exact structural equivalent in AceView, from the first to the last intron (1,191/1,691 = 70% sensitivity, 67% specificity); an additional 12% (206) have a best match. AceView misses 210 of the Gencode variants but sees 225 new variants in other Gencode genes. Finally, AceView misses some Gencode genes containing a total of 88 transcripts with introns, but sees some genes missed by Gencode, containing 170 transcripts with introns.

The second most sensitive track is ECgene (60% sensitivity; 26% specificity), which has twice as many spliced models as Gencode to choose from, but has fewer identical transcripts than AceView. The next best, ExonWalk and 'Known Genes', drop sharply to 23% exact matches. RefSeq sees only 17% of the Gencode transcripts and covers 53% of the unique nucleotides and 56% of the unique introns: despite its well recognized quality, RefSeq does not provide a comprehensive representation of the transcriptome.

Intronless transcripts

By definition, the above analysis only evaluated transcripts with introns. Transcripts and genes without introns are in fact rare in the Gencode annotation (2.3% of the models, plus 1.7% due to clipping at the boundaries of the ENCODE regions). To better mimic Gencode, most were filtered by the EGASP participants, including AceView, which normally annotates a large number of single exon genes (Table 1; Additional data file 4).

Although rare on the entire UCSC browser, unspliced transcripts appear to be an important part of the human transcriptome, and they are an order of magnitude more frequent in mammals than in simpler Metazoa. Indeed, we compared the high quality full-length cDNA libraries made by Sugano using the oligo-capping method in either the nematode *Caenorhabditis elegans* or human (all sequences are in GenBank [7,9]). We removed 1.1% of clones that may be genomic contaminants (331/29,562 aligned in an intron-less gene ending on an A-rich region in the genome) and found that the percentage of fully sequenced intronless clones is 10 times greater in human than in worm (36% in human (10,578/29,562 FLJ clones) versus 3.5% in worm (2,010/56,671 worm yk capped clones)). These clones also map in 11 times more unspliced genes (with no spliced variants) in human than in worm (25% of these FLJ-containing genes in human (4,261/17,214 genes) versus 2%

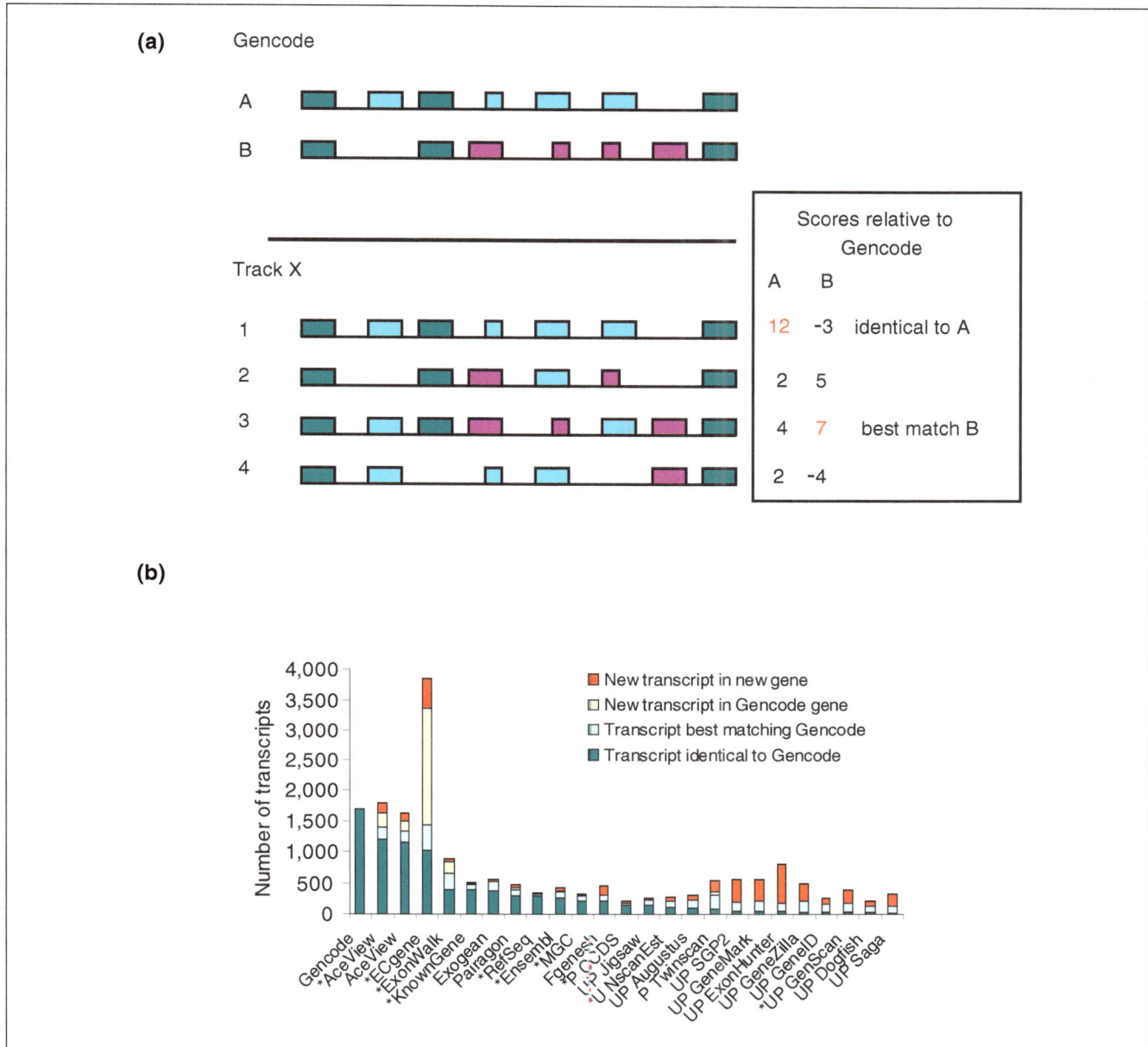

Figure 2
Comparison of whole transcripts. **(a)** Strategy for selecting the best one to or e matching pairs. **(b)** Comparison of whole transcripts through their intron signatures. The number of transcripts identical to Gencode, best-matching but different from Gencode, new transcripts in Gencode genes and new transcripts in new genes are represented.

in worm (155/7,223 genes)). However, the level of possibly immature transcripts (unspliced, but from a gene with introns) is similar in the libraries from both species (57% (6,043/10,578) in human, 59% (1,187/2,010) in the worm).

According to our analysis, the huge increase in intronless genes is a major difference between the worm and human transcriptomes. It may have co-evolved with the increased usage of alternative splicing, increased intron length or other transcriptional features. An intronless transcript is not

associated with an exon junction complex, so it is expected to be translated and degraded less efficiently in human [10-12], and it might be submitted to less evolutionary pressure: unspliced genes could be where new functions arise. Indeed, some of the unspliced genes potentially encode small proteins (18,385 intronless genes have hypothetical CDS of more than 100 amino acids; note that most are human or mammal specific), some may be regulatory non-coding RNAs, and some may just be transcription by-products. We do annotate these genes in the public AceView [7] and expect that in the future the role of intronless genes will be better apprehended.

Are Gencode mRNAs fully validated and complete?

We compared the introns of Gencode to those of other tracks (Additional data file 2.4); 189 unique projected introns (5% of their 3,618) are seen exclusively in the Gencode transcripts but in no other track. We expect those to be supported by Gencode experimental validation, although the evidence was not submitted to GenBank as of August 2005. Another possibility is that some might be supported only by homology to non-human transcripts, as described in the standard HAVANA procedure [4], in which case we hope they are not exported to UniProt. Conversely, 681 'consensual' introns are seen by three tracks or more, or 340 are seen by the strictly cDNA-supported AceView that are not seen by Gencode. Actually, Gencode sees 8% of the consensual introns missed by AceView, but AceView sees 72% of the consensual introns missed by Gencode.

Overall, we find that the Gencode transcript annotation is quite comprehensive except for the quasi-absence of intronless genes. We confirm its high quality: it missed or annotated as pseudogene only a few expressed genes seen by AceView, and it did not exploit in the order of 15% of the introns represented in cDNAs from the public databases.

In general, we especially appreciate the fact that they report all observed transcripts with good alignments and standard introns, without filtering those that are structurally candidates for nonsense mediated RNA decay (NMD) [12]. According to our estimates [7], putative nonsense transcripts represent about 13% of the fully supported transcripts with predicted proteins of more than 100 amino acids in human (12,855 of 101,877 have introns larger than 60 base-pairs (bp) and with standard boundaries lying at least 55 bp downstream of the Stop codon), in contrast to 4% in worm (671 of 15,119 using the same criteria as above, except that minimal intron size is 30 bp). This increase may parallel the evolution of NMD into an essential multifunctional mechanism in mammals [10-12], or it may indicate that our cells have a more tolerant life style than the worm. Over the past few years, evidence that mRNA may be functionally active beyond its protein coding ability has accumulated [13]; human transcripts have a complex life, with mechanisms present to protect the mRNA, modulate its accessibility to the ribosome or to specific modification or processing enzymes, monitor its aging, its position in the cell, or its stability [10]. A comprehensive uninterpreted catalog of observed transcripts is needed to help understand all this complexity, and Gencode or AceView aim at this goal.

Selecting protein coding transcripts and regions can only be an educated guess

A glimpse at the ENCODE regions of the UCSC genome browser [3] shows that Gencode and AceView transcripts

look quite different from the other tracks, but so similar to one another that it would be hard to guess which is which if the names of the tracks were masked. However, Guigo and Reese [5] state that there is no clear winner at finding the Gencode coding regions. These two observations are in fact not contradictory: annotation of transcripts is based on large amounts of experimental cDNA evidence, so Gencode and AceView can agree almost perfectly. In contrast, we diverge on protein annotation, because there is almost no experimental protein sequence data available today. For this reason, most protein annotation remains hypothetical. Even proteins from UniProt/SwissProt are now contaminated by CDS predictions derived from transcriptome annotation, which makes new predictions by homology more and more circular. For instance, UniProt currently harbors 23,298 *C. elegans* proteins, but one should be aware that those are predictions: at most, 9,487 have complete cDNA support, and almost none has been sequenced. Until a substantial amount of direct protein sequences is generated, it is not surprising that different points of view coexist. For instance, Gencode annotates a CDS in only 41% of their transcripts, whereas all other tracks, including AceView, conventionally annotate a CDS in almost all their transcripts. Precisely because Gencode and AceView transcripts are so similar, the apparent specificity of AceView automatically drops by 59% when we compare CDSs, and the resemblance between the two methods is no longer striking (Additional data file 3, in agreement with [6] for the projected view).

In this context, AceView considers the CDS problem fully open, and offers no guarantee on which actual proteins are made [7]. In practice, we identify all possible CDSs, usually more than one per transcript, and annotate those larger than 50 amino acids using BlastP, PFAM, and Psort2. All hypothetical CDSs are available from our download page to help identify mass spectra. But to simplify the display, we pick a single 'best product' per transcript, knowing it may not reflect the situation *in vivo*, since it does not follow closely the rules indicated by Kozak [14]. For example, we do not necessarily choose the first CDS, which is quite often a short upstream open reading frame (uORF) [15,16], and we do not reinitiate and display multiple products per transcript. The 'best' protein is defined by considering, in a graded fashion, the presence of a Pfam protein domain, BlastP homologies, TaxBlast conservation, specific Psort annotations, maximization of introns within the CDS, position along the transcript and size of the CDS. If the mRNA is not known to be complete (if its 5' end is not defined by a capped clone) and the frame is open on the 5' side, the AceView CDS starts at the first in frame codon. But if the mRNA appears to be complete, the CDS starts at the first AUG codon, unless there is, in the correct environment, an in frame NUG or ANG codon [14,17-20] at least 180 bp upstream of the first AUG. In such a case, we annotate a predicted CDS starting at the non-AUG codon (the limit was set at 60 bp in the August 2005 release, leading to an excessive 24% complete CDS starting on an NUG start codon). *Escherichia coli* is reported to use about 17% non-AUG start, and to our surprise 7% of the human best complete products in AceView (20,616 of 293,158) actually have an alternative Start codon in the correct environment 60 amino acids upstream of the first AUG. It will be interesting to see how many are occasionally used as Start *in vivo*.

On the other hand, Gencode departs from all other programs in that it does not call a CDS unless it is conserved or already annotated in SwissProt, and it has a 'sensible' gene structure that is not a candidate for NMD. The product they annotate is almost always the same as AceView, except that the Start codon may differ. In reality, they probably do not really mean that 59% of the transcripts from protein coding genes are non-coding, but they just have to be very careful, because their proteins are poured directly in the UniProt reference database, so any annotation error will spread. We respect their attitude, yet some of their choices can be questioned.

NMD transcripts for instance are expected to produce, briefly but efficiently, truncated proteins, some of which could be functional [12]. Indeed, activation of mRNA degradation by this pathway requires a pioneering round of translation that, due to the dual role of NMD proteins in activating translation, should be very efficient on transcripts still decorated with exon junction complexes (reviewed in [10,11]). Furthermore, NMD is only shortening transcripts lifetime by 1.5 to 11 times [21], and it remains possible that it does not act in all cells and tissues at all times. The leaky behavior of the surveillance machinery is well known to geneticists: if no protein was produced from NMD candidates, the great majority of nonsense mutations would behave as complete loss of function (true nulls), but there are a number of counter-examples where a well positioned stop mutation leads to a gain of function phenotype (for example, *lin-1* allele n1790 [22]).

With respect to the choice of the initiation codon, Gencode and other groups give much weight to interspecies conservation, they do not annotate upstream ORFs, and consider only AUG codons. But it is difficult to conceive how the ribosomes would be aware of these rules, instead of following the scanning mechanism experimentally established by Kozak [14]. If the transcript is accessible, upon scanning, the ribosome subunits assemble at the first AUG (or more rarely at an alternative start), irrespective of the length and interspecies conservation of the protein. If there is a stop codon soon after the AUG, the ribosome will keep scanning rather than drop off, and may reinitiate synthesis of a second product.

In summary, protein annotation is not supported by enough direct protein sequence evidence, and large scale mass spectrometry data on proteins are badly needed to clarify what happens *in vivo*.

Validating the transcriptome through democratic consensus

Annotating the transcriptome is a difficult and dynamic task, the data and the rules do evolve, and even the most careful manual annotation cannot be expected to provide by itself an incontestable and final truth. There are strengths and weaknesses in each annotation, but good mRNA models supported by strong cDNA data should be found by more than one method. Therefore, an alternative way to select a benchmark may be to take a democratic approach: instead of considering a single track as 'reference', we propose to pool independent cDNA-supported annotations and search for consensual models.

A caveat is that some annotations, such as RefSeq, CCDS or UniProt/SwissProt, are so renowned that most methods use them as a data source in addition to the primary cDNA or protein data, so these models are sticky and will end up validated, whether or not they are correct. Manual annotation is invariably the source of second-hand annotation problems. In fact, in AceView, we do not use SwissProt for this reason, and we now explicitly label all transcripts whose structure is supported only by a RefSeq model as possibly suspect. In the same vein, we use only human cDNAs at the exclusion of any other species, even mammals or primates.

We implemented the democratic idea in two flavors in UCSCtrackCompare: we either rotate the reference from Gencode to any other track and perform a closest neighbor consensus analysis (Figure 3b); or, alternatively, we pick as reference a selected pool of tracks and the program extracts their consensus models (for instance those whose intron structure is identical in at least two independent annotation tracks) and measures, for each of the 25 tracks, the number of models exactly matched, hence their sensitivity and specificity.

Table 2 and Figure 3a show the results of the pooling analysis, where the consensual set are the 1,556 spliced transcripts seen by at least 2 of the 7 evidence-based independent methods: Gencode, UCSC Known Genes, RefSeq, Ensembl, AceView, ECgene and ExonWalk. AceView and Gencode fare best, with 84% and 81% sensitivity and 73% and 74% specificity, respectively. The next most sensitive method is ECgene, with 77% confirmed models, but its specificity is only 31%. ExonWalk, UCSC Known Genes, and Exogean provide considerably fewer models (33% to 26% sensitivity). ExonWalk and its closest neighbor ECgene suffer from low specificity, unless both are included in the reference set, probably because they allow combinatorial arrangements of the introns. It would be interesting to know how frequently these entire models are validated in RT-PCR experiments. Then come in order RefSeq, Pairagon, Ensembl, MGC, Fgenesh, and CCDS. Finally, the 12 remaining methods are less sensitive than CCDS, as shown in Figure 3a. This analysis is robust against variations in the composition of

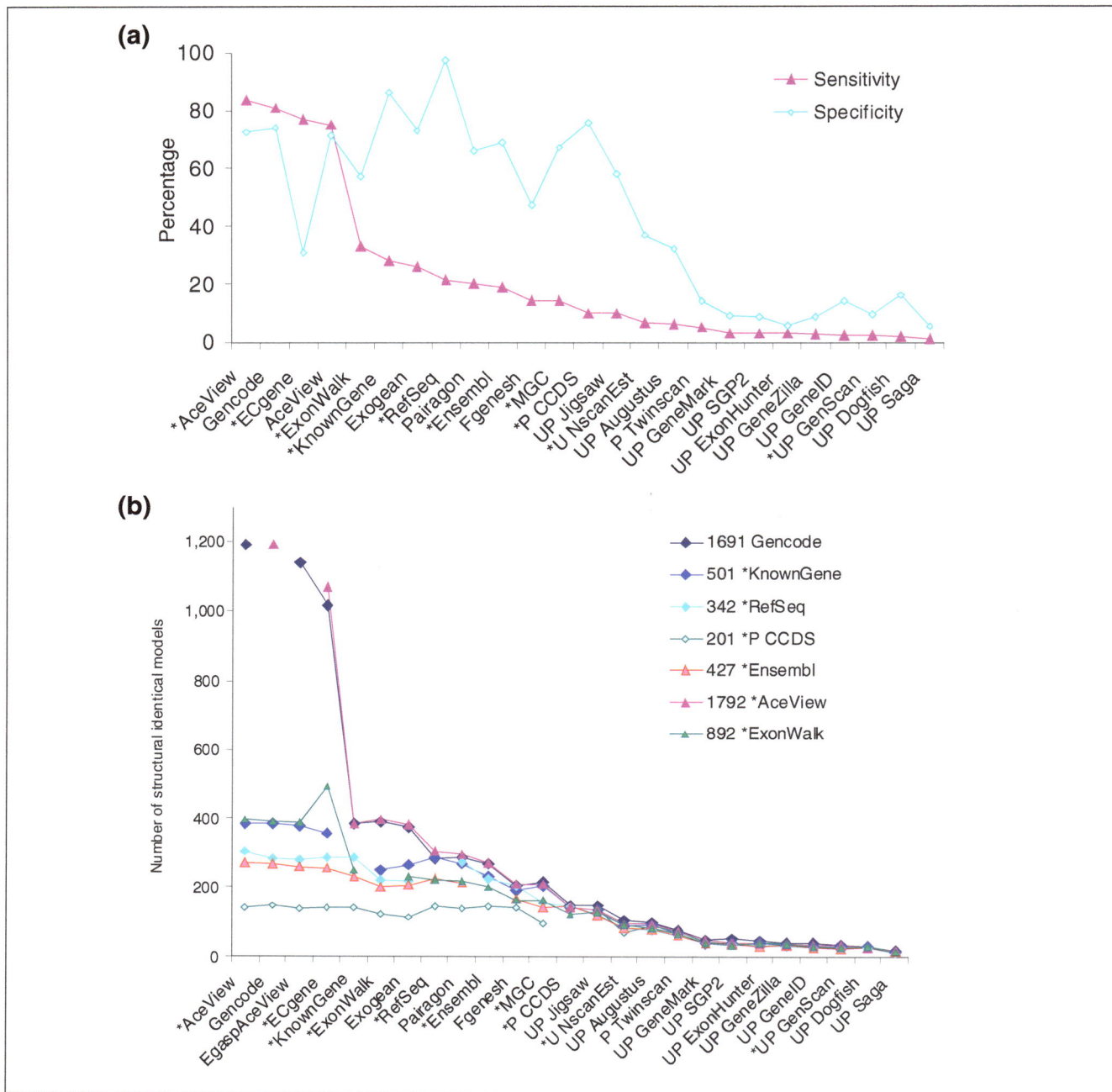

Figure 3

Consensus analysis. (a) Sensitivity and specificity at identifying 1,556 consensus transcripts from the pool of the following evidence-based tracks: RefSeq, Known Gene, Ensembl, Gencode, AceView, ECgene and ExonWalk. The sensitivity and specificity of all tracks at identifying these consensus models is plotted and listed in Table 2. (b) Closest neighbor consensus, evaluated by switching the track of reference. This figure shows the number of evidence-based models from CCDS, RefSeq, UCSC Known Genes, Gencode, or AceView, ExonWalk and Ensembl whose intron-exon structure is exactly matched by the 25 tracks. Tracks are arranged in decreasing order of averaged detection sensitivity, defined here as the sum of all evidence-based models from these seven reference tracks detected exactly.

the reference set, as long as both Gencode and AceView or ECgene are included. It is even stable if we include the 23 unrelated tracks in the reference pool, yielding 1,957 consensual models (Additional data file 2.6). But if AceView and ECgene are excluded, the consensus falls down to only

478 transcripts across NCBI RefSeq, UCSC Known Gene, EBI Ensembl and Sanger Institute Gencode. In this context, Gencode has an appalling specificity of only 26%. Fortunately, its agreement with AceView cross-validates both methods.

Table 2

Sensitivity and specificity of each method at detecting the 1,556 consensus transcripts

Track	Number of models with introns	Consensual models (of 1,556 total)	Sensitivity	Specificity
*AceView	1,792	1,302	84%	73%
Gencode	1,691	1,255	81%	74%
*ECgene	3,851	1,198	77%	31%
AceView	1,630	1,165	75%	71%
*ExonWalk	892	511	33%	57%
*Known Gene	501	432	28%	86%
Exogean	554	404	26%	73%
*RefSeq	342	332	21%	97%
Pairagon	471	310	20%	66%
*Ensembl	427	295	19%	69%
*MGC	323	217	14%	67%
Fgenesh	462	217	14%	47%
*P CCDS	201	152	10%	76%
UP Jigsaw	259	150	10%	58%
*U NscanEst	282	104	7%	37%
UP Augustus	312	100	6%	32%
P Twinscan	547	77	5%	14%
UP GeneMark	551	50	3%	9%
UP SGP2	552	48	3%	9%
UP GeneZilla	477	47	3%	10%
UP ExonHunter	807	41	3%	5%
UP GeneID	267	38	2%	14%
*UP GenScan	395	37	2%	9%
UP Dogfish	204	33	2%	16%
UP Saga	331	18	1%	5%

Sensitivity and specificity of each method at detecting the 1,556 consensus transcripts across the pool of the following evidence based tracks: RefSeq, Known Gene, Ensembl, Gencode, AceView, ECgene and ExonWalk, as in Figure 3a. Coded information has been added in front of the track name: asterisks distinguish standard gene tracks, available genome-wide, from an ENCODE only track; a U track predicts a unique model per gene; P predicts protein coding regions only.

Another view is shown in Figure 3b, which displays the closest neighbor consensus analysis (see also Additional data files 2.6 and 3.1 and 3.6). The sum over seven references of the number of exactly matching models was used to order the tracks. Interestingly, all tracks spontaneously appear to be ordered in overall sensitivity, almost independently of the chosen reference. Gencode and AceView are nearly indistinguishable. With 1,191 models in common, they are by far the most sensitive, but retain a very good specificity. In a robust way, they detect the largest number of models from all other tracks; they are the most inclusive, and three times more thorough than any other reference track (see Additional data file 2.6).

To our surprise, Ensembl, which is often used as a reference catalog, in particular to count the human genes [23], is not consensual, and far below Gencode and AceView in both sensitivity and specificity. While it offers more models than RefSeq (427 versus 342), fewer are confirmed by at least one other method (270 versus 304 in the closest neighbor analysis, 295 versus 332 in the democratic consensus), but the caveat about the artificial increase of specificity and sensitivity of RefSeq certainly applies here. However, the quality of the RefSeq is truly higher than suggested by ExonWalk, Ensembl or Exogean, which validate only 220 to 224 of the 342 RefSeqs: in the EGASP AceView version, we purposely did not use the RefSeqs as a source, yet we confirmed the intron-exon structure of 82% of them (279). However, RefSeq is far from comprehensive. The even smaller CCDS collection is equally well matched by many tracks, but at the protein level (Additional data file 3.6), they are perfectly matched only by Ensembl (201), and not quite by the other members of the CCDS collaboration: RefSeq (197), Gencode (182) and UCSC (189). It is definitely difficult to agree on any standard for protein annotation.

AceView summarizes GenBank and dbEST into a comprehensive evidence-based gene annotation by performing hand-supervised automatic annotation

The fact that the manually curated Gencode and automatic AceView transcripts are so similar shows that the critical information for the intron-exon structure of a Gencode-like validated annotation is almost entirely contained in the combination of human ESTs, mRNAs and the genome. It appears that AceView is now able to automatically extract this information, with little more noise than a team of careful human experts. The resemblance also indicates that we have the same, possibly biased, way of looking at the data and that we apply similar filters when annotating transcripts.

To reconstruct the genes, AceView considers all cDNA sequences submitted to the public databases, and stringently co-aligns them at their single best position on the genome [7]. Its cDNA to genome alignment algorithms are finely tuned to clip vectors and poly-A and to filter away 3% of the cDNAs because of insufficient quality of their best alignment (especially if they map in multiple genomic locations), and 2.2% because of suspected structural defects; 3.7% of the cDNAs are strand-inverted. The alignments are seeded on exact matches of 15 bp and extended using a finite automaton able to switch from normal to insertion or deletion mode when the EST fasta file starts calling bases at the wrong frequency. Missing exons are researched aggressively, seeding on 6 bp words. Short hits are counted not in base-pairs but in entropy, assuming that each base is statistically independent from its neighbors. This is an over-simplification, but the advantage is that, for instance, an AT rich region is penalized and at the extreme a pure poly-A hit counts zero. The intron-exon boundaries are then refined by co-alignment. Finally, we reject the very long introns unless they are bounded by strong exon support. Aligned cDNAs are then clustered into the minimal set of transcripts (that is, a gene) consistent with their complete intron-exon structure. Most of the gene models with multiple cDNAs have alternative variants, but since September 2004, in order to limit combinatorial expansion of variants, we minimize concatenation by using each cDNA in one and only one transcript, favoring a silent merge in a known compatible transcript, so that only cDNAs containing a specific alternative feature are singled out. As a result, some variants are partial, but 70% of all AceView transcripts have their predicted CDS entirely supported by a single identified cDNA. The remaining models require concatenation of rarer

forms, and will possibly be split into multiple alternative variants when additional data become available.

We then name the gene by physical contact to an NCBI Entrez gene model, else by alignment of a RefSeq or GenBank mRNA assigned to an Entrez gene, else by a Pfam-containing name, else by a nickname. The nickname is a number encoded in decodable pseudo English or pseudo Japanese by using a set of phonemes as basic digits. All names and previous aliases are tracked from release to release, and *de facto* AceView closely follows the official HUGO and Entrez gene nomenclature.

However, genome annotation cannot be fully automatic. We must often look at the genes, and take significant decisions to resolve the irregularities. The difficulty is to maintain this hand annotation over the years, as new data become available. For example, the hand annotations of the first *Drosophila* jamboree and of the initial version of the Celera human genome were nearly entirely lost. In AceView, we have limited manpower, just the two of us, so we had to devise an efficient cumulative methodology. Rather than hand annotating the final report of a gene, we only provide hints that are incorporated automatically in the context of the most recent data. For example, if the program seems to merge two genes in an unreasonable way, we do not create a permanent wall between them; instead, we hand annotate a few cDNA clones as having a 'real 3' end' or 'real 5' end', and then port these annotations from build to build. These hints will probably induce the program to split the gene, but if tomorrow a new mRNA sequence strongly bridges the two genes, they will automatically be reunified. When a significant number of genes need the same kind of manual hints, we add a piece of code that performs the same task and then drop the manual annotations once they are automatically reproduced, and even often enhanced. For example, at the EGASP meeting, we learnt that non-standard introns (neither GT/AG, nor GC/AG) are not usually validated by RT-PCR: we now discard any cDNA variant with a non-standard intron, unless it also brings a novel alternative intron with standard boundaries. As a result, we reduce the noise, but if many clones in a gene use a particular non-standard intron boundary, for instance because of an error in the genome sequence or because the intron is truly non-standard, this intron will naturally sift through our tolerant rule and be kept in AceView.

AceView is a service to the community, it does not provide a final answer, but rather some rated proposals aimed at stimulating confirmatory experiments. By using the genome as a guide, it automatically rectifies the sequencing errors in the cDNAs and brings these sequences in line with the excellent quality of the genome itself. But it only provides a partial view of the entire transcriptome, because we are still far from saturation in cDNAs. From release to release, we improve the models by incorporating the latest cDNA data, but also by refining the rules. For example, we recently redefined the gene as a set of transcripts sharing at least one intron boundary, instead of a simple sequence contact. This disentangled the numerous contiguous genes with 3' 5' untranslated region (UTR) overlap and separated the unspliced variants, improving the gene annotation in directions wished for by the users.

Because Gencode annotation is manual, it may prove more difficult for them to include new data or to implement a change in strategy. For example, one can read on the HAVANA guideline site: "Occasionally a short two exon product is supported by Fgenesh and Genscan, in which case the object can be translated. It is then annotated as "believable CDS"." This rule was recently abandoned, but it will be labor-intensive to hand revise all previous models accordingly.

The puzzle of gene counts
Gencode annotates 3,618 distinct introns. But this is possibly only the tip of the iceberg, since 10,241 other introns in coding regions are predicted in EGASP, mostly by *ab initio* methods. If a proportion of those were correct, we might have mRNA or EST support for maybe only half of the introns and, by extension, we might be missing an appreciable fraction of the genes.

The parallel with the nematode *C. elegans* is interesting. In "So many genes, such a little worm" [24], Hillier *et al.* count 19,735 coding genes in WormBase. But when we analyze all available cDNA sequences in AceView WormGenes [7], we find only 16,094 worm genes with direct experimental evidence, of which about 700 are not annotated in the current WormBase (WS150). There are 12,083 genes supported by cDNAs, mainly from the large scale libraries from Kohara and collaborators (all sequences are in GenBank). An additional 4,011 genes or gene fragments are supported only by the systematic RT-PCR amplification of predicted ORFs from the Vidal ORFeome project [25]. So we conclude that the authors of [24] are confident that, in addition to the cDNA supported genes and the 4,011 gene fragments supported by RT-PCR amplification, close to 4,400 genes that remain pure *ab initio* predictions really exist. Indeed, they exported them to SwissProt/UniProt.

On the other hand, these authors and their collaborators [23] claim that the human genome contains a maximum of 25,000 protein coding genes. They consider that nearly all of them are already known, that the numerous cDNAs that map outside of their official gene list possibly 'reflect reproducible transcriptional noise', and they do not expect any reliable gene to come from *ab initio* predictions. However, AceView unambiguously reconstructs from the readily available human cDNAs about 40,000 genes potentially encoding more than 100 amino acids (22,280 spliced and 18,385 intronless in the August 2005 version), in addition to 13,133

spliced genes encoding shorter proteins or non-coding. Moreover, as we see in EGASP, many *ab initio* predictions can be proposed in between cDNA supported genes and, unlike in the worm, no intense RT-PCR experiments have yet been launched in human. Preliminary EGASP results [6] only provide a lower bound on their existence, because only a fraction of the cDNA supported introns of Gencode and AceView have been validated and *ab initio* predictions are expected to be less expressed. If we apply a uniform method to count genes, we are forced to conclude that human has at least 3 times as many coding genes as the worm, and at least 10 times as many protein isoforms.

Conclusions
EGASP [5,6] and the availability of the excellent Gencode/ HAVANA models have helped us to significantly refine the AceView pipeline. The structure of the AceView transcripts is extremely similar to the Gencode benchmark, so AceView appears to provide today the most comprehensive and accurate representation of the entire human transcriptome. On the other hand, due to a profound lack of experimental protein evidence, annotation of coding sequences remains controversial. We hope that this situation will rapidly improve with the current progress of mass spectrometry and a new understanding of the complex regulation of the translation machinery in vertebrates.

There are currently at least three times more protein coding genes in human than in worm, but the human transcriptome is still far from saturation; 23% of the standard introns observed today in Gencode or AceView are still only supported by a single cDNA. Consistently, the number of variants and alternative introns keeps increasing almost linearly with new cDNA sequences: the addition to GenBank in January 2006 of close to two million 5' complete capped ESTs by the Japanese FLJ group [26] proportionately increased by 26% the number of alternative variants, and added 7% new spliced genes to the AceView collection [7]. In line with these observations, *ab initio* methods propose a wide variety of new models, and suggest that we may currently know only a fraction of the protein coding genes. To learn more about the genes expressed at low level, we depend on future technological improvements, in particular in the microarray domain, and on the continued acquisition of new data. We hope to integrate this flow of information seamlessly in the AceView hand-supervised automatic pipeline.

Materials and methods
The UCSCtrackCompare program used in this analysis can be downloaded from the AceView web page [7] and can be compiled on any properly configured Unix, Mac or Windows machine. A few precompiled executables and the whole source code are available. A description of the program and

relevant examples of the analyses it generates are provided in the four Additional data files.

Additional data files

The following additional data are available with the online version of this paper. Additional data file 1 is the UCSCtrackCompare package: a user's guide. The UCSCtrackCompare code is written in C. This document explains how to use the code, which options can be specified, and which results and analyses are generated. Additional data file 2 provides analyses of the entire transcript models. This document includes notes on the data and the selected tracks, comparisons to the Guigo *et al.* tables [6], and a series of diagrams showing comparisons, over the 25 tracks, of nucleotides, exons, spliced transcripts and regions of the transcripts annotated as coding. Both qualitative (projected on the genome, each object counts only once) and quantitative (multiple alternative variants contribute separately) analyses are described graphically. Gencode is evaluated, and results from a consensus analysis, depicting the transcript agreement across all tracks, are presented. Additional data file 3 provides analyses of the regions annotated as protein-coding. This document is similar to Additional data file 2, but the analysis is limited to the regions annotated as CDS. When alternative variants are considered separately, the resemblance between Gencode and AceView remains apparent. But when all alternative variants are collapsed, as done in [6], the amazing coherence between Gencode and AceView is obscured. As shown in our consensus analysis, this bias is due to the fact that Gencode annotates proteins in only 41% of their mRNAs. Additional data file 4 includes direct results of the UCSCtrackCompare program arranged in seven excel sheets: Regions, Summary stats, Nucleotides, Introns, Exons, Complete models (includes comparisons to Gencode and to 14 other tracks taken as reference), and Consensus analysis.

Acknowledgements
We thank Martin Reese for convincing us to write this paper and for his gentle comments, Yuji Kohara and Sumio Sugano for sharing their knowledge of transcription, and our former collaborators Michel Potdevin, Mark Sienkiewicz and Vahan Simonyan. We are grateful to Jim Kent for several insightful suggestions and to the referees for their guidance. We thank David Lipman, Donna Maglott and our NCBI colleagues for the excellent work environment. This research was supported by the Intramural Research Program of the NIH, National Library of Medicine.

This article has been published as part of *Genome Biology* Volume 7, Supplement 1, 2006: EGASP '05. The full contents of the supplement are available online at http://genomebiology.com/supplements/7/S1.

References
1. ENCODE Project Consortium: **The ENCODE (ENCyclopedia Of DNA Elements) Project.** *Science* 2004, **306:**636-640.
2. Hinrichs AS, Karolchik D, Baertsch R, Barber GP, Bejerano G, Clawson H, Diekhans M, Furey TS, Harte RA, Hsu F, *et al.*: **The UCSC Genome Browser Database: update 2006.** *Nucleic Acids Res* 2006, **34:**D590-598.
3. **UCSC Genome Browser: ENCODE Regions** [http://genome.ucsc.edu/ENCODE/encode.hg17.html]
4. **HAVANA** [http://www.sanger.ac.uk/HGP/havana/havana.shtml]
5. Guigo R, Reese MG: **EGASP: collaboration through competition to find human genes.** *Nat Methods* 2005, **2:**575-577.
6. Guigo R, Flicek P, Abril JF, Reymond A, Lagarde J, Denoeud F, Antonarakis S, Ashburner M, Bajic VB, Birney E, *et al.*: **EGASP: The ENCODE Genome Annotation Assessment Project.** *Genome Biology* 2006. **7(Suppl 1):**S2.
7. **The AceView genes** [http://www.ncbi.nlm.nih.gov/IEB/Research/Acembly]
8. Suzuki Y, Taira H, Tsunoda T, Mizushima-Sugano J, Sese J, Hata H, Ota T, Isogai T, Tanaka T, Morishita S, *et al.*: **Diverse transcriptional initiation revealed by fine, large-scale mapping of mRNA start sites.** *EMBO Rep* 2001, **2:**388-393.
9. Ota T, Suzuki Y, Nishikawa T, Otsuki T, Sugiyama T, Irie R, Wakamatsu A, Hayashi K, Sato H, Nagai K, *et al.* **Complete sequencing and characterization of 21,243 full-length human cDNAs.** *Nat Genet* 2004, **36:**40-45.
10. Moore MJ: **From birth to death: the complex lives of eukaryotic mRNAs.** *Science* 2005, **309:**1514-1518.
11. Wilkinson MF: **A new function for nonsense-mediated mRNA-decay factors.** *Trends Genet* 2005, **21:**143-148.
12. Maquat LE: **Nonsense-mediated mRNA decay in mammals.** *J Cell Sci* 2005, **118:**1773-1776.
13. Mendes Soares LM, Valcarcel J: **The expanding transcriptome: the genome as the 'Book of Sand'.** *EMBO J* 2006 **25:**923-931.
14. Kozak M: **Pushing the limits of the scanning mechanism for initiation of translation.** *Gene* 2002, **299:**1-34.
15. Suzuki Y, Sugano S: **Transcriptome analyses of human genes and applications for proteome analyses.** *Curr Protein Pept Sci* 2006, **7:**147-163.
16. Oyama M, Itagaki C, Hata H, Suzuki Y, Izumi T, Natsume T, Isobe T, Sugano S: **Analysis of small human proteins reveals the translation of upstream open reading frames of mRNAs.** *Genome Res* 2004, **14:**2048-2052.
17. Kozak M: **Context effects and inefficient initiation at non-AUG codons in eucaryotic cell-free translation systems.** *Mol Cell Biol* 1989, **9:**5073-5080.
18. Riechmann JL, Ito T, Meyerowitz EM: **Non-AUG initiation of AGAMOUS mRNA translation in Arabidopsis thaliana.** *Mol Cell Biol* 1999, **19:**8505-8512.
19. Chang KJ, Lin G, Men LC, Wang CC: **Redundancy of non-AUG initiators. A clever mechanism to enhance the efficiency of translation in yeast.** *J Biol Chem* 2006, **281:**7775-7783.
20. Touriol C, Bornes S, Bonnal S, Audigier S, Prats H, Prats AC, Vagner S: **Generation of protein isoform diversity by alternative initiation of translation at non-AUG codons.** *Biol Cell* 2003, **95:**169-178.
21. Mendell JT, Sharifi NA, Meyers JL, Martinez-Murillo F, Dietz HC: **Nonsense surveillance regulates expression of diverse classes of mammalian transcripts and mutes genomic noise.** *Nat Genet* 2004, **36:**1073-1078.
22. Jacobs D, Beitel GJ, Clark SG, Horvitz HR, Kornfeld K: **Gain-of-function mutations in the *Caenorhabditis elegans* lin-1 ETS gene identify a C-terminal regulatory domain phosphorylated by ERK MAP kinase.** *Genetics* 1998, **149:**1809-1822.
23. International Human Genome Sequencing Consortium: **Finishing the euchromatic sequence of the human genome.** *Nature* 2004, **431:**931-945.
24. Hillier LW, Coulson A, Murray JJ, Bao Z, Suslton JE, Waterston RH: **Genomics in *C.elegans*: So many genes, such a little worm.** *Genome Res* 2005, **15:**1651-1660.
25. Reboul J, Vaglio P, Tzellas N, Thierry-Mieg N, Moore T, Jackson C, Shin-i T, Kohara Y, Thierry-Mieg D, Thierry-Mieg J, *et al.*: **Open-reading-frame sequence tags (OSTs) support the existence of at least 17,300 genes in C. elegans.** *Nat Genet* 2001, **27:**332-336.
26. Kimura K, Wakamatsu A, Suzuki Y, Ota T, Nishikawa T, Yamashita R, Yamamoto J, Sekine M, Tsuritani K, Wakaguri H, *et al.*: **Diversification of transcriptional modulation: large-scale identification and characterization of putative alternative promoters of human genes.** *Genome Res* 2006, **16:**55-65.

Research

A computational approach for identifying pseudogenes in the ENCODE regions

Deyou Zheng* and Mark B Gerstein*†‡

Addresses: *Department of Molecular Biophysics and Biochemistry, Yale University, Whitney Avenue, New Haven, CT 06520, USA. †Department of Computer Science, Yale University, Prospect Street, New Haven, CT 06520, USA. ‡Program in Computational Biology and Bioinformatics, Yale University, New Haven, CT 06520, USA.

Correspondence: Mark B Gerstein. Email: Mark.Gerstein@yale.edu

Published: 7 August 2006

Genome *Biology* 2006, **7(Suppl 1)**:S13

The electronic version of this article is the complete one and can be found online at http://genomebiology.com/2006/7/S1/S13

Abstract

Background: Pseudogenes are inheritable genetic elements showing sequence similarity to functional genes but with deleterious mutations. We describe a computational pipeline for identifying them, which in contrast to previous work explicitly uses intron-exon structure in parent genes to classify pseudogenes. We require alignments between duplicated pseudogenes and their parents to span intron-exon junctions, and this can be used to distinguish between true duplicated and processed pseudogenes (with insertions).

Results: Applying our approach to the ENCODE regions, we identify about 160 pseudogenes, 10% of which have clear 'intron-exon' structure and are thus likely generated from recent duplications.

Conclusions: Detailed examination of our results and comparison of our annotation with the GENCODE reference annotation demonstrate that our computation pipeline provides a good balance between identifying all pseudogenes and delineating the precise structure of duplicated genes.

Background

Pseudogenes occupy a significant portion of vertebrate genomes, and are especially prevalent in the mammalian genomes [1-6]. It is estimated that the human genome may contain approximately 20,000 pseudogenes and pseudogene fragments [1,4]. These are copies of functional genes that have lost their potential as DNA templates for functional products (for example, proteins). Usually, they have accumulated various detrimental sequence mutations (for example, nonsense mutation) during evolution. Based on the processes of their formations, pseudogenes are often separated into: processed pseudogenes, which have been retrotransposed back into a genome from mRNA intermediates; and non-processed pseudogenes [2,5,6].

Pseudogenes have traditionally been recognized as an important resource for exploring dynamics and evolutionary history of genes and genomes. The common wisdom is that pseudogenes are non-functional and evolve neutrally. Therefore, they are often used for calibrating the parameters in various models of molecular evolution. However, some pseudogenes are transcribed and a few of them have been indicated to be involved in biological processes [5,7-9]. While the functional roles of pseudogenes are yet to be elucidated with more studies, the prevalence of pseudogenes in mammalian genomes has been problematic for gene annotation [10,11]. Because of high sequence similarity with functional genes, pseudogenes can sometimes be mistakenly annotated as genes, especially in an automated annotation pipeline. The task of distinguishing real genes from duplicated pseudogenes (a subtype of non-processed) is even more challenging. Therefore, the correct identification of pseudogenes is not only essential for subsequent pseudogene studies *per se* but also important for the overall accuracy of gene annotation [11].

Several computational algorithms have been described previously for annotating human pseudogenes [1,4,9,10,12-16]. All of them identify pseudogenes based on their two key sequence properties: similarity to genes and non-functionality. In practice, the former is often characterized by the sequence similarity between a pseudogene and its closest functioning gene relative (referred to as the 'parent gene') in the present-day genome. The latter is somewhat more elusive but is most commonly manifested by the occurrence of disablements (that is, premature stop codons, frameshifts and indels) in the 'putative coding region' of a pseudogene. Using such features as a pseudogene signature, Zhang *et al.* [1] identified approximately 8,000 processed pseudogenes in the human genome. The total number of human pseudogenes has been estimated to be between 10,000 and 20,000 according to this study and analyses from other groups [1,4,12,13].

Here we describe our pseudogene annotation for the ENCODE regions in the human genome. Our current computational pipeline contains various modifications and improvements from previous methods [1,9,16], with a new emphasis on delineating the precise structures of pseudogenes arising from recent gene duplication. Unlike their processed counterparts, duplicated pseudogenes arise from gene duplication and usually have intron-exon like structures inherited from their gene ancestors. This structure is also present in a unitary pseudogene, which has no functional relative in the same genome. The constituents of these structures may be called 'pseudo-introns' and 'pseudo-exons', terms that will be used in this paper. Previously, such 'introns' were inferred by aligning a pseudogene's nucleotide sequence to its parent gene's protein [1,4,13,14,16]. They were then used to distinguish duplicated from processed pseudogenes. As a result, processed pseudogenes with insertions (for example, transposons) could be incorrectly classified as duplicated unless extra care had been taken [16]. Our current method examines the preservation of a parent gene's intron-exon structure in a pseudogene and uses it as direct evidence for identifying duplicated pseudogenes. Applying this approach to the ENCODE regions found 164 pseudogenes (note that this number refers to the status in August 2005), which overlap very well with a reference set of manually curated pseudogenes from the GENCODE research group [17]. In addition, we found that 16 duplicated pseudogenes have their 'introns' and 'exons' arranged in the same patterns as those of their parent genes. These results demonstrate that our pipeline can identify pseudogenes correctly, and, as importantly, can delineate the precise structures of duplicated pseudogenes.

Results
Overview of our pipeline and number of pseudogenes in ENCODE regions

Gene prediction usually starts with the building of gene models from a specific training set of genes [11]. These models are subsequently applied to predict genes in unannotated genomic sequences. Many algorithms are presented in accompanying papers in this special ENCODE Genome Annotation Assessment Project (EGASP) issue [18]. The special characteristics of pseudogenes, on the other hand, have led researchers to adopt rather different strategies for their prediction. A homology-based approach like ours scans a genome for DNA sequences similar to a set of query genes. The resulting gene-like sequences are then scrutinized and those possessing pseudogene features are extracted. It is obvious that such a method requires a good set of known genes that is as complete and accurate as possible. After evaluating several data sources (data not shown), we decided to use the annotation from the ENSEMBL [19]. To be precise, we used version 29.35e (released in March 2005), which contained 24,194 genes (including 1,978 pseudogenes) and 28,479 proteins (composed of 292,306 non-redundant exons).

One criterion commonly used for separating processed and non-processed pseudogenes is based on the occurrence of pseudo-intron(s). Processed pseudogenes should have no pseudo-introns as they are the consequence of retrotransposition, but the non-processed ones typically retain introns or at least parts of them. In order to explore such a difference, we implemented a computational pipeline composed of two routines, with one (routine P) focusing on processed pseudogenes and the other (routine D) on duplicated ones (Figure 1). The major difference of these two routines lies in: the homology search step, where D uses individual exons while P uses full length proteins as queries; the step of assembling BLAST [20] hits into putative pseudogenes, where only D explicitly uses the intron-exon information of query genes (see Materials and methods for details). Putative pseudogenes from the P and D routines were combined and further inspected before they were finally classified. In our work, we specifically separated non-processed pseudogenes further into duplicated pseudogenes and fragments. The former have recognizable 'intron-exon' arrangements nearly identical to that of their parent genes whereas the latter do not.

In the end, the above pipeline identified a total of 211 pseudogenes (provided to EGASP/2005 in May 2005) in the ENCODE regions. Of these, 27 turned out to be LINE/SINE fragments after cross-reference with an updated version of RepeatMasker library. Excluding them, we identified 184 pseudogenes (Table 1), of which 93 were classified as processed, 19 as duplicated and 72 as pseudogene fragments. We also found one instance of a partially processed pseudogene; it is located at ENm011:80704-81919. The parent gene β-actin contains five exons. The processed

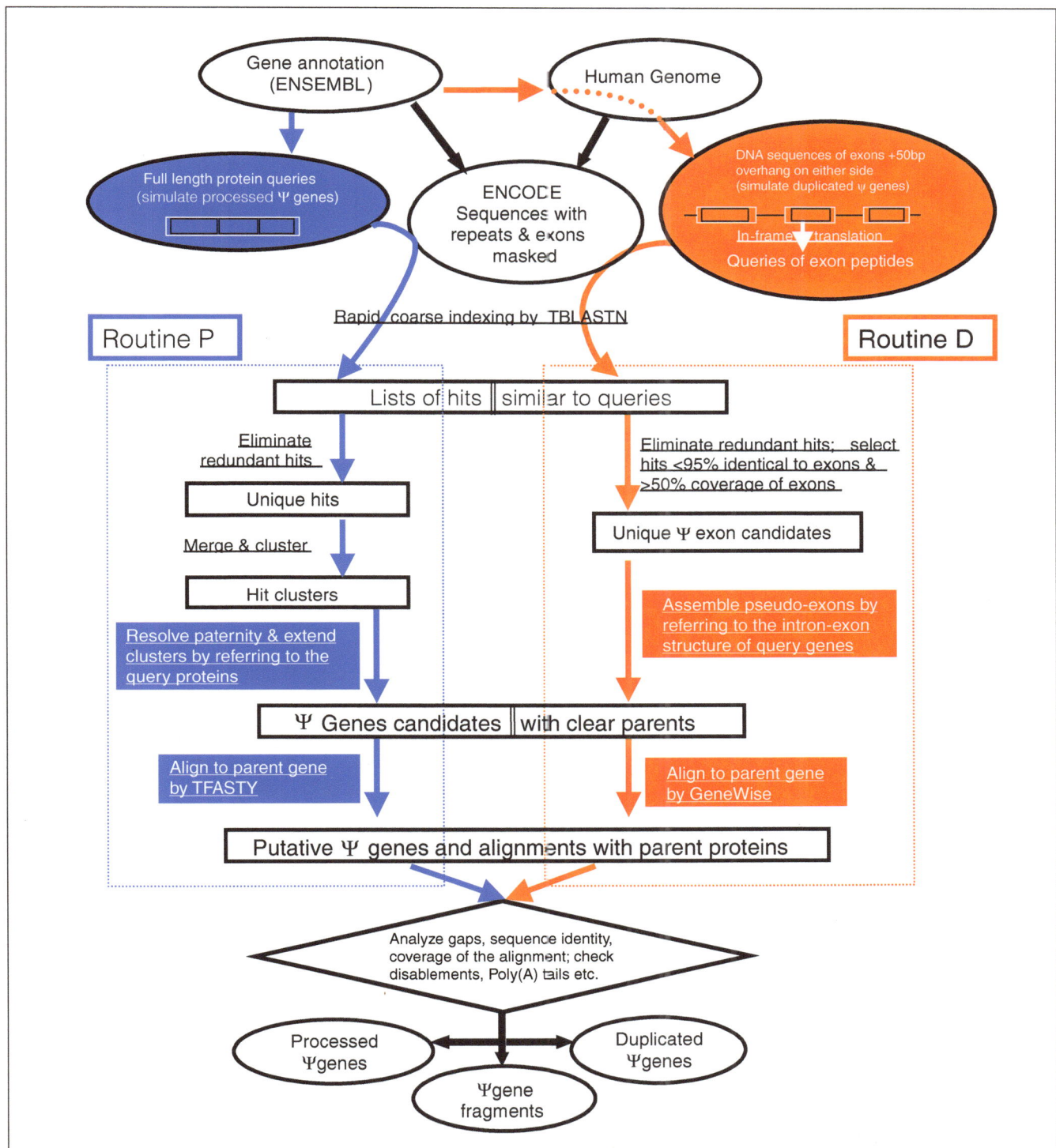

Figure 1
A flow chart of our computational pipeline for identifying pseudogenes. It contains two parallel procedures, one on the left (routine P) is mainly for processed pseudogenes and the other on the right (routine D) is for duplicated pseudogenes. The steps common to both are shown at the top and in the bottom. Both procedures searched the ENCODE regions for DNA sequences similar to human genes as annotated by the ENSEMBL. The two routines differ in how to perform the search and how to process the search results. The key differences are highlighted with blue in P and orange in D. At the end, an alignment between a known gene and a pseudogene candidate was constructed either by TFASTY or GeneWise. Information in this alignment and the computational path taken by a pseudogene were used together to separate pseudogenes into three classes: duplicated, processed and fragment.

Table 1

Separation of 184 pseudogenes in ENCODE regions identified in this study

Final pseudogene type*	Detected only by routine P	Detected by both routines	Detected only by routine D
Processed	60	30	3
Non-processed			
Duplicated	3	13†	3
Fragment	60	1	11

*The types are the final classification after information from routines P and D was combined. They could be different from a pseudogene's initial type labeled in either routine P or D. †In routine P, two were annotated as processed and two as fragments and another four were identified partially.

pseudogene retains only the third intron (95 base-pairs (bp)) while the remaining three introns have been spliced out.

The 44 ENCODE regions were picked with a variety of gene densities and conservation [21]. As shown in Figure 2, the number of pseudogenes varies in different regions. Many have only one or two pseudogenes, but two (ENm007 and ENm009) contain more than 20. Both of these two regions are also gene dense [17]. ENm009 contains the well characterized β-globin locus and is known to have many olfactory receptor (OR) pseudogenes [22]. In fact, 24 of our 29 pseudogenes in ENm009 were identified with olfactory receptor genes as their parent genes. Since the coding region of an OR gene is intronless, all but one OR pseudogene were put into the group of pseudogene fragments in our pipeline. Overall, the number of pseudogenes appears to correlate well with the number of genes in individual regions ($r^2 = 0.65$) (Figure 2).

Duplicated pseudogenes

Duplicated pseudogenes are an important evolutionary residue of a genome's past activity. It is generally thought that gene duplication is one of the main driving forces for creating genes with novel functions [23] . Therefore, the accurate identification of duplicated pseudogenes is valuable both for understanding the process of gene duplication and for studying the subsequent evolutionary fate of duplicated genes, which can either lead to gene death (that is, becoming a pseudogene or deleting a gene) or gene birth (that is,

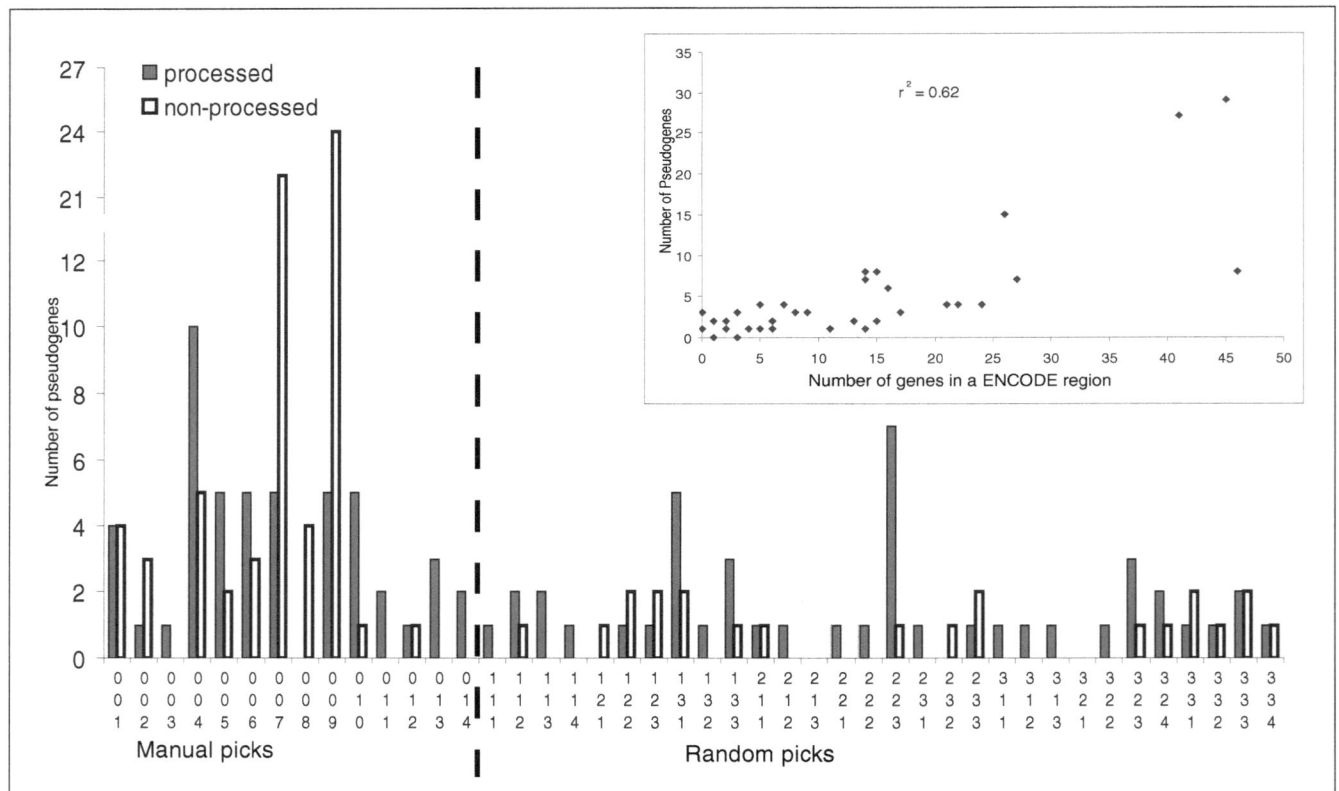

Figure 2
Distribution of 184 pseudogenes in ENCODE regions. Pseudogenes were first grouped into processed and non-processed (duplicated and fragments). Their numbers in the 44 ENCODE regions are plotted. The inserted panel shows that the number of pseudogenes is approximately correlated to that of genes within individual regions.

arising of a gene with new function). Only 19 of the 91 non-processed pseudogenes retain clear evidence of duplication, as supported by the preservation of intron-exon structures matching to their parent genes. The longest one contains 10 pseudo-exons spanning about 19,000 nucleotides. However, most of them (16) have lost at least one exon from their ancestors based on comparisons with their modern gene relatives. Notably, five of our duplicated pseudogenes are on the same chromosome as their parent genes, suggesting that they may have arisen from local gene duplication.

The majority of non-processed pseudogenes did not contain a pseudo-intron and, therefore, were classified as fragments. With the exception of OR pseudogenes, which originate from single-exon genes, most of them only match a short fragment of their parent proteins. They may represent single-exon duplication of their parent genes or have entirely lost their original intron-exon signatures. In this sense, it is appropriate to say that the duplicated pseudogenes identified by us arise from recent events of gene duplication.

Brief comparison of data from routines P and D

We merged pseudogenes from our two computational routines. We have examined how pseudogenes were identified and labeled by these two routines. As shown in Table 1, nearly all processed pseudogenes were detected (and labeled correctly; data not shown) in routine P. Routine D is intended for duplicated pseudogenes, but we allow it to pick up processed ones as well (see Materials and methods for details). In fact, it recognized one-third of our final 93 processed pseudogenes with an additional three not detected in routine P. These three were quite short and shared rather weak sequence similarity with their parent genes, so they were filtered out in routine P.

As mentioned above, approximately 80% of non-processed pseudogenes did not have a pseudo-intron and in many cases could be reliably aligned to only a fraction (<70%) of their parent genes. These were classified as pseudogene fragments [1,16]. Since they did not contain detectable pseudo-introns, they look like 'processed pseudogenes' and were mainly identified from routine P as expected. Most (25) OR pseudogenes were in this class; they actually result from gene duplication but were classified as pseudogene fragments in our computational scheme.

Most final duplicated pseudogenes were discovered by both D and P routines. It might appear that this defeats the whole purpose of routine D. However, detecting the presence of a pseudogene is one thing but recovering its full structure with accurate pseudo-intron-exon boundaries is another. The goal of routine D is really the latter. For six cases, only part of the pseudogene was identified in routine P while the entire structure with pseudo-exons and pseudo-introns was correctly annotated in routine D. Furthermore, two of these six were labeled as processed in routine P. These inaccuracies

would not have been corrected without information from routine D. On the other hand, to our surprise, three duplicated pseudogenes were missed in routine D. Further manual inspection showed that one in ENm001 (1092641-1094417) was more likely to be a processed pseudogene disrupted by a 1.2 kb DNA insertion; the other two (ENm006: 796815-805109; ENm008: 4095-8064) were almost identical (>95% sequence identity) to their parent genes. In routine D, we did not analyze genomic sequences with the latter feature.

In conclusion, the above results indicate that by combining two routines our computation pipeline provides a good balance between detecting all pseudogenes and identifying the exact structure of duplicated pseudogenes.

Comparison with GENCODE/HAVANA annotation

The GENCODE group and the HAVANA team have produced a high quality manual annotation for ENCODE regions, including 521 genes and 167 pseudogenes [17]. These served as the gold standard for evaluating other prediction methods in the EGASP/2005 workshop [18]. Table 2 summarizes the comparison between our pseudogenes and HAVANA annotation; 136 of our 184 pseudogenes overlapped with 135 of their pseudogenes. One-quarter of pseudogenes was unique to each method. This is a very promising result since pseudogenes annotated by different methods often did not agree very well [9,14].

In addition, 95 and 20 of our pseudogenes intersected with introns and exons from GENCODE, respectively. The overlapping between our pseudogenes and exons raises an important issue. Our method uses annotated genes for two purposes: as queries to search for similar genomic sequences; and as filters to eliminate exon sequences (that is, remove known genes). In our work, the gene annotation was obtained from the ENSEMBL, which contained 576 predicted genes in the ENCODE regions. Any discrepancy between ENSEMBL and the HAVANA/GENCODE annotation would be carried over to our annotation. For example, six GENCODE pseudogenes overlapped with ENSEMBL exons and thus could not be found by us. Conversely, some pseudogenes in our list could be components of genes missed by the ENSEMBL. Of the 20 overlapping with GENCODE exons, 11 were classified as pseudogene fragments, suggesting that they probably are real exons missed in our gene collection.

In order to illustrate the difficulty of gene/pseudogene annotation, we present two cases of discrepancy between our pseudogene annotation and GENCODE's gene prediction. First, in ENr122, we predicted a duplicated pseudogene at 359245-366200. There is a frame shift mutation in this pseudogene as shown in its alignment with an ENSEMBL protein ENSP00000331368 (Figure 3a). The parent gene (Serpin B8 or CAP-2) is in very close vicinity at

Table 2

Overlapping of our 184 pseudogenes with GENCODE annotations

Annotation in this study	GENCODE annotation			
	Processed	Non-processed	Not annotated	Exons
Processed	70	7	13	3
Non-Processed	15	44	17	17
Not Annotated	15	18	-	-

ENr122:375942-395286 (chr18: 59788243-59807587). This pseudogene contains three pseudo-exons corresponding to the first three exons of the parent gene, which has seven exons. It overlaps a GENCODE gene whose transcript (ID: 'AC009802.2-001') contains our three pseudo-exons and one extra untranslated exon at the 5' end. The disablement is in the first pseudo-exon (Figure 3a). However, if this disablement is skipped and an internal ATG is used as an alternative translation start site, a truncated protein can be produced. Without further experiments, the contradictory annotations for this region can not be resolved convincingly.

In the second case, we predicted a pseudogene at ENm005:200473-211501. Again, a frameshift mutation was found in the fourth pseudo-exon as shown in its alignment with an ENSEMBL protein ENSP00000283507 (Figure 3b). The parent gene (TCP-10) is in a different chromosome at chr6: 167554536-167579329. The pseudogene retains the intact structure of its parent gene with six exons and five introns. The first four pseudo-exons were included in a five-exon GENCODE transcript (ID: 'AP000274.7-001'). There is a full length cDNA (H-Inv: HIT000014684) matching this transcript, suggesting that this is likely a gene instead of pseudogene. However, transcription alone cannot be used as exclusive evidence to disapprove a pseudogene annotation because some pseudogenes are transcribed [9].

Comparison with known pseudogenes

We compared our annotation to a few available known pseudogenes in the ENCODE regions. We began with four duplicated pseudogenes. Previously, a transcribed β-globin pseudogene (HBBP1) [24] in ENm009 was discovered with a substitution mutation in the start codon, a nonsense mutation in codon 15 and frameshift mutations in the second and third exons. It was detected by us and GENCODE annotators, but both predicted a shorter version (Table 3). We did not observe sequence similarity for the 1 kb sequence at the 3' end. Two α-globin and one ζ-globin pseudogene have also been described in ENm008 [25] (Table 3). One of the α-globin pseudogenes was present in the GENCODE annotation and ours as well. The other was missed by both groups because of its low sequence similarity to the parent

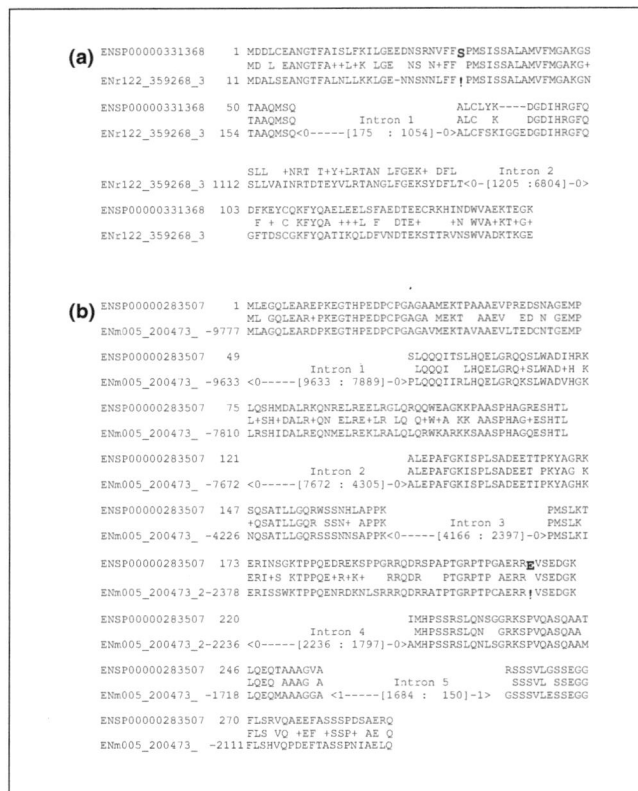

Figure 3
Two pseudogenes inconsistent with GENCODE gene annotation. **(a)** A pseudogene in ENr122: 359245-366200 (+) and its alignment with an ENSEMBL protein ENSP00000331368 (Serpin B8). This pseudogene overlaps a GENCODE gene whose transcript (ID: 'AC009802.2-001') contained the three pseudo-exons and one additional 5' exon. **(b)** A pseudogene at ENm005:200473-211501 (-) and its alignment with an ENSEMBL protein ENSP00000283507 (TCP-10 homolog). The first four pseudo-exons were included in a five-exon GENCODE transcript (ID: 'AP000274.7-001'). The frameshift mutations ('!' in the alignment) in both pseudogenes are highlighted.

gene, α-globin. (Note, we did find part of it during a homology search, but we did not pursue it because of its short sequence and no disablement.) The ζ-globin pseudogene is nearly identical to the ζ-globin gene except a single nonsense mutation in codon 7 [25]. It was identified in routine P but this stop codon was not displayed by alignment tools (see Discussion). Since this disablement was not visible and the remaining material was very similar (>95%) to the gene, we (perhaps over-cautiously) treated the sequence as a gene and did not report it.

Finally, previous studies have annotated many OR pseudogenes in ENm009 [22]. Since these are single exon pseudogenes, they are relatively easy to identify with computational pipelines. We found the majority of them and two examples are listed in Table 3.

Table 3

Examples of known pseudogenes in ENCODE regions

Name	Region	Pseudogene location	Our annotation	GENCODE annotation
β-globin	ENm009	488570-490726 (-)	489920-490351 (-)	488931-490348 (-)
α-globin	ENm008	156150-156704 (+)	NA	NA
α-globin	ENm008	158635-159503 (+)	158920-159084 (+)	158678-159333 (+)
ζ-globin	ENm008	152711-155400 (+)	NA	153121-155155 (+)
OR51H2P	ENm009	123369-124314 (+)	123353-124305 (+)	123368-124273 (+)
OR51B8P	ENm009	577399-578156 (-)	577369-578174 (-)	577403-578171 (-)

NA, not annotated.

Discussion

Genes, especially protein coding genes, have been and will remain the major focus of research on the genome. The launch of the ENCODE project, however, aims to identify all structural and functional elements in the human genome [21]. Pseudogenes are a major component of our genome and a few of them have been suggested to have functions. Nevertheless, pseudogene annotation is often considered as a side-project or by-product of gene annotation. However, most pseudogenes have traceable origins and sequence features distinct from genes, suggesting that computational strategies specific to pseudogene prediction are necessary.

In this paper, we describe our general algorithm for annotating pseudogenes. For the EGASP held in May 2005, we identified 184 pseudogenes, of which 136 overlap with the reference set of pseudogenes manually curated by the GENCODE team. About a quarter of the pseudogenes are unique to our own method. Although pseudogene prediction was not part of the official competition in EGASP [18], it was discussed extensively during the workshop. In addition, several research groups have subsequently been working together to obtain an accurate pseudogene annotation in the 44 ENCODE regions, and to improve methods that can be applied to the entire genome.

Limitation of our methods and future improvement
Gene annotation is in flux, so is pseudogene identification
Eighteen pseudogenes unique to our method were found to intersect with exons predicted by GENCODE (Table 2). Although a few of these pseudogenes may be *bone fide* pseudogenes, many of them are likely components of functional genes. A homology-based approach like ours needs gene annotation to compile a list of known genes (and proteins) as queries and as filters for eliminating genic sequences. Therefore, our result is limited by the source of gene annotation. Since annotation of the human genome is an ongoing dynamic process, our result will also be in flux. In this study, we used the ENSEMBL annotation [19]

because of its good coverage of the human genome. It is also deeper than the RefSeq collection [26] but more specific than annotation derived purely from computation prediction using software like GenScan [11,27]. Having said that, we note that the ENSEMBL gene collection in itself includes some pseudogenes, due to the complexity of gene annotation as discussed above. For example, the human genome has three GAPDH and 80 ribosomal protein genes, but harbors approximately 80 GAPDH pseudogenes and approximately 1,700 ribosomal protein pseudogenes [1,13]. Some of these pseudogenes were incorrectly annotated as genes by ENSEMBL.

Another issue in relation to the quality of the data source is the correct identification of repetitive sequences. If these sequences are not masked, they could be easily annotated as pseudogenes simply because they have the features of pseudogenes (and they are pseudogenes in some sense). As a matter of fact, we mistook 27 LINE/SINE sequences as pseudogenes. It is fair to say we would have not annotated the above 20 gene components and 27 repeats as pseudogenes if the relevant information was available to us in the beginning.

Need a better way to align a pseudogene to its parent protein sequence
The assessment of a genomic sequence as a pseudogene depends on the correct identification of its parent gene and the alignment between them. Currently, we assume that the most similar gene in the present-day genome represents the parent. This assumption may introduce unexpected artifacts into the alignment between a gene and its pseudogene relative, as both are descendents of an ancestral functional gene. Another practical issue is how to construct a 'correct' alignment. Stop codons and frameshifts are accommodated by programs like GeneWise [28] and TFASTY [29], but such disablements can break an alignment and leave it incomplete. Fundamentally, these programs are developed for genes so disablements are strongly disfavored in

constructing alignments. This is the reason for our failure to identify the ζ-globin pseudogene in ENm008. This pseudogene contains a nonsense mutation in codon 7, but the rest of the sequence is identical to ζ-globin gene. Both GeneWise and TFASTY constructed an alignment starting from codon 8 that totally ignored the first seven codons. As a result, we overlooked this pseudogene. This case clearly indicates that a better tool is needed to align a pseudogene to its parent gene. An algorithm specifically designed for aligning pseudogenes [30] appears promising and a new program, GeneMapper [31], may be useful for addressing this problem.

Strength and limitation of our computational pipeline
Processed pseudogenes are derived from processed mRNA. They are usually not disrupted by large indels and thus easier to be identified than duplicated pseudogenes. Our pipeline, especially through routine P, is very good at identifying these pseudogenes. In routine P, the presence of pseudo-intron is inferred if an insertion (relative to its parent protein in the alignment) larger than a threshold (for example, 60 bp) is found in a pseudogene. As shown in Table 2 and discussed above, this parameter is sufficient for detecting most duplicated pseudogenes even though it may not lead to the identification of the full length pseudogenes. However, it will misclassify disrupted processed pseudogenes as duplicated ones. To overcome this limitation, we developed routine D, which explicitly uses the intron-exon structure of a parent gene to classify duplicated pseudogenes. This idea appears very reasonable but it assumes that the intron-exon structure of a gene is at least partially preserved in its pseudogene relatives. Further investigation will be required to validate this assumption. Nevertheless, the combination of routines P and D provides a good balance between discovering all pseudogenes and identifying the exact structure of duplicated pseudogenes.

Conclusions
Using a homology-based approach, we have identified 184 pseudogenes in the ENCODE regions. The majority of them (74%) overlap with high quality pseudogenes annotated by the GENCODE group and the HAVANA team, an indication that our method worked successfully. Excluding the 20 pseudogenes overlapping with GENCODE exons, we would obtain a set of 164 pseudogenes, of which 91 are processed, 16 are duplicated, and the rest are tentatively classified as fragments. The list of our final 164 pseudogenes and the two ambiguous cases described above (Figure 3) can be found at [32]. This work also provides some insights for improving our approach in the future. At the current stage, there are not enough experimentally reported pseudogenes to establish a gold standard dataset for evaluating different prediction methods. However, several groups have worked together to reach a consensus and reliable list of pseudogenes for the ENCODE regions. Relevant information of that project is available at [33].

Materials and methods
Genomic sequence and annotated genes
The human genome sequence (build 35) was downloaded from the ENSEMBL [19] and sequences of the 44 ENCODE regions were extracted to serve as targets of our pseudogene annotation. Gene annotation was also obtained from the ENSEMBL. It included a set of known genes (as defined by ENSEMBL), with their intron, exon positions and their corresponding protein sequences.

Our computational pipeline contains two routines, with each focusing on a special type of pseudogene (Figure 1). In both routines, repetitive and exonic sequences in the ENCODE regions were masked.

Routine P focuses on processed pseudogenes
Processed pseudogenes are generated by retrotransposition, the process of reverse transcription of a processed mRNA into DNA and its subsequent insertion into a genome. As a consequence, these pseudogenes usually do not contain pseudo-introns. Although some of them may contain indels, they generally can be reliably aligned to their parent proteins, and the alignments often expand the full parent sequences. In recognition of this, routine P (for 'processed') uses human protein sequences in their full lengths as queries to search for pseudogenes. The steps (Figure 1) involved in this routine have been described previously [1,16]. In brief, intergenic and intronic sequences similar (>40% sequence identity) to human proteins are identified. Putative pseudogenes covering >70% of their parent proteins without an insertion longer than 60 nucleotides are then labeled as processed pseudogenes, and those with a gap as duplicated. Pseudogenic sequences aligned to only part (<70%) of proteins are classified as pseudogene fragments.

Routine D focuses on duplicated pseudogenes
A duplicated pseudogene usually contains pseudo-introns. To exploit this, we revised routine P and developed a new computational scheme, routine D (for 'duplicated'), which is more suitable for identifying duplicated pseudogenes (Figure 1). As described in detail below , there are two major changes from routine P: one during the homology search and the other in assembling search results into putative pseudogenes. In the homology search step, we used individual exons (of a gene) as our queries instead of the full length protein. When assembling pseudo-exons into pseudogene candidates, we referred to the intron-exon structures of our query genes to distinguish duplicated from processed pseudogenes.

Exon based BLAST
For each exon, we retrieved its DNA sequence and an extra 50 nucleotides adjacent to both ends of its exon (Figure 1). The extra 100 nucleotides are important for separating duplicated from processed pseudogenes because they enable our queries to span the pseudo-intron-exon boundaries of

duplicated pseudogenes. All 'exonic' DNA queries were first in-frame translated to peptides and then used by the program TBLASTN [20] to search for similar sequences in the ENCODE regions.

Assemble BLAST hits

BLAST hits (that is, the sequence regions found by TBLASTN) were assembled into pseudogene candidates based on the intron-exon organization of their query genes. A hit (putative pseudo-exon) was skipped if it covered less than half of its query exon, or if it was nearly identical (>95% sequence identity) to its query. Based on their genomic coordinates, two neighboring hits were joined together and labeled as 'Dup' if: they were similar to adjacent exons of the same genes; and the distance between them was within the size (plus an extra 500 nucleotides) of the intron separating the two parent exons. They were otherwise labeled as 'NonDup' if they were separated by less than 50 nucleotides. This step noticeably considered the alignment running across the intron-exon boundary because the distance of two 'pseudo-exons' in a processed pseudogene would presumably be 0 or at least less than 50 nucleotides. This labeling step ran through all BLAST hits. In the end, neighboring hits were assembled into pseudogene candidates.

Identify the parent gene for a pseudogene

As expected, many pseudogenic regions shared sequence similarity to more than one gene. For such cases, we enforced a one-to-one relationship by picking the gene most similar (defined by the smallest e-value) to this region as its parent.

Align a pseudogene candidate to its parent protein

After the relationship between a pseudogene candidate and its parent was established, the pseudogenic DNA sequence and the gene's 'protein' sequence were retrieved from databases and then aligned using the program GeneWise [28]. We chose GeneWise because it allows frameshift mutations and can accommodate very large insertions. GeneMapper [31] will be an alternative to explore in the future. Furthermore, we also used information from the alignment to adjust the start and end positions of a pseudogene. From the final alignments, we defined the locations of pseudo-exons and pseudo-introns in reference to the parent proteins. Disablements in the aligned regions were also used as criteria for our pseudogene assignment. Finally, we compared a duplicated pseudogene to its parent gene's DNA sequence in order to validate our pseudogene classification and to refine the genomic locations of a duplicated pseudogene.

Acknowledgements

The authors would like to thank Zhaolei Zhang and Nicholas Carriero for their valuable assistance in this study, and financial support from the NIH (1U01HG003156-01: Transcriptional and Regulatory Elements in the ENCODE Region).

This article has been published as part of Genome Biology Volume 7, Supplement 1, 2006: EGASP '05. The full contents of the supplement are available online at http://genomebiology.com/supplements/7/S1.

References

1. Zhang Z, Harrison PM, Liu Y, Gerstein M: **Millions of years of evolution preserved: a comprehensive catalog of the processed pseudogenes in the human genome.** *Genome Res* 2003, 13:2541-2558.
2. Zhang Z, Carriero N, Gerstein M: **Comparative analysis of processed pseudogenes in the mouse and human genomes.** *Trends Genet* 2004, 20:62-67.
3. Zhang Z, Gerstein M: **Large-scale analysis of pseudogenes in the human genome.** *Curr Opin Genet Dev* 2004, 14:328-335.
4. Torrents D, Suyama M, Zdobnov E, Bork P: **A genome-wide survey of human pseudogenes.** *Genome Res* 2003, 13:2559-2567.
5. Balakirev ES, Ayala FJ: **Pseudogenes: are they "junk" or functional DNA?** *Annu Rev Genet* 2003, 37:123-151.
6. Mighell AJ, Smith NR, Robinson PA, Markham AF: **Vertebrate pseudogenes.** *FEBS Lett* 2000, 468:109-114.
7. Korneev SA, Park JH, O'Shea M: **Neuronal expression of neural nitric oxide synthase (nNOS) protein is suppressed by an antisense RNA transcribed from an NOS pseudogene.** *J Neurosci* 1999, 19:7711-7720.
8. Hirotsune S, Yoshida N, Chen A, Garrett L, Sugiyama F, Takahashi S, Yagami K, Wynshaw-Boris A, Yoshiki A: **An expressed pseudogene regulates the messenger-RNA stability of its homologous coding gene.** *Nature* 2003, 423:91-96.
9. Zheng D, Zhang Z, Harrison PM, Karro J, Carriero N, Gerstein M: **Integrated pseudogene annotation for human chromosome 22: evidence for transcription.** *J Mol Biol* 2005, 349:27-45.
10. van Baren MJ, Brent MR: **Iterative gene prediction and pseudogene removal improves genome annotation.** *Genome Res* 2006, 16:678-685.
11. Brent MR, Guigo R: **Recent advances in gene structure prediction.** *Curr Opin Struct Biol* 2004, 14:264-272.
12. Harrison PM, Hegyi H, Balasubramanian S, Luscombe NM, Bertone P, Echols N, Johnson T, Gerstein M: **Molecular fossils in the human genome: identification and analysis of the pseudogenes in chromosomes 21 and 22.** *Genome Res* 2002, 12:272-280.
13. Ohshima K, Hattori M, Yada T, Gojobori T, Sakaki Y, Okada N: **Whole-genome screening indicates a possible burst of formation of processed pseudogenes and Alu repeats by particular L1 subfamilies in ancestral primates.** *Genome Biol* 2003, 4:R74.
14. Khelifi A, Duret L, Mouchiroud D: **HOPPSIGEN: a database of human and mouse processed pseudogenes.** *Nucleic Acids Res* 2005, 33(Database issue):D59-66.
15. Bischof JM, Chiang AP, Scheetz TE, Stone EM, Casavant TL, Sheffield VC, Braun TA: **Genome-wide identification of pseudogenes capable of disease-causing gene conversion.** *Hum Mutat* 2006, 27:545-552.
16. Zhang Z, Carriero N, Zheng D, Karro J, Harrison PM, Gerstein M: **PseudoPipe: an automated pseudogene identification pipeline.** *Bioinformatics* 2006. 22 1437-1439
17. Harrow J, Denoeud F, Frankish A, Reymond A, Chao-Kung C, Chrast J, Lagarde J, Gilbert JGR, Storey R, Swarbreck D, *et al*: **GENCODE: producing a reference annotation for ENCODE.** *Genome Biol* 2006, 7(Suppl 1):S4.
18. Guigó R, Flicek P, Abril JF, Reymond A, Lagarde J, Denoeud F, Antonarakis SE, Ashburner M, Bajic VB, Birney E, *et al*: **EGASP: the human ENCODE genome annotation assessment project.** *Genome Biol* 2006, 7(Suppl 1):S2.
19. Hubbard T, Andrews D, Caccamo M, Cameron G, Chen Y, Clamp M, Clarke L, Coates G, Cox T, Cunningham F, *et al*: **Ensembl 2005.** *Nucleic Acids Res* 2005, 33(Database issue):D447-453.
20. Altschul SF, Madden TL, Schaffer AA, Zhang J, Zhang Z, Miller W, Lipman DJ: **Gapped BLAST and PSI-BLAST: a new generation of protein database search programs.** *Nucleic Acids Res* 1997, 25:3389-3402.
21. ENCODE Project Consortium: **The ENCODE (ENCyclopedia Of DNA Elements) Project.** *Science* 2004, 306:636-640.
22. Glusman G, Yanai I, Rubin I, Lancet D: **The complete human olfactory subgenome.** *Genome Res* 2001, 11:685-702.
23. Zhang J: **Evolution by gene duplication: an update.** *Trends Ecol Evol* 2003, 18:292-298.
24. Chang LY, Slightom JL: **Isolation and nucleotide sequence analysis of the beta-type globin pseudogene from human, gorilla and chimpanzee.** *J Mol Biol* 1984, 180:767-784.
25. Proudfoot NJ, Gil A, Maniatis T: **The structure of the human zeta-globin gene and a closely linked, nearly identical pseudogene.** *Cell* 1982, 31:553-563.

26. Pruitt KD, Tatusova T, Maglott DR: **NCBI Reference Sequence (RefSeq): a curated non-redundant sequence database of genomes, transcripts and proteins.** *Nucleic Acids Res* 2005, **33(Database issue):**D501-504.

27. Burge C, Karlin S: **Prediction of complete gene structures in human genomic DNA.** *J Mol Biol* 1997, **268:**78-94.

28. Birney E, Clamp M, Durbin R: **GeneWise and Genomewise.** *Genome Res* 2004, **14:**988-995.

29. Pearson WR, Wood T, Zhang Z, Miller W: **Comparison of DNA sequences with protein sequences.** *Genomics* 1997, **46:**24-36.

30. Csuros M, Miklos I: **Statistical alignment of retropseudogenes and their functional paralogs.** *Mol Biol Evol* 2005, **22:**2457-2471.

31. Chatterji S, Pachter L: **Reference based annotation with GeneMapper.** *Genome Biol* 2006, **7:**R29.

32. **Supplemental Data** [http://www.pseudogene.org/pseudoPipeD/]

33. **Updated Data for ENCODE Pseudogenes** [http://www.pseudogene.org/ENCODE/]